Elementary Introduction to Quantum Geometry

This graduate textbook provides an introduction to quantum gravity, when spacetime is two-dimensional. The quantization of gravity is the main missing piece of theoretical physics, but in two dimensions it can be done explicitly with elementary mathematical tools, but it still has most of the conceptional riddles present in higher dimensional (not yet known) quantum gravity.

It provides an introduction to a very interdisciplinary field, uniting physics (quantum geometry) and mathematics (combinatorics) in a non-technical way, requiring no prior knowledge of quantum field theory or general relativity.

Using the path integral, the chapters provide self-contained descriptions of random walks, random trees and random surfaces as statistical systems where the free relativistic particle, the relativistic bosonic string and two-dimensional quantum gravity are obtained as scaling limits at phase transition points of these statistical systems. The geometric nature of the theories allows one to perform the path integral by counting geometries. In this way the quantization of geometry becomes closely linked to the mathematical fields of combinatorics and probability theory. By counting the geometries, it is shown that the two-dimensional quantum world is fractal at all scales unless one imposes restrictions on the geometries. It is also discussed in simple terms how quantum geometry and quantum matter can interact strongly and change the properties both of the geometries and of the matter systems.

It requires only basic undergraduate knowledge of classical mechanics, statistical mechanics and quantum mechanics, as well as some basic knowledge of mathematics at undergraduate level. It will be an ideal textbook for graduate students in theoretical and statistical physics and mathematics studying quantum gravity and quantum geometry.

Key features:

- Presents the first elementary introduction to quantum geometry
- Explores how to understand quantum geometry without prior knowledge beyond bachelor level physics and mathematics
- Contains exercises, problems and solutions to supplement and enhance learning

Elementary Introduction to Quantum Geometry

Jan Ambjørn

CRC Press
Taylor & Francis Group
Boca Raton London New York

CRC Press is an imprint of the
Taylor & Francis Group, an **informa** business

First edition published 2022
by CRC Press
4 Park Square, Milton Park, Abingdon, Oxon, OX14 4RN

and by CRC Press
6000 Broken Sound Parkway NW, Suite 300, Boca Raton, FL 33487-2742

© 2023 Jan Ambjørn

CRC Press is an imprint of Informa UK Limited

The right of Jan Ambjørn to be identified as author of this work has been asserted in accordance with sections 77 and 78 of the Copyright, Designs and Patents Act 1988.

All rights reserved. No part of this book may be reprinted or reproduced or utilised in any form or by any electronic, mechanical, or other means, now known or hereafter invented, including photocopying and recording, or in any information storage or retrieval system, without permission in writing from the publishers.

For permission to photocopy or use material electronically from this work, access www.copyright.com or contact the Copyright Clearance Center, Inc. (CCC), 222 Rosewood Drive, Danvers, MA 01923, 978-750-8400. For works that are not available on CCC please contact mpkbookspermissions@tandf.co.uk

Trademark notice: Product or corporate names may be trademarks or registered trademarks, and are used only for identification and explanation without intent to infringe.

British Library Cataloguing-in-Publication Data
A catalogue record for this book is available from the British Library

---

Library of Congress Cataloging-in-Publication Data

Names: Animasaun, Isaac Lare, author. | Shah, Nehad Ali, author. | Wakif, Abderrahim, author. | Mahanthesh, Basavarajappa, author. | Sivaraj, Ramachandran, author. | Koríko, Olubode Kolade, author.
Title: Ratio of momentum diffusivity to thermal diffusivity : introduction, meta-analysis, and scrutinization / Isaac Laare Animasaun, Nehad Ali Shah, Abderrahim Wakif, Basavarajappa Mahanthesh, Ramachandran Sivaraj, Olubode Kolade Koriko.
Description: First edition. | Boca Raton : Chapman & Hall/CRC Press, 2023. | Includes bibliographical references and index. |
Identifiers: LCCN 2022005066 (print) | LCCN 2022005067 (ebook) | ISBN 9781032108520 (hbk) | ISBN 9781032310893 (pbk) | ISBN 9781003217374 (ebk)
Subjects: LCSH: Materials--Thermal properties. | Thermal diffusivity.
Classification: LCC TA418.52 .A55 2023 (print) | LCC TA418.52 (ebook) | DDC 620.1/1296--dc23/eng/20220401
LC record available at https://lccn.loc.gov/2022005066
LC ebook record available at https://lccn.loc.gov/2022005067

ISBN: 978-1-032-33555-1 (hbk)
ISBN: 978-1-032-34100-2 (pbk)
ISBN: 978-1-003-32056-2 (ebk)

DOI: 10.1201/9781003320562

Typeset in Nimbus
by KnowledgeWorks Global Ltd.

Publisher's note: This book has been prepared from camera-ready copy provided by the authors.

Contents

Preface ix

Author xi

Chapter 1   Preliminary Material Part 1: The Path Integral 1

    1.1   The Classical Action 1
    1.2   Statistical Mechanics 2
    1.3   Classical to Quantum 5
    1.4   The Feynman Path Integral in Quantum Mechanics 7
    1.5   The Feynman-Kac Path Integral and Imaginary Time 12
    1.6   Problem Sets and Further Reading 14

Chapter 2   The Free Relativistic Particle 15

    2.1   The Propagator 15
    2.2   The Path Integral 17
    2.3   Random Walks and Universality 21
    2.4   ProblemSets and Further Reading 23

Chapter 3   One-Dimensional Quantum Gravity 25

    3.1   Scalar Fields in One Dimension 25
    3.2   Hausdorff Dimension and Scaling Relations 30
    3.3   Problem Sets and Further Reading 34

Chapter 4   Branched Polymers 35

    4.1   Definitions and Generalities 35
    4.2   Rooted Branched Polymers and Universality 37
    4.3   The Two-Point Function 39
    4.4   Intrinsic Properties of Branched Polymers 42
    4.5   Multicritical Branched Polymers 44
    4.6   Global and Local Hausdorff Dimensions 46
    4.7   Problem Sets and Further Reading 47

Chapter 5   Random Surfaces and Bosonic Strings 49

    5.1   The Action, Green Functions and Critical Exponents 49
    5.2   Regularizing the Integration over Geometries 55
    5.3   Digression: Summation over Topologies 63

	5.4	Scaling of the Mass 68
	5.5	Scaling of the String Tension 76
	5.6	Problem Sets and Further Reading 79

Chapter 6   Two-Dimensional Quantum Gravity 81

	6.1	Solving 2D Quantum Gravity by Counting Geometries 81
	6.2	Counting Triangulations of the Disk 83
	6.3	Multiloops and the Loop-Insertion Operator 92
	6.4	Explicit Solution for Bipartite Graphs 93
	6.5	The Number of Large Triangulations 96
	6.6	The Continuum Limit 99
	6.7	Other Universality Classes 102
	6.8	Appendix 104
	6.9	Problem Sets and Further Reading 105

Chapter 7   The Fractal Structure of 2D Gravity 107

	7.1	Universality and the Missing Correlation Length 107
	7.2	The Two-Loop Propagator 107
	7.3	The Two-Point Function 114
	7.4	The Local Hausdorff Dimension in 2D Gravity 117
	7.5	Problem Sets and Further Reading 120

Chapter 8   The Causal Dynamical Triangulation model 121

	8.1	Lorentzian Versus Euclidean Set Up 121
	8.2	Defining and Solving the CDT Model 122
	8.3	GCDT: Showcasing Quantum Geometry 131
	8.4	GCDT Defined as a Scaling Limit of Graphs 137
	8.5	The Classical Continuum Theory Related to 2D CDT 140
	8.6	Problem Sets and Further Reading 142

Appendix A   Preliminary Material Part 2: Green Functions 145

Appendix B   Problem Sets 1–13 161

	B.1	Problem Set 1 161
	B.2	Problem Set 2 164
	B.3	Problem Set 3 167
	B.4	Problem Set 4 170
	B.5	Problem Set 5 175
	B.6	Problem Set 6 178
	B.7	Problem Set 7 185
	B.8	Problem Set 8 191
	B.9	Problem Set 9 195
	B.10	Problem Set 10 204

		B.11 Problem Set 11 210
		B.12 Problem Set 12 217
		B.13 Problem Set 13 219
Appendix C		Solutions to Problem Sets 1–13 225
	C.1	Solutions to Problem Set 1 225
	C.2	Solutions to Problem Set 2 229
	C.3	Solutions to Problem Set 3 232
	C.4	Solutions to Problem Set 4 235
	C.5	Solutions to problem set 5 238
	C.6	Solutions to Problem Set 6 243
	C.7	Solutions to Problem Set 7 247
	C.8	Solutions to Problem Set 8 250
	C.9	Solutions to Problem Set 9 252
	C.10	Solutions to Problem Set 10 257
	C.11	Solutions to Problem Set 11 260
	C.12	Solutions to Problem Set 12 263
	C.13	Solutions to Problem Set 13 267

References 273

Index 275

Preface

These Lecture Notes and the related Problem Sets have been used for some years in a course in theoretical physics given to Master and PhD students at the Niels Bohr Institute, Copenhagen and at Radboud University, Nijmegen, the Netherlands. The idea has been to provide a non-technical introduction to what can be called *the statistical theory of geometries*. The theory of General Relativity is at present our best attempt to formulate a classical theory of geometry, but it has been difficult to quantize the theory, and even now, it is not entirely clear how to proceed when spacetime is four-dimensional. However, two-dimensional spacetime provides an interesting playground for trying to understand what a quantum theory of geometry might entail. If one rotates from Lorentzian signature to Euclidean signature of spacetime and use the path integral formalism one arrives at a statistical theory of two-dimensional geometries. This theory can be solved by quite elementary methods, and we will do that. The solution will also provide us with a beautiful illustration of the Wilsonian view on quantum field theories as associated with universality classes of statistical theories at their critical points. If matter fields live on these two-dimensional geometries one easily ends up with string theory, so also the "surrounding" of two-dimensional quantum geometry is quite rich and important. We will also discuss some elementary aspects of matter fields coupled to two-dimensional geometries.

As mentioned the notes will often be descriptive rather than providing proofs of the statements. For those interested in more technical details, I can refer to the book *Quantum Geometry, a statistical field theory approach* [1] by B. Durhuus, T. Jonsson and myself. However, when first preparing the lectures, I realized that the book was not really suited for students who have just completed their bachelor/undergraduate degree, so the idea of these notes is that the only requirement is some basic knowledge of classical analytic mechanics, quantum mechanics and statistical mechanics at bachelor/undergraduate level, and similarly, only simple mathematics, like contour integration in the complex plane, is used. Of course, sometimes more advanced concepts are mentioned, but hopefully never in a way that is essential for an understanding of the topic in question. No knowledge of quantum field theory or General Relativity is needed (although it does not harm, of course). Admittedly, a few concepts from Riemannian geometry are used, like the concept of a metric that describes the geometry and the concept of curvature. However, since we only discuss geometry related to at most two-dimensional surfaces, these concepts, in the context we use them, can be introduced in an intuitive fashion, and in particular, they can be defined for so-called piecewise linear geometries where no coordinates are needed for the description.

Since not much prior knowledge is assumed, the notes have sections of "preliminary material". Chapter 1, called *Preliminary material, part 1*, is a reminder of the absolute basics in classical mechanics, statistical theory and quantum mechanics, and (this is the primary reason it is written) an introduction to the path integral, since the path integral is not standard material in elementary quantum mechanics courses. This part of the preliminary material is put in front of the real lectures and I usually start the lecture series discussing this part. *Preliminary material, part 2* reminds the student of the use of Green functions in (classical) physics. It is a good starting point to know what a classical Green function is, as we will basically be calculating (quantum) Green functions of geometry. It is included as an Appendix at the end of the lectures, intended for self-study for the students for whom the concept of a Green function has become a little hazy.

The Problem Sets are an important integral part of the course. Most of the exercises are not really meant to provide the student with specific technical skills, but rather to supplement or explain in more detail some of the topics discussed in the notes. For the same reason detailed solutions are included. Both the Lecture Notes and the Problem Sets existed for a number of years as handwritten notes, and I am thankful to Joren Brunekreef, who was my teaching assistant for two years at Radboud University and who started to tex the Problem Sets and the solutions, despite my advice not to bother to do so. This eventually motivated me to also get the Lecture Notes themselves into a more readable form and to expand the Problem Sets.

While writing these notes I have benefitted from the insight and skills of my numerous collaborators over the years, and if anything deep or ingenious is present in the notes, they are the ones to be credited. However, when it comes to mistakes or conceptual blunders the blame is entirely on me. In relation to the topics covered in these notes I am, apart of course for the co-authors of the book "Quantum Geometry", Begfinnur Durhuus and Thordur Jonsson, particularly indebted to Jerzy Jurkiewicz, Renate Loll, Yuri Makeenko, Charlotte Kristjansen, Yoshiyuki Watabiki, Kostas Anagnostopoulos, Timothy Budd, Leonid Chekhov, Yuki Sato, Stephan Zohren, Willem Westra, Andrzej Görlich, Lisa Glaser, Asger Ipsen, Gudmar Thorleifsson and Zdzislaw Burda.

Author

Jan Ambjørn is a Danish physicist regarded as one of the founders of the statistical theory of geometries. The formalism has been applied to bosonic strings and quantum gravity in two and higher dimensions, and it was developed as a tool to study string theory and quantum gravity non-perturbatively. A later development, especially designed to study quantum gravity, is known as Causal Dynamical Triangulation Theory. During his career, Ambjørn has done research in numerous other areas, including quantum field theory and QCD, lattice gauge theories, the baryon asymmetry of the universe, matrix models, non-commutative field theory, string theory as well as the statistical theories of random paths and random surfaces. He is currently a professor at the Niels Bohr Institute, Copenhagen and Radboud University, Nijmegen.

1 Preliminary Material Part 1: The Path Integral

1.1 THE CLASSICAL ACTION

Consider a non-relativistic particle with mass m moving in one dimension in a potential $V(x)$. The simplest Hamiltonian and the corresponding equations of motion (eom) are then, p denoting the momentum of the particle,

$$H(x,p) = \frac{p^2}{2m} + V(x) \qquad \dot{x} = \frac{\partial H}{\partial p}, \quad \dot{p} = -\frac{\partial H}{\partial x}. \tag{1.1}$$

The Lagrangian $L(x,\dot{x})$ is defined as

$$\dot{x}p - H(x,p) = \frac{1}{2}m\dot{x}^2 - V(x) \equiv L(x,\dot{x}) \tag{1.2}$$

and the corresponding eom

$$\frac{d}{dt}\frac{\partial L(x,\dot{x})}{\partial \dot{x}} - \frac{\partial L(x,\dot{x})}{\partial x} = 0. \tag{1.3}$$

The so-called *action* $S[x]$ will play a central role in the course. Given a (particle) path $x(t)$, it is defined as

$$S[x] = \int_{t_1}^{t_2} dt\, L(x(t),\dot{x}(t)) = \int_{t_1}^{t_2} dt \left[\frac{m}{2}\left(\frac{dx}{dt}\right)^2 - V(x(t))\right]. \tag{1.4}$$

The action should be viewed as a *functional* on the set of paths $x(t)$, $t \in [t_1, t_2]$. Its relation to the eom is that the eom is an extremum of $S[x]$:

$$\frac{\delta S[x]}{\delta x(t)} = 0, \quad \left(\delta x(t_1) = \delta x(t_2) = 0\right) \quad \Rightarrow \quad \frac{d}{dt}\frac{\partial L(x,\dot{x})}{\partial \dot{x}} - \frac{\partial L(x,\dot{x})}{\partial x} = 0. \tag{1.5}$$

More precisely we consider a path $x(t), t \in [t_1, t_2]$ and an infinitesimal variation $\delta x(t)$ away from $x(t)$, with the boundary conditions that the variations at the endpoints of the path are zero as illustrated in Fig. 1.1. We then define

$$\begin{aligned}
\delta S[x] &\equiv S[x(t) + \delta x(t)] - S[x(t)] \tag{1.6}\\
&= \int_{t_1}^{t_2} dt \left[\left(\frac{\partial L(x,\dot{x})}{\partial x}\delta x + \frac{\partial L(x,\dot{x})}{\partial \dot{x}}\delta \dot{x}\right) + O\left(\delta x^2, \delta x \delta \dot{x}, (\delta \dot{x})^2\right)\right]\\
&= \int_{t_1}^{t_2} dt \left[\left(\frac{\partial L(x,\dot{x})}{\partial x} - \frac{d}{dt}\frac{\partial L(x,\dot{x})}{\partial \dot{x}}\right)\delta x(t) + O\left(\delta x^2, \delta x \delta \dot{x}, (\delta \dot{x})^2\right)\right].
\end{aligned}$$

Thus demanding that $\delta S[x] = 0$ for all infinitesimal variations (where also the time derivative of $\delta x(t)$ can be viewed as infinitesimal of the same order) leads to the classical eom for $x(t)$ as indicated in eq. (1.5).

DOI: 10.1201/9781003320562-1

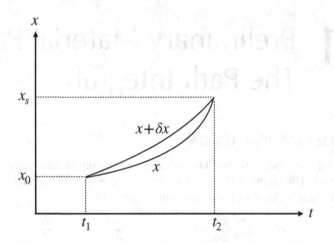

Figure 1.1 A generic path, denoted $x(t)$, starting from x_0 at time t_1 and ending at x_s at time t_2. In addition, one sees another path, denoted $x(t) + \delta x(t)$, which differs from the path $x(t)$ by a small deformation $\delta x(t)$ which vanishes at the endpoints.

1.2 STATISTICAL MECHANICS

Given a statistical system where the possible energy states s have energies E_s, we define the partition function as a function of the temperature T by

$$Z(T) = \sum_s e^{-\beta E_s}, \qquad \beta = \frac{1}{kT}, \tag{1.7}$$

where k denotes the Boltzmann constant. The summation is over all states s, counting also degeneracies. An important example is a classical ferromagnetic spin system. We are in d dimensions and consider a hyper-cubic lattice (Fig. 1.2 shows a two-dimensional such lattice), where the spins are located at the vertices, which we denote with integer coordinates $n = (n_1, \ldots, n_d)$. The classical spin at site n is then represented as a k-dimensional vector $\vec{S}(n)$, and a spin state is then the set $\{\vec{S}(n)\}$ of spins assigned to all sites n. A model for spin-spin interactions in a crystal assigns the following classical energy to a spin-state:

$$E(\{\vec{S}(n)\}) = -J \sum_{\text{neighboring } n,n'} \vec{S}(n) \cdot \vec{S}(n') + \sum_n a\vec{S}(n)^2 + b\left(\vec{S}(n)^2\right)^2, \tag{1.8}$$

where J, a, b are coupling constants and $J > 0$ for ferromagnetic system. The partition function is then

$$Z(T, J, a, b) = \sum_{\{\vec{S}(n)\}} e^{-\beta E(\{\vec{S}(n)\})} = e^{-\beta F(T, J, a,)}, \tag{1.9}$$

where $F(T)$ denotes the free energy of the system. For the classical spin system, the formal summation over the spin states is actually an integration. If the lattice has an

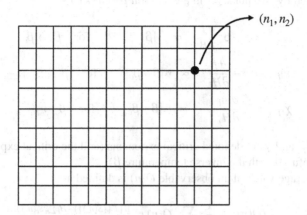

Figure 1.2 A two-dimensional square lattice and in this lattice a vertex with coordinates (n_1, n_2).

extension N_i in direction i, we have

$$\sum_{\{\vec{S}(n)\}} = \int \prod_{i=1}^{d} \prod_{n_i=1}^{N_i} \prod_{a=1}^{k} dS_a(n_1,\ldots,n_d) \qquad (1.10)$$

When we have an infinite lattice, i.e. $N_i \to \infty$, such a statistical system can have a phase transition as a function of the temperature. In this case, if the phase transition is of order n, the nth derivative of $Z(T)$ will be discontinuous at the critical temperature T_c where the phase transition takes place. The phase transitions are characterized by certain critical exponents, which we will now define and discuss since they and the associated critical behavior will be important for our understanding of quantum geometry.

Let us impose an external magnetic field H. In the simplest such ferromagnetic system the spins $\vec{S}$ have d components S_i, $i = 1,\ldots,d$, i.e. the same number of components as $\vec{H}$, and we have a partition function which now depends on H:

$$Z(T, H) = e^{-\beta F(T,H)} = \sum_{\{\vec{S}(n)\}} e^{-\beta \left(E(\{\vec{S}(n)\}) - \vec{H}\cdot\sum_n \vec{S}(n) \right)}, \qquad (1.11)$$

We can define a number of thermodynamical observables associated with the spin system. Let us write $F(T, H) = V f(T, H)$, where $V = \prod_{i=1}^{d} N_i$ is the volume of the lattice, and $f(T, H)$ thus the free energy density. As stated above, we have to take V to infinity to have a genuine phase transition in the system. Denote the assumed critical temperature of the system by T_c and the corresponding β by β_c. The specific heat c_v per volume, the magnetization m_i per volume and the susceptibility χ_{ij} are then defined by (we put $k=1$ in the relation $\beta = 1/kT$)

$$c_v = -\beta^2 \frac{\partial f}{\partial \beta} \propto |\beta - \beta_c|^{-\alpha} \quad \text{for} \quad \beta \to \beta_c \tag{1.12}$$

$$m_i = -\frac{\partial f}{\partial H_i} \propto |\beta - \beta_c|^{\beta} \quad \text{for} \quad \beta > \beta_c \; \beta \to \beta_c \tag{1.13}$$

$$\chi_{ij} = \frac{\partial m_i}{\partial H_j} \propto |\beta - \beta_c|^{-\gamma} \quad \text{for} \quad \beta \to \beta_c \tag{1.14}$$

where α, β and γ are denoted critical exponents (and the critical exponent β should not be confused with the inverse temperature β).

The average value of an observable $O(n)$ is defined as

$$\langle O(n) \rangle = \frac{1}{Z} \sum_{\{\vec{S}(n')\}} O(n) \, e^{-\beta \left(E(\{\vec{S}(n')\}) - H_i \sum_n S_i(n') \right)}, \tag{1.15}$$

and we can write

$$m_i = \frac{1}{V} \sum_n \langle S_i(n) \rangle, \quad \chi_{ij} = \frac{1}{V} \sum_{n,n'} \left\langle \left(S_i(n) - \langle S_i(n) \rangle \right) \left(S_j(n') - \langle S_i(n') \rangle \right) \right\rangle. \tag{1.16}$$

We define the spin-spin correlator as

$$G_{ij}(n,n') = \left\langle \left(S_i(n) - \langle S_i(n) \rangle \right) \left(S_j(n') - \langle S_i(n') \rangle \right) \right\rangle, \tag{1.17}$$

and thus

$$\chi_{ij} = \frac{1}{V} \sum_{n,n'} G_{ij}(n,n') = \sum_{n'} G_{ij}(n-n'), \tag{1.18}$$

where we in the last equation have assumed translational invariance of the system (i.e. for a finite lattice periodic boundary conditions). Thus we have

$$\boxed{\chi_{ij}(\beta) = \sum_n G_{ij}(n-n')} \tag{1.19}$$

Away from β_c the spin-spin correlation function (1.17) will be short ranged, and it falls off exponentially over a few lattice spacings and thus $\chi_{ij}(\beta)$ will be finite. If the phase transition at β_c is a second-order transition (i.e. the first derivative of $F(\beta, H=0)$ wrt β is finite at β_c, but the second derivative diverges when $\beta \to \beta_c$), the correlation length of $G_{ij}(n,n')$ will diverge when $\beta \to \beta_c$ and $\chi_{ij}(\beta)$ will also diverge as indicated in (1.14) (assuming the critical exponent $\gamma > 0$). Denote by $\xi(\beta)$ the correlation length of $G_{ij}(n,n')$, defined by the asymptotic exponential fall off of G:

$$\frac{-\ln G_{ij}(n,n')}{|n-n'|} \to \xi(\beta) \quad \text{for} \quad |n-n'| \to \infty. \tag{1.20}$$

Generically the long- and short-distance behavior of $G(n,n')$ close to the critical (inverse) temperature β_c is characterized by:

$$G(n,n') \propto |n-n'|^{2-d-\eta}, \quad \text{for } 1 \ll |n-n'| \ll \xi(\beta), \quad (1.21)$$
$$G(n,n') \propto e^{-|n-n'|/\xi(\beta)+\mathcal{O}(\ln|n-n'|)} \quad \text{for } \xi(\beta) \ll |n-n'| \quad (1.22)$$

where

$$\xi(\beta) \propto |\beta - \beta_c|^{-\nu} \quad \text{for } \beta \to \beta_c. \quad (1.23)$$

In (1.21) and (1.23) we have introduced two critical exponents, η and ν, in addition to α, β and γ. For many systems they are not independent since there exist so-called hyperscaling relations:

$$\alpha = 2-\nu d, \qquad \alpha + 2\beta + \gamma = 2, \qquad \gamma = \nu(2-\eta). \quad (1.24)$$

Thus there are only two independent exponents. The last relation is called *Fisher's scaling relation*. Since we will meet it again in our quantum geometry theories, let us just show how it can be derived heuristically from the definitions already given. We suppress the indices i, j and assume translational invariance of G. Approximate G by the asymptotic form (1.21) for $|n| < \xi(\beta)$ and put it to zero for $|n| > \xi(\beta)$, since it, according to (1.22), is more or less exponentially suppressed in that region. We thus have, replacing summation by integration:

$$\chi(\beta) = \int d^d n \, G(n) \propto \int_{|n|<\xi(\beta)} \frac{d^d n}{n^{d-2+\eta}} \propto \left(\xi(\beta)\right)^{2-\eta} \propto \frac{1}{|\beta-\beta_c|^{\nu(2-\eta)}}, \quad (1.25)$$

which implies that $\gamma = \nu(2-\eta)$ according to the definition (1.14) of γ.

The importance of these exponents is that they are *universal*. Different spin systems can have the same exponents even if their local spin-spin interactions and their critical temperatures can be quite different. While the local detail of the spin-spin interactions might be unimportant for the system's critical exponents, the symmetries are important. Thus spin systems with different symmetries can have different exponents. When the dimension of space $d \geq 4$, all spin systems will have the same exponents, the so-called *mean-field exponents*. They are

$$\nu = \frac{1}{2}, \quad \eta = 0, \quad \text{and from hyperscaling} \quad \gamma = 1, \quad \text{etc.} \quad (1.26)$$

In Problem Set 4 we will calculate the critical exponents of the above spin system using a so-called mean-field approximation, and we will (not surprisingly....) find the mean field values (1.26).

1.3 CLASSICAL TO QUANTUM

In the transition from classical physics to quantum physics for the simple one-dimensional classical system we considered above, we first introduce the Hilbert space $\mathcal{H}$ of square integrable functions $\mathcal{H} = L^2(\mathbb{R})$ on the real axis. Next we promote the classical variable x, p to operators $\hat{x}, \hat{p}$ in the following way:

$$\hat{x}: \psi(x) \to (\hat{x}\psi)(x) = x\psi(x), \qquad \hat{p}: \psi(x) \to (\hat{p}\psi)(x) = \frac{\hbar}{i}\frac{d\psi}{dx} \qquad (1.27)$$

where $\psi \in L^2(\mathbb{R})$. Both $\hat{x}$ and $\hat{p}$ are unbounded but Hermitian operators. The same is true for the quantum Hamiltonian, which is obtained by replacing x, p in the classical Hamiltonian with the operators $\hat{x}, \hat{p}$:

$$\hat{H} = \frac{1}{2m}\hat{p}^2 + V(\hat{x}) = -\frac{\hbar^2}{2m}\frac{d^2}{dx^2} + V(\hat{x}). \qquad (1.28)$$

Any vector $|\psi\rangle$ in $\mathcal{H}$ can be expanded in any orthonormal basis $|e_n\rangle$ in $\mathcal{H}$:

$$|\psi\rangle = \sum_n |e_n\rangle\langle e_n|\psi\rangle, \qquad \sum_n |e_n\rangle\langle e_n| = \hat{1}. \qquad (1.29)$$

The eigenvectors $|x\rangle$ and $|p\rangle$ of the operators $\hat{x}$ and $\hat{p}$, corresponding to the eigenvalues x and p, respectively, are defined by

$$\hat{x}|x\rangle = x|x\rangle, \qquad \hat{p}|p\rangle = p|p\rangle. \qquad (1.30)$$

These eigenvectors do not belong to $L^2(\mathbb{R})$. Nevertheless, we can still expand the vectors in $\mathcal{H}$ on these vectors, as in eq. (1.29):

$$|\psi\rangle = \int dx\, |x\rangle\langle x|\psi\rangle, \qquad \int dx\, |x\rangle\langle x| = \hat{1}. \qquad (1.31)$$

$$= \int dp\, |p\rangle\langle p|\psi\rangle, \qquad \int dp\, |p\rangle\langle p| = \hat{1}. \qquad (1.32)$$

$\langle x|\psi\rangle \equiv \psi(x)$ is denoted the wave function of the state $|\psi\rangle$ and it is when expanding the states $|\psi\rangle$ on the vectors $|x\rangle$ that we in (1.27) defined the operators $\hat{x}$ and $\hat{p}$. In particular we have for the state $|p\rangle$:

$$\langle x|p\rangle = \frac{e^{ipx/\hbar}}{\sqrt{2\pi\hbar}}, \quad \text{i.e.} \quad |p\rangle = \int dx\, |x\rangle\langle x|p\rangle = \int dx\, |x\rangle \frac{e^{ipx/\hbar}}{\sqrt{2\pi\hbar}} \qquad (1.33)$$

The eigenstates and eigenvalues of $\hat{H}$ are of course of particular interest. Denote an eigenstate $|E\rangle$ where E is the corresponding eigenvalue of $\hat{H}$:

$$\hat{H}|E\rangle = E|E\rangle, \quad \psi_E(x) = \langle x|E\rangle, \quad \left(-\frac{\hbar^2}{2m}\frac{d^2}{dx^2} + V(x)\right)\psi_E(x) = E\psi_E(x). \qquad (1.34)$$

The spectrum (the eigenvalues of the Hamiltonian) can be discrete, as when the potential $V(x) = \omega^2 x^2/2$ is that of the harmonic oscillator and where the eigenvalues are $E_n = \hbar\omega(n + \frac{1}{2})$. It can also be continuous as when $V(x) = 0$, i.e. the free particle case, where the eigenstates of the Hamiltonian are just the states $|p\rangle$ and the corresponding eigenvalues of the Hamiltonian are $p^2/2m$.

We can now define the *quantum partition function* as in (1.7), just by replacing the classical energies by the quantum energy calculated from (1.34):

$$Z(T) = \sum_E e^{-\beta E} = \sum_E \langle E|e^{-\beta \hat{H}}|E\rangle = \text{tr}\, e^{-\beta \hat{H}} = \int dx\, \langle x|e^{-\beta \hat{H}}|x\rangle, \quad (1.35)$$

where $\sum_E$ is a summation if the eigenvalues are discrete and a suitable integration if they are continuous.

The *time evolution in quantum mechanics* is simplest described by the Schrödinger equation:

$$i\hbar \frac{\partial}{\partial t}|\psi(t)\rangle = \hat{H}|\psi(t)\rangle, \quad (1.36)$$

$$i\hbar \frac{\partial}{\partial t}\psi(x,t) = \left(-\frac{\hbar^2}{2m}\frac{d^2}{dx^2} + V(x)\right)\psi(x,t), \quad \psi(x,t) \equiv \langle x|\psi(t)\rangle. \quad (1.37)$$

The formal solution to (1.36) is

$$|\psi(t)\rangle = e^{-i\hat{H}t/\hbar}|\psi(0)\rangle. \quad (1.38)$$

The basic question asked in quantum mechanics is the following: given a state $|\psi_0\rangle$ at time $t=0$, what is the probability amplitude for finding the system in the state $|\psi_s\rangle$ at time t? The answer is:

$$\langle \psi_s|e^{-i\hat{H}t/\hbar}|\psi(0)\rangle. \quad (1.39)$$

Using the expansion (1.31) in terms of wave functions we can write

$$\langle \psi_s|e^{-i\hat{H}t/\hbar}|\psi(0)\rangle = \int dx_s \int dx_0\, \psi_s^*(x_s)\psi_0(x_0)\langle x_s|e^{-i\hat{H}t/\hbar}|x_0\rangle. \quad (1.40)$$

In principle we can then answer all questions if we can only calculate

$$\langle x_s|e^{-i\hat{H}t/\hbar}|x_0\rangle, \quad (1.41)$$

i.e. the probability amplitude for a particle which at time 0 is located at x_0 to be found at x_s at time t. This probability amplitude can be represented as a *Feynman path integral*. The Feynman path integral formalism of quantum mechanics will be central in the following text.

1.4 THE FEYNMAN PATH INTEGRAL IN QUANTUM MECHANICS

$$\boxed{\langle x_s|e^{-i\hat{H}t/\hbar}|x_0\rangle = \int_{\substack{x(0)=x_0 \\ x(t)=x_s}} \mathcal{D}x(\tilde{t})\, e^{\frac{i}{\hbar}S[x(\tilde{t})]}} \quad (1.42)$$

The "integration" on the rhs of this formula is over the set of continuous path $x(\tilde{t})$, $\tilde{t} \in [0,t]$ where $x(0) = x_0$ and $x(t) = x_s$. For a given path $x(\tilde{t})$, the weight of

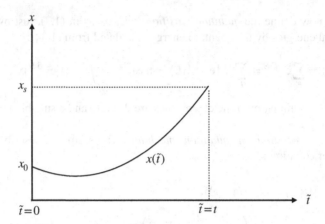

Figure 1.3 A path $x(\tilde{t})$, $\tilde{t} \in [0,t]$ contributing to the path integral (1.42), i.e. it starts at x_0 when time is 0 and ends at x_s when time is equal t. However, as we will discuss below, the path is much too "nice" to represent a typical path appearing in the path integral.

the integrand is $e^{iS[x]/\hbar}$ where $S[x]$ is the classical action of the path $x(\tilde{t})$. Fig. 1.3 shows a "typical" such path (well, as we will see, it is not so typical. A "typical" continuous path is much more "wild") . Before discussing how the path integral can be derived in a simple way from the standard time evolution of quantum mechanics, let us discuss how the path integral provides an intuitive link between quantum and classical physics. Let us consider the limit $\hbar \to 0$ and ask which paths contribute to (1.42) in that limit. For small $\hbar$ a little change of a given path $x(\tilde{t}) \to x(\tilde{t}) + \delta x(\tilde{t})$ will introduce a large phase shift in $e^{iS[x+\delta x]/\hbar}$ compared to $e^{iS[x]/\hbar}$ and the contributions from neighboring paths will cancel unless $S[x+\delta x] \approx S[x]$. We have

$$S[x+\delta x] = S[x] + \int_0^t d\tilde{t}\, \frac{\delta S}{\delta x(\tilde{t})} \delta x(\tilde{t}) + \mathcal{O}((\delta x)^2). \tag{1.43}$$

Thus we can obtain $S[x+\delta x] \approx S[x]$ provided that

$$\frac{\delta S}{\delta x(\tilde{t})} = 0 \quad \text{i.e. according to (1.5)} \quad x(\tilde{t}) = x_{\mathrm{cl}}(\tilde{t}), \tag{1.44}$$

and therefore the paths which contribute the most in the $\hbar \to 0$ limit to the path integral will be the solutions to the classical eom (1.5). The larger $\hbar$ is compared to the constants which appear in the Lagrangian (the mass of the particle, the cyclic frequency ω in the case of a harmonic oscillator, etc.), the more "non-classical" paths, i.e. paths which do not satisfy the classical eom, will be important when calculating the path integral.

Let us now give a precise meaning to the path integral appearing in (1.42). First, we note the following mathematical facts:

Preliminary Material Part 1: The Path Integral

(1) Let a, b be complex numbers. Then

$$e^{-a-b} = e^{-a}e^{-b} = \lim_{n \to \infty} \left(e^{-a/n} e^{-b/n} \right)^n. \tag{1.45}$$

(2) Let $\hat{A}$ and $\hat{B}$ be $N \times N$ matrices. Then in general $e^{-\hat{A}-\hat{B}} \neq e^{-\hat{A}}e^{-\hat{B}}$ when $[\hat{A}, \hat{B}] \neq 0$. However, it is still true that

$$e^{-\hat{A}-\hat{B}} = \lim_{n \to \infty} \left(e^{-\hat{A}/n} e^{-\hat{B}/n} \right)^n, \quad \text{(the Lie-Trotter product formula)} \tag{1.46}$$

(3) Let $\hat{A}$ and $\hat{B}$ be Hermitian operators on a Hilbert space $\mathscr{H}$ with eigenvalues that are bounded from below. They can be unbounded operators, but have to be such that $D_A \cap D_B$ is dense in $\mathscr{H}$, where D_A and D_B denote the domains of definition of $\hat{A}$ and $\hat{B}$. Then $e^{-\hat{A}}$, $e^{-\hat{B}}$ and $e^{-\hat{A}-\hat{B}}$ are bounded operators with a well-defined norm, and one has the equivalent of (1.46):

$$e^{-\hat{A}-\hat{B}} = \lim_{n \to \infty} \left(e^{-\hat{A}/n} e^{-\hat{B}/n} \right)^n, \quad \text{(the Kato-Trotter theorem)}, \tag{1.47}$$

where the convergence is in operator norm, i.e. a quite strong convergence. The same relation is true if we replace $\hat{A}$ by $i\hat{A}$, and similarly for $\hat{B}$ and $\hat{A}+\hat{B}$. In that case $e^{i\hat{A}}$ etc. become unitary operators with norm 1.

With these remarks in mind, we now apply (1.47) to

$$i\frac{t}{\hbar}\hat{H} = i\hat{A} + i\hat{B}, \quad \hat{A} = \frac{t}{2m\hbar}\hat{p}^2, \quad \hat{B} = \frac{t}{\hbar}V(\hat{x}). \tag{1.48}$$

$$e^{-it\hat{H}/\hbar} = \lim_{n \to \infty} \left(e^{-i\hat{A}/(n+1)} e^{-i\hat{B}/(n+1)} \right)^{n+1} = \lim_{n \to \infty} \left(\hat{O}_\varepsilon \right)^{n+1} \tag{1.49}$$

where

$$\hat{O}_\varepsilon = e^{-i\varepsilon \hat{p}^2/(2m\hbar)} e^{-i\varepsilon \hat{V}(\hat{x})/\hbar}, \quad \varepsilon = \frac{t}{n+1}. \tag{1.50}$$

We can calculate the matrix element $\langle x | \hat{O}_\varepsilon | y \rangle$:

$$\boxed{\langle x | \hat{O}_\varepsilon | y \rangle = \left(\frac{m}{2\pi i \varepsilon \hbar} \right)^{\frac{1}{2}} e^{i(x-y)^2 m/(2\varepsilon \hbar)} e^{-i\varepsilon V(y)/\hbar}} \tag{1.51}$$

Here is the calculation: we use (1.32) and (1.33) to write

$$\begin{aligned}
\langle x | \hat{O}_\varepsilon | y \rangle &= \langle x | e^{-i\varepsilon \hat{p}^2/(2m\hbar)} \left(\int dp \, |p\rangle \langle p| \right) e^{-i\varepsilon \hat{V}(\hat{x})/\hbar} | y \rangle \\
&= e^{-i\varepsilon V(y)/\hbar} \int dp \, \langle x | p \rangle \langle p | y \rangle \, e^{-i\varepsilon p^2/(2m\hbar)} \\
&= e^{-i\varepsilon V(y)/\hbar} \int \frac{dp}{2\pi \hbar} \, e^{ip(x-y)/\hbar} e^{-i\varepsilon p^2/(2m\hbar)} = \text{rhs of } (1.51),
\end{aligned}$$

where the last equality follows from a Gaussian integral, which will be discussed in Problem Set 1.

We can now write

$$\hat{O}_\varepsilon^{n+1} = \hat{O}_\varepsilon \hat{I} \hat{O}_\varepsilon \hat{I} \cdots \hat{I} \hat{O}_\varepsilon, \quad \hat{I} = \int dx\, |x\rangle\langle x| \qquad (1.52)$$

Thus we can write the matrix elements of $\hat{O}_\varepsilon^{n+1}$ as standard "matrix" multiplication, where we define $x_{n+1} = x_s$:

$$\langle x_{n+1}|\hat{O}_\varepsilon^{n+1}|x_0\rangle = \int \prod_{i=1}^n dx_i\, \langle x_{n+1}|\hat{O}_\varepsilon|x_n\rangle\langle x_n|\hat{O}_\varepsilon|x_{n-1}\rangle\cdots\langle x_1|\hat{O}_\varepsilon|x_0\rangle \qquad (1.53)$$

$$= \left(\frac{m}{2\pi i\varepsilon\hbar}\right)^{\frac{n+1}{2}} \int \prod_{i=1}^n dx_i\, \exp\left(\frac{i}{\hbar}\sum_{i=0}^n \varepsilon\left[\frac{m}{2}\left(\frac{x_{i+1}-x_i}{\varepsilon}\right)^2 - V(x_i)\right]\right)$$

We finally have, with $\varepsilon = t/(n+1)$,

$$\boxed{\langle x_s|e^{-i\hat{H}t/\hbar}|x_0\rangle = \lim_{n\to\infty}\left(\frac{m}{2\pi i\varepsilon\hbar}\right)^{\frac{n+1}{2}} \int \prod_{i=1}^n dx_i\, \exp\left(\frac{i}{\hbar}\sum_{i=0}^n \varepsilon\left[\frac{m}{2}\left(\frac{x_{i+1}-x_i}{\varepsilon}\right)^2 - V(x_i)\right]\right)}$$

(1.54)

This gives a precise meaning to the rhs of eq. (1.42) via the Kato-Trotter theorem. However, the x_i's which enter in the formula are at this point merely integration variables. The BCH (Baker, Campbell, Hausdorff) formula tells us that[1]

$$\hat{O}_\varepsilon = e^{-i\varepsilon\hat{H}/\hbar}\, e^{\mathcal{O}(\varepsilon^2)}. \qquad (1.55)$$

It is clear that we can trivially write

$$\langle x_s|e^{-i\hat{H}t/\hbar}|x_0\rangle = \int \prod_{i=1}^n dx_i\, \langle x_{n+1}|e^{-i\frac{\varepsilon\hat{H}}{\hbar}}|x_n\rangle\langle x_n|e^{-i\frac{\varepsilon\hat{H}}{\hbar}}|x_{n-1}\rangle\cdots\langle x_1|e^{-i\frac{\varepsilon\hat{H}}{\hbar}}|x_0\rangle.$$

(1.56)

In (1.56), the x_i's, $i = 1,\ldots,n$ are also independent integration variables, but we can now associate the time $t_i = i\varepsilon$ to the index i since $e^{-i\varepsilon\hat{H}/\hbar}$ precisely is the evolution operator during a time-interval ε, i.e. it makes sense to write $x(t_i) \equiv x_i$ in (1.56). In view of (1.55), we will do the same (1.54), although strictly speaking there is no strict link of the i in (1.54) to the time $t_i = i\varepsilon$ for $1 \leq i \leq n$. With this assignment $x_i \to x(t_i)$, it is tempting to assign a "path" to the sequence of points $x(t_i)$ by joining $x(t_i)$ and $x(t_{i+1})$ by a straight line such we can write $x(\tilde{t})$, $\tilde{t} \in [0,t]$. This is illustrated in Fig. 1.4. With this assignment, we can view (1.54) as a certain limit of the integration over the class of "piecewise linear" paths shown in Fig. 1.4.

---

[1] One version of the BCH formula useful in this context is $e^{t(\hat{A}+\hat{B})} = e^{t\hat{A}}e^{t\hat{B}}e^{-t^2[\hat{A},\hat{B}]+O(t^3)}$, where the term $O(t^3)$ can be expressed in terms of two or more commutators of $\hat{A}$ and $\hat{B}$. Note that the Lie-Trotter (or as it is also called, the Suzuki-Trotter) product formula (1.46) follows from this BCH formula.

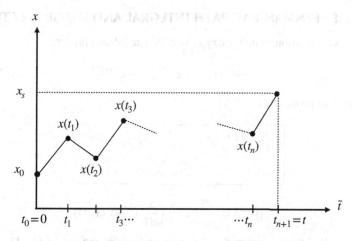

Figure 1.4 The points $\{x(t_0), x(t_1), \ldots, x(t_{n+1})\}$ shown in the figure are connected into a piecewise linear path $x(\tilde{t})$, $\tilde{t} \in [0, t]$. The first point at time t_0 is the starting point of the path integral. The last point at t_{n+1} is the endpoint of the path integral. The other points at time t_1 to time t_n are x-variables with respect to which one integrates in eq. (1.54) or eq. (1.56).

With the above "path"-interpretation it is now tempting to perform the following identifications:

$$\frac{x(t_{i+1}) - x(t_i)}{\varepsilon} \approx \frac{dx(\tilde{t})}{d\tilde{t}}, \quad \tilde{t} \in]t_i, t_{i+1}[, \quad \text{(wrong!)}, \quad (1.57)$$

$$\sum_{i=0}^{n} \varepsilon \left[\frac{m}{2} \left(\frac{x_{i+1} - x_i}{\varepsilon} \right)^2 - V(x_i) \right] \xrightarrow[n \to \infty]{} \int_0^t d\tilde{t} \left[\frac{m}{2} \left(\frac{dx}{d\tilde{t}} \right)^2 - V(x(\tilde{t})) \right] = S[x(\tilde{t})] \quad (1.58)$$

$$\mathscr{D}x(\tau) = \lim_{n \to \infty} \mathscr{N}(n) \prod_{i=1}^{n} dx(t_i), \quad \mathscr{N}(n) = \left(\frac{m}{2\pi i \varepsilon \hbar} \right)^{\frac{n+1}{2}}. \quad (1.59)$$

With these identifications the rhs of (1.42) will be precisely the rhs of (1.54) which has a well-defined limit. However, a number of points should be made clear. First, as explicit written in eq. (1.57), it is somewhat misleading to make the identification of the lhs and the rhs of eq. (1.57), even if it is, of course, correct for a piecewise linear path like the one shown in Fig. 1.4. The reason is that when $\varepsilon \to 0$ the numerical value of the derivative goes to infinity as we will discuss shortly (see eq. (1.68)). Thus (1.58) does not really make much sense for a typical path of the kind shown in Fig. 1.4. Rather it should be used moving from the rhs to the lhs when one wants an assign a meaning to the rhs of (1.42). Nevertheless, the formal use of the classical action in (1.42) is extremely useful in a number of formal manipulations one can make with the path integral as we will also see later. Finally, one could get the idea from the notation (1.59) that $\mathscr{D}x(\tau)$ is (apart from the factor i in $\mathscr{N}(n)$) a kind of generalization of the finite-dimensional Lebesgue measure. However, this is not the case. There exists no Lebesgue measure on $\mathbb{R}^\infty$. Below, we will discuss what kind of measure one can associate with the path integral.

1.5 THE FEYNMAN-KAC PATH INTEGRAL AND IMAGINARY TIME

Let us make a rotation to imaginary (so-called Euclidean) time τ:

$$t \to -i\tau, \qquad e^{-i\hat{H}t/\hbar} \to e^{-\hat{H}\tau/\hbar} \qquad (1.60)$$

As before one proves that

$$\boxed{\langle x_s|e^{-\hat{H}\tau/\hbar}|x_0\rangle = \int_{\substack{x(0)=x_0 \\ x(\tau)=x_s}} \mathscr{D}x(\tilde{\tau})\, e^{-\frac{1}{\hbar}S_E[x(\tilde{\tau})]}} \qquad (1.61)$$

$$S_E[x(\tilde{\tau})] = \int_0^\tau d\tilde{\tau}\left[\frac{m}{2}\left(\frac{dx}{d\tilde{\tau}}\right)^2 + V(x(\tilde{\tau}))\right] \qquad (1.62)$$

which is called the Feynman-Kac formula and where the rhs of (1.61) should be understood as

$$\lim_{n\to\infty}\left(\frac{m}{2\pi\varepsilon\hbar}\right)^{\frac{n+1}{2}}\int\prod_{i=1}^n dx_i\,\exp\left(-\frac{1}{\hbar}\sum_{i=0}^n \varepsilon\left[\frac{m}{2}\left(\frac{x_{i+1}-x_i}{\varepsilon}\right)^2 + V(x_i)\right]\right) \qquad (1.63)$$

If we choose $\tau = \beta\hbar$ in (1.61) and $x_s = x_0 = x$ and integrate wrt x we obtain a path integral representation of the quantum partition function (1.35):

$$\boxed{Z(\beta) = \int dx\,\langle x|e^{-\beta\hat{H}}|x\rangle = \int dx\int_{\substack{x(0)=x \\ x(\beta)=x}} \mathscr{D}x(\tilde{\tau})\, e^{-\frac{1}{\hbar}S_E[x(\tilde{\tau})]}.} \qquad (1.64)$$

This formula is also denoted the Feynman-Kac formula and we see that one, loosely speaking, obtains the partition function by integrating over all paths, which are periodic with period β.

Note the difference of sign for the potential in (1.62) compared to (1.4). This can formally be understood by making the replacement $t \to -i\tau$ in the classical action (1.4), whereby

$$iS = i\int d\tilde{t}\left[\frac{m}{2}\left(\frac{dx}{d\tilde{t}}\right)^2 - V(x(\tilde{t}))\right] \to -\int d\tilde{\tau}\left[\frac{m}{2}\left(\frac{dx}{d\tilde{\tau}}\right)^2 + V(x(\tilde{\tau}))\right] = -S_E. \qquad (1.65)$$

However, this is the wrong way to think about it. In general a curve $x(t)$ will have no analytic continuation when $t \to -i\tau$. As mentioned one arrives at $S_E[x]$ by following the same steps as before, but since there is no "i" in front of the Hamiltonian one obtains instead of (1.51)

$$\langle x|\hat{O}_\varepsilon^{(E)}|y\rangle = \langle x|e^{-\frac{\varepsilon\hat{p}^2}{2m\hbar}}e^{-\frac{\varepsilon\hat{V}(\hat{x})}{\hbar}}|y\rangle = \left(\frac{m}{2\pi\varepsilon\hbar}\right)^{\frac{1}{2}}e^{-\frac{(x-y)^2 m}{2\varepsilon\hbar}}e^{-\frac{\varepsilon V(y)}{\hbar}}. \qquad (1.66)$$

Preliminary Material Part 1: The Path Integral

which explains the change of sign. $S_E[x]$ *is called the Euclidean action* because the replacement $t \to -i\tau$, when viewed in a Minkowskian spacetime formally corresponds to changing the signature of the metric to that of Euclidean space:

$$ds_M^2 = dx^2 - c^2 dt^2 \to ds_E^2 = dx^2 + c^2 d\tau^2. \tag{1.67}$$

Let us finally return to (1.57) and explain why it is wrong and what kind of "paths" we are "integrating" over in the path integral. In this discussion, it is more convenient to use the Euclidean version (1.61)–(1.63) since one can then actually talk about a measure on a suitable set of path. When we look at the integrand in (1.63) it is clear that the dominant terms in the limit $\varepsilon \to 0$ will be the kinetic terms proportional $(x_{i+1} - x_i)^2/\varepsilon$, and they will kill any contribution to the integrals unless these terms are $\mathcal{O}(1)$. On the other hand, they are not really suppressed any further. Thus we typically expect for $\varepsilon \to 0$:

$$\frac{|x_{i+1} - x_i|}{\varepsilon} \approx \frac{1}{\sqrt{\varepsilon}} \sqrt{\frac{2\hbar}{m}}, \quad \text{i.e.} \quad \frac{|x_{i+1} - x_i|}{\varepsilon} \to \infty. \tag{1.68}$$

The picture of the piecewise linear path shown in Fig. 1.4 is, therefore, somewhat misleading. In the limit $\varepsilon \to 0$ the derivative of the curve diverges everywhere, and it should be viewed as a continuous curve that is nowhere differentiable. We can also estimate the length of such a curve when $\varepsilon \to 0$ (recall $\varepsilon = \tau/(n+1)$):

$$\ell(x(\tilde{\tau})) = \sum_{i=0}^{n} |x_{i+1} - x_i| \approx \sqrt{\frac{2\hbar}{m}} n \sqrt{\varepsilon} \approx \sqrt{\frac{2\hbar\tau}{m}} \sqrt{n}. \tag{1.69}$$

The fact that the length $\ell(x(\tilde{\tau})) \propto \sqrt{n}$ for large n, rather than going to a constant (the length ℓ of the given nice continuous curve $x(\tilde{\tau})$) signifies that the curve $x(\tilde{\tau})$ is *not* "nice", but actually *fractal* with a so-called *Hausdorff dimension* equal to 2 (we will discuss this in detail later).

Finally, does it make mathematical sense to view (1.61)–(1.63) as an integration over continuous path from x_0 to x_s? The answer is yes. We will not go into any detail but just mention a few things. We assume the parameter range $[0, \tau]$ is the same for all curves $x(\tilde{\tau})$. One can now define the distance between curves $x_1(\tilde{\tau})$ and $x_2(\tilde{\tau})$ as

$$d(x_1, x_2) = \sup_{\tilde{\tau} \in [0,\tau]} |x_1(\tilde{\tau}) - x_2(\tilde{\tau})|. \tag{1.70}$$

The existence of this distance turns the space of continuous curves from x_0 to x_s into a metric space on which one can define a measure, which again allows us to define the integration of functions of curves. It turns out that, loosely speaking, this integration measure is just our $\mathcal{D}x(\tilde{\tau})$ multiplied by the action of a free particle. This integration measure is denoted the *Wiener measure on the set of continuous paths*. Putting $\hbar = 1$ and $m = 1$, we can write

$$\mathcal{D}\mu(x(\tilde{\tau})) = \mathcal{D}x(\tilde{\tau}) e^{-\frac{1}{2} \int d\tilde{\tau} \left(\frac{dx}{d\tilde{\tau}}\right)^2} = \lim_{n \to \infty} \frac{1}{\sqrt{2\pi\varepsilon}} \prod_{i=1}^{n} \frac{dx_i}{\sqrt{2\pi\varepsilon}} e^{-\frac{1}{2} \sum_{i=0}^{n} \varepsilon \left(\frac{x_{i+1} - x_i}{\varepsilon}\right)^2} \tag{1.71}$$

Contrary to $\mathscr{D}x(\tilde{\tau})$, $\mathscr{D}\mu(x(\tilde{\tau}))$ can be shown to be well defined, and one can now integrate functions defined on the set of continuous curves (what we usually call functionals, since the arguments of such a function are itself a function (the curve)). Let $F[x(\tilde{\tau})]$ be such a function (functional). The integral of $F[x]$ can now formally be written as

$$\int \mathscr{D}\mu(x(\tilde{\tau})) \, F[x(\tilde{\tau})], \tag{1.72}$$

and the precise meaning is obtained by sub-dividing the $\tilde{\tau}$ parameter range $[0, \tau]$ by n points $\tau_i = i\varepsilon$, as in (1.71), also for the functional $F[x(\tilde{\tau})]$. In particular the Feynman-Kac path integral (1.61) is now the integral (1.72) with the functional

$$F[x(\tilde{\tau})] = e^{-\int d\tilde{\tau} V(x(\tilde{\tau}))} = e^{-\varepsilon \sum_{i=0}^{n} V(x_i)}. \tag{1.73}$$

The Wiener measure acts as a probability measure on the set of continuous functions and one can then ask interesting questions like: what is the probability that a continuous function is differentiable in a single point? Maybe not surprising from our discussion above the probability is zero! Nevertheless, the set of C^∞ functions is dense in the set of continuous functions (much in the same way as the set of rational numbers have measure zero but still is dense in the set of real numbers).

There are not many potentials $V(x)$ where one can calculate the path integral (1.54) for finite n and then take the limit $n \to \infty$ and in this way obtain $\langle x_s | e^{-i\hat{H}t/\hbar} | x_0 \rangle$. One is $V(x) = 0$, i.e. the potential for the free particle, where the corresponding path integral is discussed in Problem Set 1. Another one is $V(x) = m\omega x^2/2$, i.e. the harmonic oscillator potential, where the path integral is calculated in Problem Set 2. These path integrals can be performed because they only involve Gaussian integrations. From a calculational point of view, it is in general easier to solve the Schrödinger equation directly. However, there will be another class of path integrals that involve geometries, where the action $S[x]$ is "geometric" and where we will be able to perform the path integral simply by counting geometries. These are the path integrals and we will discuss in the following chapters.

1.6 PROBLEM SETS AND FURTHER READING

As already mentioned, Problem Sets 1 and 2 address the calculation of simple path integrals. Also in Problem Set 2, the concept of a *generating function* associated with a sequence of numbers is introduced. Such generating functions will be used again and again in the following Chapters.

A more mathematical rigorous definition of the path integral, tailored to the use here in this book, can be found in [1]. If one, in addition, is interested in how one can use the path integral in all areas of physics (and other areas like economics), one can consult the book [2]. A book that focuses on the path integral applications in quantum mechanics is [3].

2 The Free Relativistic Particle

2.1 THE PROPAGATOR

We will now discuss how the Green function for the free *relativistic* particle, via the path integral, can be described as a scaling limit of a statistical ensemble of paths and we will encounter the first example of universality of the scaling limit of geometries. In the following we will use units where $\hbar = c = 1$. These constants will then be left out of equations, which will simplify the notation. They can of course be reinserted at any point if needed.

In Chapter 1, we discussed the Schrödinger equation for a non-relativistic particle, its rotation to "Euclidean" time (the heat- or diffusion-equation), as well as the corresponding Green functions, represented via path integrals. Let us just recapitulate, now writing the formulas in d space dimensions. The Schrödinger equation reads:

$$\left(i\frac{\partial}{\partial t} + \frac{1}{2m}\frac{\partial^2}{\partial x_i^2}\right)\psi(x,t) = 0, \tag{2.1}$$

where i can take values from 1 to d, and where a summation over index i is understood in eq. (2.1). Rotating to Euclidean time $t \to -i\tau$ leads to the diffusion equation:

$$\left(-\frac{\partial}{\partial \tau} + \frac{1}{2m}\frac{\partial^2}{\partial x_i^2}\right)\psi(x,\tau) = 0. \tag{2.2}$$

The solution to (2.2) which is zero for $\tau < \tau_0$ and starts out as

$$\psi(x,\tau_0) = \delta^d(x-x_0) \quad \text{(i.e. } |\psi(\tau_0)\rangle = |x_0\rangle\text{)} \tag{2.3}$$

can be written as

$$G(x-x_0, \tau-\tau_0) = \langle x|e^{-(\tau-\tau_0)\hat{H}}|x_0\rangle, \quad \hat{H} = -\frac{1}{2m}\frac{\partial^2}{\partial x_i^2}. \tag{2.4}$$

$G(x-x_0, \tau-\tau_1)$ is the Green function of the differential operator $\frac{\partial}{\partial \tau} - \frac{1}{2m}\frac{\partial^2}{\partial x_i^2}$:

$$\left(\frac{\partial}{\partial \tau} - \frac{1}{2m}\frac{\partial^2}{\partial x_i^2}\right)G(x-x_0, \tau-\tau_1) = \delta(\tau-\tau_0)\delta^d(x-x_0) \tag{2.5}$$

We want to generalize from a non-relativistic particle to a relativistic particle:

$$\left(-i\frac{\partial}{\partial t} - \frac{1}{2m}\frac{\partial^2}{\partial x_i^2}\right)\psi(x,t) = 0 \quad \to \quad \left(\frac{\partial^2}{\partial t^2} - \frac{\partial^2}{\partial x_i^2} + m^2\right)\phi(x,t) = 0. \tag{2.6}$$

We perform again the analytic rotation to Euclidean "time" τ:

$$t = -i\tau \equiv -ix_D, \quad D = d+1, \quad ds^2 = dx_i^2 - dt^2 \to dx_i^2 + dx_D^2 \equiv dx_i^2. \quad (2.7)$$

where we, with an abuse of notation, have included the last index D in the sum dx_i^2. Thus

$$\left(-\frac{\partial^2}{\partial x_i^2} + m^2\right)\phi(x) = 0, \quad D = d+1. \quad (2.8)$$

Again the "propagation" of the "Euclidean particle" is described by a Green function

$$\left(-\frac{\partial^2}{\partial x_i^2} + m^2\right) G(x-y) = \delta^D(x-y). \quad (2.9)$$

Introducing the Fourier transformed $\hat{G}$ by

$$G(x-y) = \int \frac{d^D k}{(2\pi)^D} e^{ik_i x_i} \hat{G}(k_i), \quad (2.10)$$

eq. (2.9) can be written as

$$(k_i^2 + m^2)\hat{G}(k_i) = 1 \quad \text{i.e.} \quad \boxed{\hat{G}(k) = \frac{1}{k_i^2 + m^2}}. \quad (2.11)$$

From (2.10) one now obtains

$$\boxed{G(x-y) = \frac{1}{(2\pi)^{\frac{D}{2}}} \left(\frac{m}{|x-y|}\right)^{\frac{D}{2}-1} K_{\frac{D}{2}-1}(m|x-y|)} \quad (2.12)$$

where $K_\nu(x)$ denotes the second modified Bessel function with index ν. The asymptotic behaviors of $G(x-y)$ are:

$$G(x-y) \approx \frac{\Gamma(\frac{D}{2}-1)}{4\pi^{\frac{D}{2}}} \frac{1}{|x-y|^{D-2}}, \quad \text{for} \quad m|x-y| \ll 1 \quad (2.13)$$

$$G(x-y) \approx \frac{1}{(2\pi)^{\frac{D}{2}}} \sqrt{\frac{\pi}{2}} \frac{m^{\frac{D}{2}-\frac{3}{2}}}{|x-y|^{\frac{D}{2}-\frac{1}{2}}} e^{-m|x-y|} \quad \text{for} \quad m|x-y| \gg 1. \quad (2.14)$$

As already mentioned when we discussed the spin-spin correlator on a lattice, this is a generic behavior for our correlators or Green functions: *a power like behavior for small distances and an exponential fall off (with power corrections) for large distances*, measured relative to a parameter, here the mass, which defines an "intrinsic" scale of the physical "system". Here we are discussing a free particle, but as we will see the correlator or Green function or propagator (many names for the same object!) will be described by a statistical ensemble of path, and it will be in this "system" that we will define the scale.

The Free Relativistic Particle

There are infinitely many Green functions for a given differential equation and we fix this ambiguity by imposing appropriate boundary conditions. Here the (Euclidean) boundary condition is:

$$G(x) \to 0 \quad \text{for} \quad |x| \to \infty, \quad (D > 2) \tag{2.15}$$

As discussed in Appendix A *Preliminary Material Part 2: Green Functions*, one obtains by analytic continuation $x_D \to it$ the so-called Feynman Green function in Minkowski spacetime. From now on we will stay in Euclidean spacetimes. As long as we deal with quantum field theories in flat spacetimes this procedure is well understood and well defined. However, its status is less clear if we consider quantum field theories in curved spacetime, partly because the concept of an analytic continuation between geometries with Euclidean signatures and geometries with Lorentzian signatures is not well understood or even always well defined. And the status of such a rotation becomes even less clear when we start discussing systems where the geometry itself is the object of quantization: has a quantum theory of geometries with Euclidean signatures any relation to a quantum theory of geometries with Lorentzian signature? It is a very interesting question and the full answer to this question is presently unknown. Here we will perform all calculations using geometries with Euclidean signature and we will not discuss the connection to a similar theory of geometries with Lorentzian signatures.

2.2 THE PATH INTEGRAL

We now want to reproduce the Euclidean Green function (2.11) or (2.12) from a path integral, using a beautiful *geometric* action for the classical free particle:

$$S[P(x,y)] = m_0 \, \ell[P(x,y)]. \tag{2.16}$$

In this formula x and y denote *spacetime* points in $\mathbb{R}^D$. After rotation to Euclidean spacetime we do no longer work with a separate time coordinate. $P(x,y)$ denotes a *geometric* path from y to x (see Fig. 2.1). Let us choose a parametrization of the path:

$$P(x,y) : \xi \to x(\xi), \quad \xi \in [0,1], \quad x(0) = y, \quad x(1) = x. \tag{2.17}$$

Having chosen such a parametrization we can calculate the length of the path:

$$\boxed{S_E[P(x,y)] = m_0 \ell[P(x,y)] = m_0 \int_0^1 d\xi \sqrt{\left(\frac{dx_i}{d\xi}\right)^2},} \tag{2.18}$$

and we can find the eom:

$$\frac{\delta S}{\delta x_i(\xi)} = -\frac{d}{d\xi}\left[\frac{dx_i}{d\xi} \bigg/ \sqrt{\left(\frac{dx_i}{d\xi}\right)^2}\right] = 0 \tag{2.19}$$

Clearly we can find the minimum of $S[P(x,y)]$ without solving (2.19), since the minimal length of a path from y to x is just $|x-y|$. This result is clearly independent of the

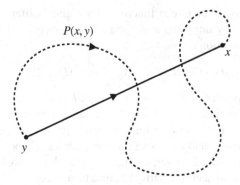

Figure 2.1 A path $P(x,y)$ (the dashed curve) between spacetime point y and spacetime point x. In addition the shortest path between y and x is shown.

chosen parametrization of the paths, in agreement with the fact that the action (2.18) is *reparametrization invariant*: for any twice differentiable function $f : \xi \to f(\xi)$ where $f(0)=0$, $f(1)=1$ and $f'(\xi) > 0$, the action will be unchanged if we replace ξ by $f(\xi)$. In particular, if $x_{cl}(\xi)$ is a solution to (2.19), then $x_{cl}(f(\xi))$ will also be a solution to (2.19), and they will both represent geometrically the straight line from y to x, just with different parametrizations. If we use the action (2.18) in the path integral, it is then natural that we sum only over geometric paths. We thus write

Free non-relativistic particle:

$$G(x-y, \tau_1-\tau_0) = \int_{\substack{x(\tau_0)=y \\ x(\tau_1)=x}} \mathcal{D}x(\tilde{\tau})\, e^{-S_E[x(\tilde{\tau})]} = \int_{\substack{x(\tau_0)=y \\ x(\tau_1)=x}} \mathcal{D}x(\tilde{\tau})\, e^{-\frac{m}{2}\int_{\tau_0}^{\tau_1} d\tilde{\tau}\, (\frac{dx}{d\tilde{\tau}})^2} \quad (2.20)$$

Free relativistic particle:

$$\boxed{G(x-y) = \int \mathcal{D}P(x,y)\, e^{-S_E[P(x,y)]} = \int \mathcal{D}P(x,y)\, e^{-m_0 \ell[P(x,y)]}.} \quad (2.21)$$

Note the difference between the two expressions. In (2.20) x refers to a spatial point, i.e. it has coordinates x_i, $i = 1,\ldots,d$ and the (Euclidean) time τ appears as the parameter in the curve $x(\tau)$ from $x(\tau_0)$ to $x(\tau_1)$. However, the "parameter" τ has the physical meaning of (Euclidean) time. Thus, if we made a reparametrization $\tau' = f(\tau)$, $f(\tau_0) = \tau_0$ and $f(\tau_1) = \tau_1$, the interpretation of $x(\tau') = x(f(\tau))$ would be that it represents a different path $x'(\tau)$, $x' = x \circ f$ of the particle, and both the path $x(\tau)$ and $x'(\tau)$ are indeed included in the path integral (2.20). In (2.21) the time-coordinate x_D, $D = d+1$ is treated on equal footing with the spatial coordinates. To specify a curve from y to x we might have to introduce an artificial parameter ξ and write $x(\xi)$. This can clearly be done in many ways, but our summation is independent of how we choose such a parametrization since we sum only over geometric paths.

The Free Relativistic Particle

We can label the paths $P(x,y)$ from y to x according to their lengths $\ell[P(x,y)]$. Since all paths of the same length have the same action we can formally split the path integral in an integration over paths of a given length ℓ followed by an integration over ℓ:

$$G(x-y) = \int_0^\infty d\ell\, e^{-m_0 \ell} \int_{\ell[P(x,y)]=\ell} \mathscr{D}P(x,y) \cdot 1 = \int_0^\infty d\ell\, e^{-m_0 \ell}\, \mathcal{N}_{x,y}(\ell), \qquad (2.22)$$

where $\mathcal{N}_{x,y}(\ell)$ denotes the number of paths of length ℓ between x and y. Thus *the propagator of the free particle is entirely determined by the entropy of paths.* We will encounter the same when we study the path integral of higher dimensional geometries than the paths. The "propagators" we can define for such ensembles will be entirely determined by the entropy, i.e. the number of such geometries, and the amazing conclusion is that we can quantize geometries, i.e. gravity, and calculate propagators, if we can only count geometries.

Of course $\mathcal{N}_{x,y}(\ell) = \infty$.

In order to make a meaningful counting, we have to introduce a cut-off in the same way as we introduced a discretization of the time interval $[\tau_0, \tau_1]$ in pieces of length ε for the non-relativistic particle, and considered piecewise linear paths in these time intervals. In the case of our (Euclidean) relativistic particle, we have a formulation where time has no special role. It is thus natural instead to consider paths in $\mathbb{R}^D$ which are piecewise linear and where the *length* of the individual linear pieces in $\mathbb{R}^D$ is a (we use a rather than ε when referring to distances in $\mathbb{R}^D$). In this way our "cut-off" a will be independent of a possible chosen parametrization of the path. Let us now consider a path P_n from y to x which consists of n pieces. We then have

$$\ell[P_n] = n \cdot a, \qquad S[P_n] = m_0 \ell[P_n] = m_0 a n, \qquad (2.23)$$

and the propagator (2.22), calculated by summing over all such piecewise linear paths, is:

$$\begin{aligned} G_a(x-y) &= \sum_{n=1}^\infty e^{-m_0 a n} \int_{\{P_n\}} \mathscr{D}P_n \cdot 1 & (2.24) \\ &= \sum_{n=1}^\infty e^{-m_0 a n} \int \prod_{j=1}^n d\hat{e}(j)\, \delta\!\left(a\sum_{j=1}^n \hat{e}(j) - (x-y)\right), & (2.25) \end{aligned}$$

where $\hat{e}(j)$ denotes a unit vector along the jth linear segment of a path P_n consisting of n such segments (see Fig. 2.2).

Note that we have to allow for all n in order not to limit the length of the paths. This is in contrast to our regularization used for the non-relativistic particle, where n was linked to ε and we for a given n could have paths of arbitrary length. The propagator now depends on the cut-off a and we want to show how one can obtain the continuum propagator in the limit $a \to 0$.

$$\hat{G}_a(p) = \int d^D x\, e^{-i p_i(x_i - y_i)}\, G_a(x-y) = \sum_{n=1}^\infty e^{-m_0 a n} \int \prod_{j=1}^n d\hat{e}(j)\, e^{-i a p_i \hat{e}_i(j)}, \qquad (2.26)$$

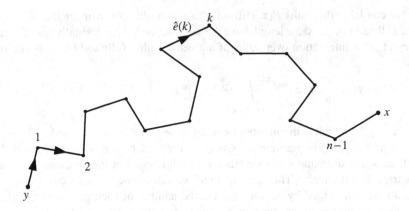

Figure 2.2 A piecewise linear curve from spacetime point y to spacetime point x, consisting of n pieces of length a. Unit vectors $\hat{e}(k)$, $k = 1,\ldots n$, are attached to each of the linear pieces, being tangent to the curve along the given piece.

and the integration over the unit vectors $\hat{e}$ can be performed[1]:

$$\int d\hat{e}\, e^{-iap_i\hat{e}_i} = f(|p|a) = f(0)\left(1 - \frac{1}{2}\sigma^2 a^2 p^2 + \mathcal{O}(a^3)\right). \quad (2.27)$$

The only important property of f is that there is no linear term in a and that the a^2-coefficient is negative, statements which follow trivially by expanding the exponential in the integrand in powers of a. Thus we obtain from (2.26)

$$\hat{G}_a(p) = \sum_{n=1}^{\infty}\left(e^{-m_0 a}f(|p|a)\right)^n = \frac{e^{-m_0 a}f(|p|a)}{1 - e^{-m_0 a}f(|p|a)}. \quad (2.28)$$

We are interested in the limit $a \to 0$. If we can arrange

$$e^{-m_0 a}f(|p|a) \to 1 - \frac{1}{2}a^2\sigma^2(p^2 + m_{ph}^2) + \mathcal{O}(a^3) \quad (2.29)$$

then we obtain

$$\boxed{\hat{G}_a(p) \to \frac{2}{a^2\sigma^2}\frac{1}{p^2 + m_{ph}^2} = \frac{2}{a^2\sigma^2}G_{cont}(p)}, \quad (2.30)$$

which shows that except for a divergent factor in front (which has a clear interpretation, as we will discuss later), we obtain precisely the desired continuum result. Is it possible to arrange (2.29)? Yes, by treating m_0 as an adjustable parameter not directly related to the *physical* mass of our particle, which we now denote m_{ph}:

$$m_0 = \frac{\ln(f(0))}{a} + \frac{1}{2}a\sigma^2 m_{ph}^2. \quad (2.31)$$

---

[1] The integral is equal $2\pi^{\frac{D}{2}}J_{\frac{D-1}{2}}(|p|a)/(|p|a/2)^{\frac{D-1}{2}}$, where $J_\nu(x)$ is the Bessel function with index ν.

The Free Relativistic Particle

Note that m_0 actually never appeared in the classical eom, so it seems not disastrous to change it, and (2.31) is the simplest example of *mass renormalization* in quantum field theory.

Is this result accidental? At first, a procedure like the one outlined above might seem rather arbitrary. Note however that the result did not depend on the detailed form of the function f, and we will now show that the *scaling limit* (2.28)–(2.30) can indeed be viewed as natural, and it will be our first example of *universality*.

2.3 RANDOM WALKS AND UNIVERSALITY

Let us consider a simple model of so-called random walks (RW) in $\mathbb{R}^D$. The nth step in the walk is characterized by the initial position $x(n-1)$ and a probability distribution $\mathscr{P}(v) = \mathscr{P}(|v|)$ such that $x(n) = x(n-1)+v$, where v is selected to be between v and $v+dv$ with probability $\mathscr{P}(v)d^D v$. Such a stochastic process is called Markovian, meaning that the step is independent of any $x(k)$, $k < n-1$. Further, let $1-e^{-\mu}$ be the probability that the process stops at $x(n)$ and $e^{-\mu}$ the probability that it continues. The probability that the process will bring us from y to x and then stop is

$$\mathscr{G}(x-y) = (1-e^{-\mu}) \sum_{n=0}^{\infty} e^{-\mu n} \int \prod_{i=1}^{n} d^D x(i) \prod_{k=0}^{n} \mathscr{P}(x(k+1))-x(k)) \quad (2.32)$$

where $x(n+1)=x$ and $x(0)=y$. In discussions related to the path integral it is often convenient to work with "unnormalized" probabilities $P(v)$, i.e. we have

$$P(v) = e^{\mu_c} \mathscr{P}(v), \quad \int d^D v\, P(v) = e^{\mu_c}. \quad (2.33)$$

and we then write

$$\boxed{G(x-y) = \sum_{n=0}^{\infty} e^{-\mu n} \int \prod_{i=1}^{n} d^D x(i) \prod_{k=0}^{n} P(x(k+1))-x(k))} \quad (2.34)$$

In the same way as $\mu \geq 0$ in order for (2.32) to make sense, $\mu \geq \mu_c$ in order for (2.34) to make sense. Now $G(x-y)$ is no longer a normalized probability because P is no longer normalized and because we have chosen to drop the factor $1-e^{-\mu}$ in (2.32), but we trivially get back to $\mathscr{G}(x-y)$ with μ replaced by $\mu-\mu_c$ by dividing $G(x-y)$ by

$$\chi(\mu) = \int d^D x\, G(x-y) = e^{\mu_c} \sum_{n=0}^{\infty} (e^{\mu_c} e^{-\mu})^n = \frac{e^{\mu_c}}{1-e^{-(\mu-\mu_c)}}. \quad (2.35)$$

We denote $\chi(\mu)$ *the susceptibility* because of the obvious analogy with (1.19). Note that

$$\chi(\mu) \propto \frac{1}{\mu-\mu_c} \quad \text{for} \quad \mu \to \mu_c. \quad (2.36)$$

This singular behavior of the susceptibility will play an important role later.

Let us analyze the Fourier transform of $G(x-y)$. We denote the Fourier transform of $\mathscr{P}(v)$, the so-called characteristic function of the probability distribution, by $\hat{\mathscr{P}}(k)$. Since the Fourier transforms of convolutions of functions are the product of the Fourier transforms of the functions (see Problem Set 1 for discussions of this), we have

$$\hat{\mathscr{P}}(k) = \int d^D x\, e^{-ik_j x_j}\, \mathscr{P}(x), \qquad \hat{G}(k) = \int d^D x\, e^{-ik_j x_j}\, G(x), \qquad (2.37)$$

and

$$\hat{G}(k) = e^{\mu_c}\, \hat{\mathscr{P}}(k) \sum_{n=0}^{\infty} \left(e^{-(\mu-\mu_c)}\, \hat{\mathscr{P}}(k) \right)^n = \frac{e^{\mu_c}\, \hat{\mathscr{P}}(k)}{1 - e^{-(\mu-\mu_c)}\, \hat{\mathscr{P}}(k)}. \qquad (2.38)$$

From the assumption $\mathscr{P}(x) = \mathscr{P}(|x|)$ and the assumption that the second and fourth moment of the probability distribution exists, the characteristic function has the expansion

$$\hat{\mathscr{P}}(k) = 1 - \frac{1}{2}\sigma^2 k^2 + \mathcal{O}(|k|^4), \qquad (2.39)$$

a result which is trivial if we are allowed to expand $e^{-ik_j x_j}$ in powers of k in (2.37).

In our RW model, we have viewed all variables as dimensionless. Let us now introduce a *scaling parameter* a with the dimension of length. Then we obtain precisely the results (2.28)–(2.31) when making the following identifications

$$\mu = a m_0, \qquad k = a p, \qquad e^{\mu_c}\, \hat{\mathscr{P}}(k) = f(k). \qquad (2.40)$$

In fact our earlier results corresponded to the choice where the general $P(v)$ used here was chosen to be proportional to $\delta(|v|-1)$.

Can we give a simple interpretation of the specific scaling we are using to obtain the propagator of a free particle:

$$k = a p, \qquad \mu - \mu_c = \frac{1}{2}\sigma^2 a^2 m_{ph}^2. \qquad (2.41)$$

Yes, it corresponds precisely to the scaling dictated by the central limit theorem for probability distributions. For our purpose we can state a simple version of the central limit theorem as follows: under the assumption (2.39), we have

$$\hat{\mathscr{P}}(k/\sqrt{n})^n \underset{n \to \infty}{\to} e^{-\frac{1}{2}\sigma^2 k^2}, \qquad (2.42)$$

which follows from (2.39) by using the formula $e^z = \lim_{n \to \infty}(1+z/n)^n$. The rhs of (2.42) is just the characteristic function of a Gaussian distribution with variance σ (see again Problem Set 1). Now apply this to the first equality in (2.38) and assume that

The Free Relativistic Particle

we can substitute (2.42) for all terms[2], not only for large n:

$$\hat{G}(k) = e^{\mu_c} \mathscr{P}(k) \sum_{n=0}^{\infty} e^{-(\mu-\mu_c)n + \frac{1}{2}\sigma^2 k^2} = \frac{e^{\mu_c}(1+\mathscr{O}(a^2))}{\sigma^2 a^2} \sum_{n=0}^{\infty} \sigma^2 a^2 \, e^{-\frac{1}{2}\sigma^2 a^2 n(m_{ph}^2 + p^2)} \tag{2.43}$$

where we have made the substitution (2.41). In the limit $a \to 0$ the sum is converted to an integral

$$\frac{1}{2} \sum_{n=0}^{\infty} \sigma^2 a^2 e^{-\frac{1}{2}\sigma^2 a^2 n(m_{ph}^2 + p^2)} \approx \frac{1}{2} \int_0^{\infty} ds \, e^{-\frac{1}{2}s(m_{ph}^2 + p^2)} = \frac{1}{m_{ph}^2 + p^2}, \quad s = \sigma^2 a^2 n. \tag{2.44}$$

Thus, since s is a finite continuum variable we can view $a^2 \sim 1/n$ and the scaling (2.41) is basically the same as the scaling in the central limit theorem (2.42), and the reason that we obtain a universal result (2.43) can be traced back to the central limit theorem. Finally, we remark that the reason we have a divergent factor $1/a^2$ in front of the rhs of eq. (2.43) can be traced to the use of an unrenormalized $G(x-y)$ in (2.34) where we have dropped a factor $(1-e^{-(\mu-\mu_c)})$ compared to $\mathscr{G}(x-y)$ in (2.32). This normalization factor is precisely the susceptibility $\chi(\mu)$, as mentioned above, and we see from (2.36) that we indeed have $\chi(\mu) \propto 1/a^2$ when $a \to 0$.

Since the central limit theorem is valid under much more general conditions than used here, one can also obtain the free particle propagator in more general settings than the one discussed above. As an example we mention here (and discuss it in detail in Problem Set 3) that one can obtain the continuum propagator from RWs on a hypercubic lattice, when the cut-off a, in this case the length of the lattice links, is taken to zero. We have illustrated such a RW on a lattice in Fig. 2.3.

2.4 PROBLEMSETS AND FURTHER READING

Problem Set 3 discusses the lattice propagator and its continuum limit. Additional aspects can be found in [1], where there also is a discussion of RWs which lead to other type of propagators. In particular it is shown how the introduction of extrinsic curvature terms in the geometric action for the particle may lead to modified, higher derivative propagators (see also the original article [4]). In [1] it is in addition shown how it is possible also to describe the *fermionic propagator* in terms of a certain kind of RWs with a rotation matrix associated to the tangent vectors of the RW. In this way the contributions from different RWs can cancel each other even in Euclidean spacetime and these cancellations will change the Hausdorff dimension of ensemble of fermionic RWs (see the original article [4] for further deails).

---

[2] It is clear that one can start the summation in (2.38) at any finite n_0 rather than at 0, and obtain the same result in the scaling limit, the reason being that the propagator diverges in that limit. All terms up to n_0 will be finite, and thus not contribute to the scaling limit result. Of course this does not prove that one can simply replace $\mathscr{P}(k)$ by a pure Gaussian distribution, but it is a strong hint.

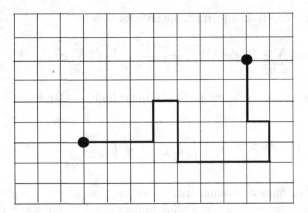

Figure 2.3 A two-dimensional cubic lattice and a random walk between two lattice points, following a connected link path between the two points.

3 One-Dimensional Quantum Gravity

3.1 SCALAR FIELDS IN ONE DIMENSION

If the RW probability distribution used in Chapter 2 is Gaussian the convolution is "exact": the convolution of two Gaussian distributions is again Gaussian (for the obvious reason that the characteristic functions are Gaussian and the product of two Gaussians again is a Gaussian, see detailed discussion in Problem Set 1) and we obtain:

$$\int \prod_{k=1}^{n-1} dx(k)\, \mathcal{P}\big(x(k)-x(k-1)\big) = \frac{e^{-\frac{|x(n)-x(0)|^2}{2\sigma^2 n}}}{(2\pi\sigma^2 n)^{\frac{1}{2}}}, \qquad \mathcal{P}(x) = \frac{e^{-\frac{|x|^2}{2\sigma^2}}}{(2\pi\sigma^2)^{\frac{1}{2}}} \qquad (3.1)$$

Somewhat surprisingly there exists an action which is Gaussian, but still *geometric* and still has the same classical eom as the geometric action $S[P(x,y)] = m_0 \ell[P(x,y)]$, where $P(x,y)$ denotes a path from y to x.

In order to describe this action we first make a little digression and describe a few aspects of *Riemannian geometry*. Consider a curved spacetime $\mathcal{M}$ of dimension M. Let $\xi = (\xi^1, \ldots, \xi^M)$ be a coordinate system on (a part of) $\mathcal{M}$. Distances on $\mathcal{M}$ are independent of the chosen coordinate system and described by a symmetric tensor $g_{ab}(\xi)$, $a,b = 1,\ldots, M$, the so-called metric tensor. $g_{ab}(\xi)$ depends on the coordinate system. Let ξ' be another coordinate system and $g'_{ab}(\xi')$ the corrsponding metric tensor. The invariant distance ds between between two points with coordinates ξ and $\xi + d\xi$ is then

$$ds^2 = g_{ab}(\xi) d\xi^a d\xi^b = g'_{a'b'}(\xi') d\xi'^{a'} d\xi'^{b'}, \qquad (3.2)$$

where the relation between $g_{ab}(\xi)$ and $g'_{ab}(\xi')$ is

$$g'_{a'b'}(\xi') = g_{cd}(\xi) \frac{\partial \xi^c}{\partial \xi'^{a'}} \frac{\partial \xi^d}{\partial \xi'^{b'}} \qquad (3.3)$$

The transformation property (3.3) defines $g_{ab}(\xi)$ as a tensor and ensures that the distance ds^2 is coordinate independent. The inverse to g_{ab} is also a tensor and we have $g^{ab} g_{bc} = \delta^a_c$. Further, the determinant g_{ab} is denoted g: $\det g_{ab} = g$. For further reference we note that the so-called Einstein-Hilbert action for the intrinsic geometry of the spacetime $\mathcal{M}$ is

$$S_{\text{E-H}}[g_{ab}] = K \int d^M \xi \, \sqrt{g(\xi)} \Big(-R(\xi) + 2\Lambda \Big), \qquad (3.4)$$

where R denotes the intrinsic scalar curvature of $\mathcal{M}$ (we will discuss the definition and meaning of R later), while $1/K$ is proportional to the gravitational coupling constant G and Λ is the cosmological coupling constant.

Let us now consider a scalar fields $X(\xi)$ defined on $\mathcal{M}$. That $X(\xi)$ is a scalar field means that under a change of coordinates $\xi \to \xi'$ it transforms as

$$X'(\xi') = X(\xi), \tag{3.5}$$

which just expresses that the *value* of a scalar field in a point $P \in \mathcal{M}$ is independent of the coordinate system. Everything we have said above is of course true also if $\mathcal{M}$ is just flat M-dimensional space, where we as a natural globally defined coordinate system can use Cartesian coordinates x^a and $g_{ab}(x) = \delta_{ab}$. The change $x \to \xi$ will then be a change from Cartesian coordinates to some curvilinear coordinates, and a scalar field will satisfy $X(x) = X'(\xi)$. The Gaussian action for a massless scalar field defined in $\mathbb{R}^M$ is

$$S[X] = \frac{\kappa}{2}\int d^M x\, \delta^{ab}\frac{\partial X(x)}{\partial x^a}\frac{\partial X(x)}{\partial x^b} = \frac{\kappa}{2}\int d^M\xi\, \sqrt{g(\xi)}\, g^{ab}(\xi)\frac{\partial X'(\xi)}{\partial \xi^a}\frac{\partial X'(\xi)}{\partial \xi^b}, \tag{3.6}$$

where κ is a coupling constant inserted for dimensional reasons if we want to assign the dimension of length to X, and where the rhs is just the action expressed in some curvilinear coordinates $x \to \xi$ and where $X'(\xi) = X(x)$. One can check that the rhs of (3.6) indeed is invariant under a coordinate change $\xi \to \xi'$ and in particular then a change $\xi \to x$ where it reduces to the lhs of (3.6). In the case where $\mathcal{M}$ with the metric $g_{ab}(\xi)$ represents a curved spacetime there exists no coordinate transformation $\xi \to x$, where x is the Cartesian coordinates in $\mathbb{R}^M$, but the rhs will then be the action for a massless free scalar field defined on the curved spacetime, where by "free" we mean that there is no self-interaction in the action, like a term X^4. If we have not only one scalar field $X(\xi)$, but D scalar fields $X_i(\xi)$, $i = 1,\ldots,D$ we can thus write down the corresponding action for these free massless fields:

$$\boxed{S[X, g_{ab}] = \frac{\kappa}{2}\int d^M\xi\, \sqrt{g(\xi)}\left[g^{ab}(\xi)\frac{\partial X_i(\xi)}{\partial \xi^a}\frac{\partial X_i(\xi)}{\partial \xi^b} + \lambda\right]} \tag{3.7}$$

Several remarks are in order here: as long as the metric $g_{ab}(\xi)$ is fixed, (3.7) just represent D independent fields X_i. The last term on the rhs of (3.7) is then irrelevant for the eom since it has no X dependence. Note that

$$\int d^M\xi\, \sqrt{g(\xi)} = \text{volume}(\mathcal{M}), \tag{3.8}$$

so the term is the cosmological term in the Einstein-Hilbert action (3.4) related to the manifold $\mathcal{M}$. In the following we are going to change the perspective on (3.7) when the dimension M of $\mathcal{M}$ is 1 or 2, by allowing g_{ab} to be a dynamical variable in addition to the scalar fields X_i. This will change everything associated with the interpretation of (3.7). The scalar fields X_i will no longer be independent since they will interact via the metric field g_{ab} and we will see that when the dimension of $\mathcal{M}$

One-Dimensional Quantum Gravity

is one or two we can view (3.7) as the complete coupled action of D scalar fields and "gravity". The reason that (3.7) can be viewed as containing also the action of gravity in these dimensions is, as we will discuss below, that in one dimension there is no intrinsic curvature while in two dimensions the part of the gravitational action (3.4) which involves the curvature term is a topological term which does not contribute to the eom, and it can consequently be left out as long as we do not consider spacetimes with changing topology. Thus, for $M = 1$ (3.7) will be the classical Lagrangian for one-dimensional gravity coupled to D scalar fields and we will show that this Lagrangian leads to precisely the same eom as the geometric action (2.18), i.e. it makes sense to view the D scalar fields $X_i(\xi)$ as the coordinates of a particle path $P : \xi \to X_i(\xi)$ in $\mathbb{R}^D$. Also, we will see that λ, the "cosmological constant" in our one-dimensional universe, will be crucial if we want to assign a mass to the free quantum particle. When the dimension of $\mathscr{M}$ is two, we will see that the action (3.7), which now describes two-dimensional gravity coupled to D scalar fields, becomes equivalent to that of a one-dimensional *string* propagating in $\mathbb{R}^D$, and where the *geometric* action of the string is given by the *area* of the worldsheet spanned by the string, in the same way as the geometric action of the particle was the length of its worldline. In the rest of this chapter, we consider the case where the dimension of $\mathscr{M}$ is $M = 1$.

If $M = 1$ we have $\xi = \xi^1$, $g_{ab} = g_{11}$ and we will suppress the index "1" and introduce the notation $g_{11}(\xi) = e^2(\xi)$ and (3.2) and (3.3) read:

$$ds = e(\xi)d\xi = e'(\xi')d\xi', \qquad e'(\xi') = e(\xi)\frac{d\xi}{d\xi'}. \tag{3.9}$$

Let us assume that ξ (and any other ξ') is normalized to be in the range $[0,1]$. The action (3.7) is then:

$$S[X,e] = \frac{\kappa}{2} \int_0^1 d\xi\, e(\xi) \left[\frac{1}{e^2(\xi)} \left(\frac{dX_i(\xi)}{d\xi} \right)^2 + \lambda \right]. \tag{3.10}$$

The eom can readily be derived:

$$\frac{\delta S}{\delta e(\xi)} = 0 \quad \Rightarrow \quad -\frac{1}{e^2(\xi)} \left(\frac{dX_i(\xi)}{d\xi} \right)^2 + \lambda = 0, \tag{3.11}$$

$$\frac{\delta S}{\delta X_i(\xi)} = 0 \Rightarrow -\frac{d}{d\xi}\left(\frac{1}{e}\frac{dX_i(\xi)}{d\xi}\right) = 0 \Rightarrow -\frac{d}{d\xi}\left[\frac{dX_i(\xi)}{d\xi} \bigg/ \left|\frac{dX_i(\xi)}{d\xi}\right|\right] = 0. \tag{3.12}$$

The rhs of (3.12) is precisely the eom (2.19), and classically (3.10) is thus equivalent to (2.18) provided that $\lambda > 0$. Note also that if we insert the value of $e(\xi)$ from (3.11) in (3.10) we obtain

$$S[X] = \kappa\sqrt{\lambda} \int_0^1 d\xi \left|\frac{dX_i(\xi)}{d\xi}\right| = m_0 \ell[X], \qquad m_0 = \kappa\sqrt{\lambda}, \tag{3.13}$$

i.e. the action (2.18) with the identification $m_0 = \kappa\sqrt{\lambda}$.

Of course it is not entirely clear that (2.18) and (3.10) will lead to the same quantum theory, since we in the latter case have two variables, $X_i(\xi)$ and $e(\xi)$. However, as we will now show, even quantum mechanically the two theories are identical.

The action (3.10) is invariant under diffeomorphisms $\xi \to \xi'(\xi)$, $\xi'(0) = 0$, $\xi'(1) = 1$, $d\xi'/d\xi > 0$. We denote the formal number of such diffeomorphisms by Vol(diff) (the number is of course infinite). Thus we define

$$\boxed{G(x-y) = \int \mathcal{D}[g_{ab}(\xi)] \int_{\substack{X_i(0)=y_i \\ X_i(1)=x_i}} \mathcal{D}X_i(\xi)\, e^{-S[X_i(\xi), g_{ab}(\xi)]}, \quad \mathcal{D}[g_{ab}] = \frac{\mathcal{D}g_{ab}}{\text{Vol(diff)}}}$$

(3.14)

For a given metric g_{ab} the path integral over the fields X_i is essential a straight forward generalization of the non-relativistic path integral for a free particle ($V(x)=0$) to D dimensions as we will discuss below. The parameter ξ will play the role of the non-relativistic time t. Many metrics g_{ab} represent the same *geometry*, which we denote $[g_{ab}]$. We should only integrate over geometries. We have formally represented this on the rhs equation in (3.14) by dividing the integration over all g_{ab} by Vol(diff). In our one-dimensional case $g_{ab}(\xi)$ has only one component which we have denoted $e^2(\xi)$ and it is not difficult to find the possible intrinsic geometries of $\mathcal{M}$. First note that the volume (3.8) of $\mathcal{M}$, which in the one-dimensional case will be called the length $\ell(\mathcal{M})$:

$$\ell(\mathcal{M}) = \int_0^1 d\xi\, \sqrt{g(\xi)} = \int_0^1 d\xi\, e(\xi), \tag{3.15}$$

is of course an invariant under diffeomorphisms. But there are no other invariants, since one can always transform $e(\xi)$ to the constant metric $e'(\xi') = \ell(\mathcal{M})$ by a suitable coordinate transformation $\xi \to \xi'(\xi)$:

$$\xi' = \frac{1}{\ell} \int_0^\xi d\tilde{\xi}\, e(\tilde{\xi}) \Rightarrow e'(\xi') = \frac{d\xi}{d\xi'}\, e(\xi) = \frac{1}{d\xi'/d\xi}\, e(\xi) = \ell. \tag{3.16}$$

In particular the existence of a constant $e(\xi)$ implies that there is no intrinsic curvature $R(\xi)$ (the expression for R involves the second derivative of $g_{ab}(\xi)$ as we will discuss later), as mentioned above: a curve has no *intrinsic* curvature. The integration over intrinsic geometries thus becomes a simple integration over ℓ, the volume of the geometry:

$$\mathcal{D}[g_{ab}] = \alpha\, dl, \tag{3.17}$$

where α is some constant which is not uniquely determined by our formal continuum arguments (it could in principle be a more complicated function $\alpha(l)$, but here we will simply assume it is just a constant independent of l). We have then already calculated the path integral in (3.14), since with the choice of $e(\xi) = \ell$ the X_i-part of the path integral just becomes the non-relativistic path integral (1.61) of a free particle (potential $V(x) = 0$) generalized from 1 to D dimensions. The result of this path integral is just (with suitable normalization) (A.59) with $t = 1$ and $b^2 = \ell/2\kappa$, and (3.14) becomes:

$$G(x-y) = \alpha \int_0^\infty d\ell \left(\frac{\kappa}{2\pi\ell}\right)^{D/2} \exp\left(-\frac{\kappa|x-y|^2}{2\ell} - \frac{1}{2}\kappa\lambda\ell\right). \quad (3.18)$$

This is the so-called Schwinger proper-time representation of the propagator. By a Fourier transformation we obtain

$$\hat{G}(p) = \int d^D x e^{-ip_j x^j} G(x-y) = \alpha \int_0^\infty d\ell\, e^{-\frac{\ell}{2\kappa}(p^2 + \lambda\kappa^2)} = \frac{1}{p^2 + m_{ph}^2}, \quad (3.19)$$

provided we choose[1] $\lambda\kappa^2 = m_{ph}^2$ and $\alpha = 1/2\kappa$. It is thus seen that $\lambda > 0$ formally seems needed in order to obtain the propagator with a mass $m_{ph} > 0$, as remarked earlier. The integral representation (3.19) is the same as we already encountered in formula (2.44).

Rather than appealing, as we did, to already derived results, when going from eq. (3.14) to (3.18), it is instructive to derive (3.18) or (3.19) (again) by introducing a cut-off ε, discretizing and taking the limit $\varepsilon \to 0$. This will bring up the question of how to discretize the space $\mathcal{M}$ (something which will play a major role later when the dimension of $\mathcal{M}$ will be larger than 1). In the case of the non-relativistic path integral we divided the (Euclidean) time interval $[0, \tau]$ into sub-intervals of length ε and at $\tau_k = k\varepsilon$ we assigned the variable $x(k) = x(\tau_k)$. We want to do the same here, but where τ had an interpretation as a physical time, the division of the coordinate ξ on $\mathcal{M}$ is not related to any physical length, and as a minimum we have to require that the cut-off introduced in $\mathcal{M}$ is invariant under reparametrization (in the limit where $\Delta\xi_k \to d\xi_k$):

$$\varepsilon = ds = e(\xi_k)\Delta\xi_k, \qquad \Delta\xi_k = \xi_{k+1} - \xi_k. \quad (3.20)$$

The discretized version of (3.10) is now:

$$\begin{aligned} S_\varepsilon[X, e] &= \frac{\kappa}{2} \sum_{k=0}^n \Delta\xi_k e(\xi_k) \left[\frac{(X_i(\xi_{k+1}) - X_i(\xi_k))^2}{e^2(\xi_k)\Delta\xi_k^2} + \lambda\right] \quad (3.21) \\ &= \frac{\kappa}{2} \sum_{k=0}^n \varepsilon \left[\frac{(X_i(\xi_{k+1}) - X_i(\xi_k))^2}{\varepsilon^2} + \lambda\right]. \end{aligned}$$

The integration over *geometries* $[g_{ab}]$ was reduced to the integration over the volume ℓ of these geometries and we have now discretized a geometry of volume ℓ into

---

[1] Note that in (3.13) we had $\lambda\kappa^2 = m_0^2$. The change from m_0 to m_{ph} comes when we perform the path integral in (3.14), which we did not actually do here. We only referred to already established results. Below we will actually perform the integral and we will see the shift to a renormalized mass.

$n+1 = \ell/\varepsilon$ pieces. In this way the integration over ℓ becomes a summation over n and we can finally write for the regularized propagator:

$$G_\varepsilon(x-y) = \sum_{n=0}^{\infty} e^{-\frac{1}{2}\lambda\kappa\varepsilon n} \int_{\substack{X(0)=y \\ X(1)=x}} \prod_{k=1}^{n} \frac{d^D X_i(k)}{(2\pi\varepsilon/\kappa)^{D/2}} e^{-\frac{\kappa}{2}\sum_{k=0}^{n} \frac{(X_i(k+1)-X_j(k))^2}{\varepsilon}}. \quad (3.22)$$

We have here chosen the normalization factor for path integral wrt the $X_i(\xi)$ variables such that we have a probability distribution $\mathscr{P}(X)$ as in (3.1). We are not forced to do that, e.g. one could have omitted the factor $\kappa^{D/2}$, in which case one would have worked with a distribution $P(X) = e^{\mu_c} \mathscr{P}(X)$, $e^{\mu_c} = \kappa^{-D/2}$. As a consequence of (3.1) we obtain by Fourier transformation:

$$\hat{G}_\varepsilon(p) = \sum_{n=0}^{\infty} e^{-\frac{\varepsilon n}{2\kappa}(\lambda\kappa^2 + p^2)}. \quad (3.23)$$

It is thus essentially the same formula as (2.43), *provided* we make the identification $\lambda\kappa^2 = m_{ph}^2$ (or more generally, for unnormalized $P(X)$, $\lambda\kappa^2 - \frac{\mu_c \kappa}{\varepsilon} = m_{ph}^2$), and

$$\boxed{a^2\sigma^2 = \frac{\varepsilon}{2\kappa}}. \quad (3.24)$$

The parameter a was a cut-off introduced in the space $\mathbb{R}^D$ where x and p live. We saw explicitly how it was related to distances in this space in the way it was introduced in (2.23) by dividing a path P there of length $\ell[P]$ in n pieces. On the other hand ε was introduced by dividing the manifold $\mathscr{M}$ of length $\ell[\mathscr{M}]$ in n pieces. However, the pieces of size ε are infinitesimal when measured in units of a, or stated differently: if $\ell[P] = n_P a$ is equal to $\ell[\mathscr{M}] = n_\mathscr{M} \varepsilon$ then $n_\mathscr{M} = n_P/(2\kappa\sigma^2 a)$. This is precisely what we discussed in (1.68)–(1.69) and is the topic which we will now study in more detail.

3.2 HAUSDORFF DIMENSION AND SCALING RELATIONS

Let us return to dimensionless units. Thus in the context of (3.22) we write $x = X\sqrt{\kappa/2\varepsilon}$, $\mu = \lambda\kappa\varepsilon/2$ and in this way (3.22) becomes a particular simple realization of the general expression (2.34), where the probability distribution $\mathscr{P}(x)$ is Gaussian. We will now use the general expression (2.34), which we replicate here for convenience:

$$G(x-y,\mu) = \sum_{n=0}^{\infty} e^{-\mu n} \int \prod_{j=1}^{n} d^D x(j) \prod_{k=0}^{n} P(x(k+1)) - x(k)), \quad \begin{array}{l} x(0) = y \\ x(n+1) = x \end{array} \quad (3.25)$$

and by Fourier transformation from eq. (2.38)

$$\hat{G}(k,\mu) = e^{\mu_c} \hat{\mathscr{P}}(k) \sum_{n=0}^{\infty} e^{-(\mu-\mu_c)n} \hat{\mathscr{P}}^n(k) = \frac{e^{\mu_c} \hat{\mathscr{P}}(k)}{1 - e^{-(\mu-\mu_c)} \hat{\mathscr{P}}(k)}. \quad (3.26)$$

Using the central limit theorem we know that $\tilde{\mathscr{P}}(k) = 1 - \frac{1}{2}\sigma^2 k^2 + \cdots$ and thus

$$\hat{G}(k,\mu) \propto \frac{1+\cdots}{m^2(\mu)+k^2+\cdots}, \quad m^2(\mu) = \frac{2}{\sigma^2}(\mu-\mu_c). \qquad (3.27)$$

where $+\cdots$ means higher order corrections in $|k|$ and $\mu - \mu_c$. By inverse Fourier transformation we obtain

$$G(x,\mu) = e^{-m(\mu)|x|+\cdots} \qquad (3.28)$$

where $+\cdots$ indicates logarithmic correction in $|x|$ for large $|x|$. Let us recall the general behavior of the spin-spin correlation function of a statistical spin system near phase transition point, which $\beta_c \sim \mu_c$ and $1/\xi(\beta) \sim m(\mu) \sim |\mu-\mu_c|^\nu$

$$G(x,\mu) \sim \frac{c}{|x|^{D-2+\eta}}, \qquad |x| \ll 1/m(\mu). \qquad (3.29)$$

$$G(x,\mu) = e^{-m(\mu)|x|+\mathcal{O}(\ln|x|)}, \qquad |x| \gg 1/m(\mu). \qquad (3.30)$$

For our free particle the central limit theorem ensures that $\nu = 1/2$, $\eta = 0$ (the mean field values). However in the following let us assume that we have an arbitrary ν, since we will later meet such cases. We now want to introduce a scaling parameter with dimension of length in $\mathbb{R}^D$, $a(\mu)$, a physical length x_{ph} and a physical mass m_{ph}:

$$m_{ph}a(\mu) = m(\mu) = c(\mu-\mu_c)^\nu, \quad x_{ph} = xa(\mu), \quad \text{i.e.} \quad m_{ph}x_{ph} = m(\mu)x. \qquad (3.31)$$

This ensures that the exponential fall off of the propagator survives in the limit $\mu \to \mu_c$ when expressed in terms of "physical" distances x_{ph} and a "physical" mass m_{ph}. A good way to think about this is to consider that propagator defined on an infinite dimensionless lattice. We now introduce the length of the lattice links as $a(\mu)$. If x is measured in number of lattice link "units", x_{ph} becomes the real physical length. When $\mu \to \mu_c$ the correlation length $\xi(\mu) = 1/m(\mu)$, measured in number of lattice links goes to infinity. However, we compensate for this by rescaling the physical length of the lattice links $a(\mu)$ such that correlation length measured in physical length x_{ph} stays fixed, namely equal to $1/m_{ph}$. This implies, from the assumed behavior of $m(\mu)$, that the length $a(\mu)$ of the lattice links scales to zero as $a(\mu) \propto (\mu-\mu_c)^\nu$. We are "scaling" the discretize lattice away and recover the continuum.

Let us now consider our ensemble of RWs (piecewise linear paths in $\mathbb{R}^D$) from 0 to x defined by (3.25). We can view $G(x,\mu)$ as the partition function for this ensemble. The expectation value of an "observable" O which takes values on the paths is then defined by

$$\langle O \rangle_\mu = \frac{1}{G(x,\mu)} \sum_{n=0}^{\infty} e^{-\mu n} \int \prod_{j=1}^{n} d^D x(j) \prod_{k=0}^{n} P\big(x(k+1)) - x(k)\big) O(\{x_j\}), \qquad (3.32)$$

We now use as an observable O the length $\ell[C(x)]$ of a piecewise linear curve $C(\{x_i\})$ from 0 to x (see Fig. 3.1) and we define the Hausdorff dimension d_H of the ensemble

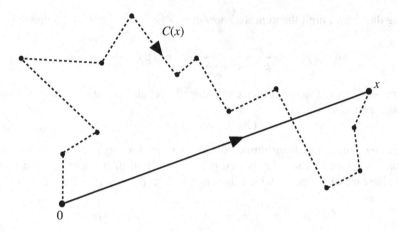

Figure 3.1 A typical piecewise linear curve $C(x)$ from spacetime point 0 to spacetime point x. Contrary to the piecewise linear curve shown in Fig. 2.2 the individual pieces here do not have the same length, but a length distribution determined by $\mathscr{P}(x)$.

of paths by:

$$\boxed{\langle \ell[C(x)]\rangle \propto |x|^{d_H}} \quad \text{for} \quad \mu \to \mu_c, \quad m(\mu)|x| = \text{const}. \tag{3.33}$$

The value of $\ell[C(x)]$ on a curve consisting of n pieces is

$$\ell[C_n(x)] = \sum_{k=0}^{n} |x(k+1)-x(k)|, \quad x(0)=0 \ \ x(n+1)=x. \tag{3.34}$$

A typical curve is shown in Fig. 3.1. When $\mu \to \mu_c$ the number n of pieces in a typical path will go to infinity and for such large n it will be approximately true that

$$\langle \ell[C(x)]\rangle \approx \sigma \langle n \rangle, \qquad \sigma = \int d^D x \, |x| \, \mathscr{P}(x), \tag{3.35}$$

σ being the average length of a step taken by the random walk, and $\langle n \rangle$ being the average number of steps. If we use the discretized action (2.23)–(2.25) this is of course exactly true, since each step in that case has a fixed length. We can easily calculate $\langle n \rangle$ since it follows directly from (3.32) that

$$\langle n \rangle_\mu = -\frac{1}{G(x,\mu)}\frac{\partial G(x,\mu)}{\partial \mu} = -\frac{\partial \ln G(x,\mu)}{\partial \mu} \approx m'(\mu)|x|, \tag{3.36}$$

where $m'(\mu)$ denotes the derivative of $m(\mu)$ wrt μ and where the rhs follows from (3.30). From (3.31) we obtain:

$$m'(\mu)|x| = \frac{\nu}{\mu-\mu_c} m(\mu)|x| = \frac{\nu m_{ph} x_{ph}}{\mu-\mu_c} \propto (m_{ph} x_{ph})^{1-\frac{1}{\nu}} |x|^{\frac{1}{\nu}} \tag{3.37}$$

One-Dimensional Quantum Gravity

We thus conclude

$$d_H = \frac{1}{\nu} \tag{3.38}$$

In the case of our RWs, we have $\nu = 1/2$ and thus $d_H = 2$.

In the case of RWs the proof that $d_H = 2$ is usually done using instead the ensemble with a fixed number of steps, and then forcing this number of steps to infinity. In this case we have from (3.25)

$$G_n(x) = \int \prod_{j=1}^n d^D x(j) \prod_{k=0}^n P\bigl(x(k+1)) - x(k)\bigr), \quad x(0) = 0, \ x(n+1) = x. \tag{3.39}$$

Before we asked about the average length of a path from 0 to x. Now we will instead ask for the distance $|x|$ travelled by a random walk of n steps (and corresponding average length $\langle \ell_n \rangle \propto n$), and we define the Hausdorff dimension by

$$\langle |x| \rangle_n \propto \langle \ell_n \rangle^{\frac{1}{d_H}}, \tag{3.40}$$

where the averages are calculated wrt $G_n(x)$. If $\mathscr{P}(x)$ in (3.39) has variance σ^2, then by the central limit theorem $G_n(x)$ will for large n be proportional to a Gaussian distribution with variance $\sigma^2 n$ and for such a distribution one readily shows that

$$\langle |x| \rangle_n \propto \sqrt{n} \propto \langle \ell_n \rangle^{\frac{1}{2}}, \tag{3.41}$$

Thus $d_H = 2$, as expected.

Let us end the section with the following remark about the significance of $d_H = 2$. As we have mentioned it is expected for spin systems that one has mean-field exponents for $D > 4$. As we have studied in Problem Set 4, mean field theory is basically the theory of Gaussian fluctuations, i.e. translated to a field theory: free fields. It is believed that one can derive the continuum scalar quantum field theories from lattice spin systems by taking the scaling limit approaching a critical point. If the corresponding critical exponents are mean field exponents it implies that the derived continuum field theory is just a free field theory. Is there a simple explanation why we cannot have interacting scalar quantum field theories in dimensions $D > 4$? Yes: $d_H = 2$. One can show that the quantum field theory of a scalar field can be formulated as a theory of particles which interact when their wold lines meet. It is not a very elegant formulation, but it shows that we cannot have interactions for $D > 4$. The particles are quantum particles, so their (quantum) paths are two-dimensional since $d_H = 2$. When $D > 4$ the probability that such two-dimensional objects meet is zero. In dimension D the intersection between a D_1- and a D_2-dimensional plane is a $(D_1 + D_2 - D \geq 0)$-dimensional plane, if they meet. Thus the paths will not meet for $D > 4$ and the particles will not interact. $D = 4$ is marginal (two planes will meet in a point), but it is believed that also here mean field prevails (see the original articles for details [8, 7].

3.3 PROBLEM SETS AND FURTHER READING

The relation with diffusion processes and fractal dimensions has been studied in great detail and $d_H = 2$ is a universal result (and by rotation of the diffusion time $\tau \to it$, also a universal result in quantum mechanics). However, there *are* diffusion processes leading to different Hausdorff dimensions (see the book [6] for a very comprehensive overview). The treatment here is slightly different since we have "integrated" over the diffusion time to obtain a relativistic invariant treatment, better suited for discussing quantum geometry. Problem Set 4 emphasizes the critical exponents of the spin-spin correlation function, which is most naturally associated with the (Euclidean) relativistic propagator.

4 Branched polymers

4.1 DEFINITIONS AND GENERALITIES

We can generalize the random process leading to the RW by enlarging the choices we have when we reach a given vertex: before we could stop or continue. Now this last choice is enlarged to branching: the RW is allowed to branch into a number of independent RWs, a process which can be repeated. The process will in this way generate a *tree-graph*, i.e. a graph which contains no loops. One can draw the abstract graph in $\mathbb{R}^2$ and we will distinguish graphs which differ by orientation as shown in Fig. 4.1 (in the actual physical systems to which such trees are approximations, this turns out to the natural thing to do). We call these graphs *planar branched polymers* or *planar trees* (but we will drop the "planar" from now on). We will mainly use the notation "branched polymers" (BP), since this was the notation used when physicists meet these objects in the study of string theory, but in general, and in particular in mathematics, the "tree" notation is used. Let v denote a vertex on the abstract BP graph G and $V(G)$ the set of vertices on G. We will always assume that G is a connected graph. We can assign points $x(v) \in \mathbb{R}^D$ to the vertices, and if v and v' are connected by a link in G we will also connect $x(v)$ and $x(v')$ in $\mathbb{R}^D$ by a straight line. In this way G is mapped to a graph $G(x)$ in $\mathbb{R}^D$ which we also denote a BP. We now associate an action with this $\mathbb{R}^D$ graph:

$$S[G(x)] = \sum_{\langle vv' \rangle} \varphi(|x(v)-x(v')|), \qquad (4.1)$$

where $\langle vv' \rangle$ denotes the link between v and v' if there is any. We are thus summing over all links in (4.1). φ is a positive function such that

$$\int d^D x \, e^{-\varphi(|x|)} = c_\varphi < \infty. \qquad (4.2)$$

The basic new aspect compared to the RW is that we assign a weight $w(v)$ to each vertex, associated with the possibility of branching. $w(v)$ will depend only on the *order* of the vertex, i.e. the number of links to which the vertex belong. Let σ_v denote the order. We will then write $w(\sigma_v)$ or w_{σ_v} instead of $w(v)$. For a RW graph (which is a special BP), one can view the factor the factor $e^{-\mu}$ as associated with vertices of order 2, while we in the case of unnormalized probabilities associated the weight 1 to (the two) vertices of order 1. For the general BP we find it more convenient to associate the weight factor $e^{-\mu}$ with the links. The final new aspect of BPs compared to RWs is that it is natural to define not only one- and two-point functions but also n-point functions, where n coordinates $\{x(i)\}$ corresponding to a

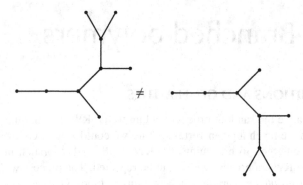

Figure 4.1 Two branched polymers (planar tree-graphs) in a plane, related by a reflection in a horizontal line. However, they cannot be made to overlap by translation and rotation in the plane where they are drawn and are thus to be considered as inequivalent.

set of n vertices $\{v(i)\}$, $i = 1, \ldots, n$, are kept fixed while we integrate over the rest:

$$G_\mu^{(n)}(x(1), \ldots, x(n)) = \sum_{B \in \mathscr{B}_n} \prod_{v \in V(B)} w(\sigma_v) \int_{\{x(v(i))\}} \prod_{v \notin \{v(i)\}} d^D x(v) \, e^{-S[B(x)] - \mu |L(B)|} \qquad (4.3)$$

In (4.3) B denotes a BP in the set $\mathscr{B}_n$ of BPs with n marked vertices $\{v(i)\}$, $V(B)$ denotes the set of vertices in B, $L(B)$ the set of links in B and $|V(B)|$ and $|L(B)|$ the number of vertices and number of links in B, respectively. For a (connected) tree-graph $|V(B)| = |L(B)| + 1$.

If we assume that vertex weights w_m are exponentially bounded, i.e. that there exist a constant c such that $w_m \leq c_w^m$, then it is relatively easy to show that $G_\mu^{(n)}(x(1), \ldots, x(n))$ exists (i.e. the sum in (4.3) is convergent for sufficiently large μ), and in addition that there exists a *critical* μ_c, such that the sum is convergent for $\mu > \mu_c$ and divergent for $\mu < \mu_c$, independent of $\{x(v(i))\}$ and n. We will not prove this here, but only outline the arguments for $n = 1$, i.e. the one-point function. The basic observation is that the number of BPs with a given number of links L is exponentially bounded. We will prove this later. Let us denote the set of BP with L links $\mathscr{B}(L)$ and the number of BP graphs with L links $\mathscr{N}(\mathscr{B}(L))$ and let us write

$$\mathscr{N}(\mathscr{B}(L)) \leq c_{bp}^{|L|}. \qquad (4.4)$$

Similarly, since each link has two vertices, $\sum_{v \in V(B)} \sigma_v = 2|L(B)|$, and we have

$$\prod_{v \in V(B)} w(\sigma_v) \leq \prod_{v \in V(B)} c_w^{\sigma_v} = c_w^{2|L(B)|}. \qquad (4.5)$$

Finally, the one-point function $G_\mu^{(1)}(x(v(1)))$ is by translational invariance of the action (4.1) independent of $x(v(1))$, and we can actually in this case perform the

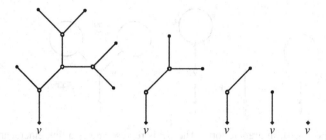

Figure 4.2 The figure illustrates how one successively can integrate with respect to the variables associated with vertices of order 1, except the variable associated to the marked vertex v. Integrating with respect to $x(v)$ where v is a vertex of order 1 implies that there will be no reference to the link connecting v to the rest of the graph. One can then "remove" the link and vertex v from the graph. Removing in this way all vertices of order 1 creates a new tree graph, where one again can integrate with respect to $x(v)$s where v are vertices of order 1. This procedure can continue until one reaches the root vertex.

integrals in (4.3) by successive integration, using the tree-nature of the graph B, see Fig. 4.2:

$$\int_{x(v(1))} \prod_{v \neq v(1)} d^D x(v)\, e^{-S[B(x)]} = \left(\int d^D x\, e^{-\varphi(|x|)} \right)^{|V(B)|-1} = c_\varphi^{|L(B)|}. \tag{4.6}$$

Thus we can write

$$G_\mu^{(1)}(x) \leq \sum_L \left(c_{bp} c_w^2 c_\varphi \right)^L e^{-\mu L}. \tag{4.7}$$

We conclude that $G_\mu^{(1)}(x)$ exists for $\mu > \ln(c_{bp} c_w^2 c_\varphi)$ and that the above mentioned $\mu_c \leq \ln(c_{bp} c_w^2 c_\varphi)$. We will later calculate μ_c more precisely.

4.2 ROOTED BRANCHED POLYMERS AND UNIVERSALITY

We now want to show that the one-point function of BPs has a universal critical behavior for $\mu \to \mu_c$, (almost) independent of $\varphi(x)$ and the weights w_m. The new aspect compared to the RW is the *universality wrt branching*. To simplify the arguments we assume that $c_\varphi = 1$ and $w_1 = 1$ (trivial assumptions), *and* we assume in addition that the marked point $v(1) = v_1$ of the one-point function has $\sigma_{v_1} = 1$. We call such one-point function the *reduced one-point function* $Z(\mu)$ and the corresponding abstract graphs *rooted BPs*. As already remarked $Z(\mu)$ is independent of $x(v_1)$. We write

$$Z(\mu) = \sum_{B \in \mathcal{B}_1'} \prod_{v \in V(B)} w(\sigma_v) \int_{x(v_1)} \prod_{v \neq v_1} d^D x(v)\, e^{-S[B(x)] - \mu |L(B)|} \tag{4.8}$$

$$= \sum_{B \in \mathcal{B}_1'} \prod_{v \in V(B)} w(\sigma_v)\, e^{-\mu |L(B)|}. \tag{4.9}$$

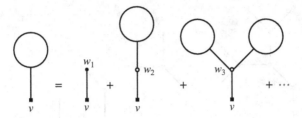

Figure 4.3 Eq. (4.10) in graphic form. The BP is represented as a link emergent from the root and connected to a "blob". The link emerging from the root can either connect to a vertex of order 1 with weight w_1 (and the graph terminates), or, if the vertex is of order $n > 1$, it can branch out in $n-1$ links. This happens with weight w_n and the vertex serves as the root for $n-1$ rooted branched polymers in which the $n-1$ links are the links in the branch polymers which are connected to the root.

where $\mathscr{B}'_1$ denotes the set of the rooted BPs. Each link is assigned a weight $e^{-\mu}$ and we see that $Z(\mu)$ satisfies the graphic equation shown in Fig. 4.3 which as an algebraic equation reads:

$$Z(\mu) = e^{-\mu} + e^{-\mu} f(Z(\mu)), \qquad f(z) = \sum_{m=2}^{\infty} w_m z^{m-1} \qquad (4.10)$$

From this we can find μ as a function of Z, shown graphically in Fig. 4.4:

$$e^{\mu} = F(Z), \qquad F(Z) = \frac{1+f(Z)}{Z}, \qquad (4.11)$$

and we can identify the critical point μ_c as the minimum of the function $F(z)$. To simplify the discussion let us assume that $w_m \geq 0$, $w_m = 0$ for $m > m_0$ and that at least one $w_m > 0$ for some $m > 2$. Also, we can assume $w_2 = 0$ since a $w_2 > 0$ simply adds w_2 to $F(Z)$, a constant which will play no role in the arguments to follow. Thus $F(z)$ has the shape shown in Fig. 4.4 and we obtain:

$$\mu - \mu_c = c\,(Z(\mu_0) - Z(\mu))^2 + \mathcal{O}\left((Z(\mu_0) - Z(\mu))^3\right), \qquad (4.12)$$

or

$$Z(\mu) \approx Z(\mu_c) - \tilde{c}\sqrt{\mu - \mu_c} = Z(\mu_c) - \tilde{c}(\mu - \mu_c)^{1-\gamma} \qquad (4.13)$$

We will show below that γ can be viewed as the susceptibility exponents for BPs, and we have derived that under the given assumptions about the branching weights w_m, *the susceptibility exponent is universal and equal 1/2*. This result is also true if we allow $w_m > 0$ for arbitrary large m, except in some special situations which we will discuss in Problem Sets 5 and 7.

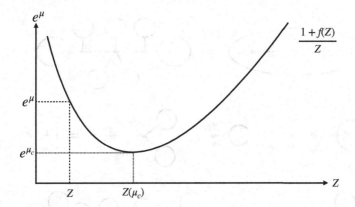

Figure 4.4 The function $e^\mu = F(Z)$ from eq. (4.11). The function behaves as $1/Z$, i.e. goes to infinity for $Z \to 0$ and like Z^n, $n > 1$ for $Z \to \infty$. It thus has a local minimum for some finite Z_c. This minimum defines μ_c by $e^{\mu_c} = F(Z_c)$.

4.3 THE TWO-POINT FUNCTION

Let us now consider the two-point function, as defined by eq. (4.3). We denote the two marked vertices v_1 and v_2, and the corresponding coordinates y and x. For any BP $B \in \mathcal{B}_2$ there is a unique shortest link-path between v_1 and v_2. We have indicated that in Fig. 4.5. At each vertex along the path, but different from v_1 and v_2, we can meet a vertex of any order $m \geq 2$, provided the weight $w_m \neq 0$. $m-2$ rooted BPs are then attached to the vertex and they can be arranged in $m-1$ ways as shown in Fig. 4.5. The total contribution from such a vertex will thus be:

$$\sum_{m=2}^{\infty} (m-1) w_m Z^{m-2} = f'(Z), \qquad (4.14)$$

where f' denotes the derivative of the function f. If the length of the shortest path is n, there will be $n-1$ such contributions. From the n links there will be a contribution $e^{-\mu n}$. Finally, there will be a contribution $(1+f(Z))$ from each of the marked vertices v_1 and v_2 from the part of the graph connected to these vertices, but not being part of the shortest path between them, as also illustrated in Fig. 4.5. Collecting this the two-point function (4.3) can be written as:

$$G_\mu^{(2)}(x-y) = \left(1+f(Z)\right)^2 \sum_{n=1}^{\infty} e^{-\mu n} \left(f'(Z)\right)^{n-1} \int \prod_{i=1}^{n-1} d^D x_i \, e^{-\sum_{i=1}^{n} \varphi(|x_i - x_{i-1}|)} \qquad (4.15)$$

where $x_0 = y$, $x_n = x$ and x_i, $i = 1, \ldots, n-1$, denote the coordinates of the $n-1$ vertices of the shortest path of length n between the marked vertices v_1 and v_2. These are the only vertices which we cannot successively integrate over in the way indicated in Fig. 4.2. It is seen that the sum in (4.15) is precisely like the sum we encounter in the

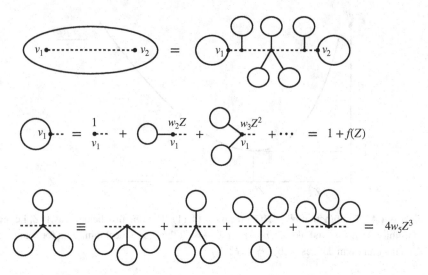

Figure 4.5 The top part shows the decomposition of a graph with two marked vertices v_1 and v_2 into a shortest link-path (the dashed line) with associated rooted branched polymers, plus the two end graphs, which again can be decomposed into rooted branched polymers as shown in the middle graph. Finally, the bottom graph shows how a contribution from a vertex of order m on the shortest path really should be understood as coming from the $m-1$ different ways $m-2$ rooted BPs can be placed relative to the shortest link path passing though the vertex.

RW analysis, namely eq. (2.34). Consequently we can write:

$$G^{(2)}_\mu(x-y) = \frac{[1+f(Z(\mu))]^2}{f'(Z(\mu))} G^{(rw)}_{\bar{\mu}}(x-y), \qquad (4.16)$$

where the superscript "rw" means the RW propagator and where $\bar{\mu}$ is a *renormalized coupling constant*:

$$e^{-\bar{\mu}} = e^{-\mu} f'(Z(\mu)) = 1 + \frac{Z(\mu)}{Z'(\mu)}, \qquad \boxed{\bar{\mu} = -\ln\left(1+\frac{Z(\mu)}{Z'(\mu)}\right)} \qquad (4.17)$$

where the expression in terms of $Z'(\mu)$ follows from (4.10) by differentiation wrt μ. From (4.13) it follows that when μ is close to μ_c, we have

$$\boxed{\bar{\mu} \propto \sqrt{\mu - \mu_c}} \qquad (4.18)$$

which tells us that the relation between μ and $\bar{\mu}$ is *non-analytically* at μ_c.

We know from our RW analysis that $G^{(rw)}_{\bar{\mu}}(x-y)$ falls off exponentially with a mass $m(\bar{\mu})$. We know that close to the RW critical point $\bar{\mu}_c$, we have $m(\bar{\mu}) \propto \sqrt{\bar{\mu} - \bar{\mu}_c}$, and we know that $\bar{\mu}_c = 0$ since we have normalized $\int d^D x\, e^{-\varphi(|x|)} = 1$. From (4.16) it is clear that $G_\mu^{(2)}(x-y)$ falls off in the same way. Let us call this mass, expressed as a function of μ, for $m_{bp}(\mu)$. We can now write:

$$m_{bp}(\mu) = m(\bar{\mu}) \propto \sqrt{\bar{\mu}} \propto (\mu-\mu_c)^{\frac{1}{4}} = (\mu-\mu_c)^{\nu_{bp}}. \quad (4.19)$$

Similarly, we define the susceptibility of our BPs as

$$\chi_{bp}(\mu) = \int d^D x\, G_\mu^{(2)}(x-y) \to \frac{c}{(\mu-\mu_c)^\gamma} \quad \text{for} \quad \mu \to \mu_c. \quad (4.20)$$

Again, from (4.16) we know it will be the same as for the random walk, but the critical behavior will be different because of the non-analytical relation (4.18) between $\bar{\mu}$ and μ:

$$\chi_{bp}(\mu) = \frac{[1+f(Z(\mu))]^2}{f'(Z(\mu))} \chi^{(rw)}(\bar{\mu}). \quad (4.21)$$

$$\chi_{bp}(\mu) \propto \chi^{(rw)}(\bar{\mu}) \propto \frac{1}{\bar{\mu}} \propto \frac{1}{\sqrt{\mu-\mu_c}}, \quad \text{i.e.} \quad \gamma_{bp} = \frac{1}{2}. \quad (4.22)$$

Note that from (4.17), we have for μ close to μ_c:

$$\chi_{bp}(\mu) \propto \frac{1}{\bar{\mu}} \propto -Z'(\mu). \quad (4.23)$$

Thus, we have shown that it was justified to use γ in formula (4.13), as promised, and it should be mentioned that this a special case of a more general relation, which we will also use when we discuss string theories. We can define a susceptibility function $\chi_{bp}^{(k)}(\mu)$ for the k-point function (4.3) by integrating over all points $x(1),\ldots,x(k)$, except one point. In this way integrals which appear for different k are actually precisely the same since we are integrating over all xs associated with vertices, except one vertex. The only difference is the way we count the graphs. Let us consider a graph with n vertices, n very large, where k of them are marked. If we want to introduce an additional marked vertex this can essentially be done in n ways (assuming $n \gg k$). Thus there will be n more graphs, but all with the same integral. We can obtain this factor n for each graph by differentiation of the n-point function from (4.3) wrt $-\mu$ since the number of links only differs from the number of vertices by 1 for connected tree graphs, and we can thus write

$$\chi_{bp}^{(k+1)}(\mu) \propto -\frac{d}{d\mu}\chi_{bp}^{(k)}(\mu), \quad \text{i.e.} \quad \gamma_{bp}^{(k+1)} = \gamma_{bp}^{(k)} + 1, \quad k \geq 2, \quad (4.24)$$

where we have defined the generalized susceptibility exponent for $G_\mu^{(k)}(x_1,\ldots,x_k)$ in an obvious way. The first formula in (4.24) is *almost* relation (4.23) for $k=1$, but not quite. $Z(\mu)$ is slightly different from $G_\mu^{(1)}(x)$ because in $Z(\mu)$ the marked vertex is of order 1. However, this different does not really change any critical behavior[1].

---

[1] If we assume there are no vertices of order 2, then it is easy to show that $V_1 = 2+V_3+2V_4+3V_5+\cdots$, where V_n denotes the number of vertices of order n. Thus more than half of the vertices are of order one, and when it comes to critical behavior depending on the number of vertices, there will be no difference if we consider vertices of order 1 or all vertices.

Finally, the short distance behavior of $G_\mu^{(2)}(x-y)$ is of course the same as that of a free particle because of (4.16):

$$G_\mu^{(2)}(x-y) \propto \frac{1}{|x-y|^{D-2}}, \qquad |x-y|\, m_{bp}(\mu) \ll 1, \tag{4.25}$$

and the exponent $\eta_{bp} = 0$. Summarizing, the BP critical exponents are

$$\boxed{\nu_{bp} = \frac{1}{4}, \quad \gamma_{bp} = \frac{1}{2}, \quad \eta_{bp} = 0, \quad d_H^{(bp)} = 4} \tag{4.26}$$

The Hausdorff dimension of BPs is $d_H^{(bp)} = 1/\nu_{bp} = 4$. A look at the top part of Fig. 4.5 makes this result quite natural. In the scaling limit the average number of vertices in each of the rooted branched polymers $Z(\mu)$ diverges as does the number of vertices in the shortest path between the two marked vertices and the divergence of total number of vertices in the BP will be determined by the product of these two numbers. More precisely, when we use the two-point function to derive the Hausdorff dimension we have $|x|^{d_H} \propto 1/(\mu-\mu_c)$ and also that the average number of vertices $\langle n \rangle_{bp} \propto 1/(\mu-\mu_c)$. At the same time the average number of vertices in a rooted BP (derived from $Z(\mu)$) is $\langle n \rangle_{rbp} \propto 1/\sqrt{\mu-\mu_c}$ and the average number of vertices in the shortest path between the two marked points $\langle n \rangle_{sp} \propto 1/(\bar\mu-\bar\mu_c) \propto 1/\sqrt{\mu-\mu_c}$. We can thus write

$$\langle n \rangle_{bp} \propto \frac{1}{\mu-\mu_c} \propto \langle n \rangle_{rbp} \langle n \rangle_{sp}. \tag{4.27}$$

4.4 INTRINSIC PROPERTIES OF BRANCHED POLYMERS

Contrary to RWs, BPs have a non-trivial "internal life", independent of the embedding in $\mathbb{R}^D$. We defined the susceptibility $\chi_{bp}(\mu)$ by integration over $x \in \mathbb{R}^D$ as in (4.20). After this integration we obtained

$$\chi_{bp}(\mu;v_1,v_2) = \frac{[1+f(Z(\mu))]^2}{f'(Z(\mu))} \sum_{r=1}^{\infty} e^{-\mu r} f'(Z(\mu))^r, \tag{4.28}$$

where we have explicitly kept the reference to the two marked vertices. We can view this as coming from a partition function for "abstract" BPs, where there is no reference to the so-called target space where the $x(v)$ live. In fact, if we define

$$\chi^{(I)}(\mu;v_1,v_2) = \sum_{B \in \mathscr{B}_2} \prod_{v \in V(B)} w(\sigma_v)\, e^{-\mu |L(B)|}, \tag{4.29}$$

where $\mathscr{B}_2$ is the set of BPs with two marked points v_1, v_2, one obtains precisely (4.28). Here we have left an explicit reference to the points v_1 and v_2 which was left out in (4.20). Similarly we would obtain our previously defined $\chi^{(n)}(\mu)$ by defining the equivalent of (4.29) with $\mathscr{B}_2$ replaced by $\mathscr{B}_n$ and keeping reference to the marked vertices $v_1,\ldots,v_n$. We now want to introduce the *intrinsic link distance* between v_1

and v_2. Denote this distance r. Then the decomposition is already given in (4.28), and (4.28) can be obtained from (4.29) by decomposing $\mathcal{B}_2$ in $\cup_{r=1}^{\infty} \mathcal{B}_2(r)$ where $\mathcal{B}_2(r)$ denotes the BPs with two marked points separated a link distance r. Thus we can write

$$G_\mu^{(I)}(r;v_1,v_2) = \sum_{B\in\mathcal{B}_2(r)} \prod_{v\in V(B)} w(\sigma_v) \, e^{-\mu|L(B)|} = \frac{[1+f(Z(\mu))]^2}{f'(Z(\mu))} e^{-r\bar{\mu}(\mu)}, \quad (4.30)$$

$$\chi^{(I)}(\mu;v_1,v_2) = \sum_{r=1}^{\infty} G_\mu^{(I)}(r;v_1,v_2) \quad (4.31)$$

Since we know from (4.18) that $\bar{\mu} \propto \sqrt{\mu-\mu_c}$ it follows that for $\mu \to \mu_c$ we have (suppressing the arguments v_1, v_2 in G)

$$G_\mu^{(I)}(r) = c e^{-m_I(\mu)r}, \quad m_I(\mu) \propto \sqrt{\mu-\mu_c}, \quad (4.32)$$

Thus

$$\boxed{\nu_I = \frac{1}{2}} \quad \text{i.e.} \quad \boxed{d_H^{(I)} = 2} \quad (4.33)$$

From (4.31) and (4.32) we find

$$\chi^{(I)}(\mu) \propto \frac{1}{\sqrt{\mu-\mu_c}}, \quad \text{i.e.} \quad \boxed{\eta_I = \frac{1}{2}} \quad (4.34)$$

It is instructive to repeat the argument which led to $d_H = 1/\nu$ in this new setting. We have our ensemble of BPs, $\mathcal{B}_2(r)$, where two marked vertices are separated a distance r, and we ask what is the average volume (i.e. the average number of links) of a graph $B \in \mathcal{B}_2(r)$. The partition function for these graphs is $G_\mu^{(I)}(r)$ and as is clear from (4.30) we obtain the average value as follows

$$\langle |L(B)|\rangle_r = -\frac{1}{G_\mu^{(I)}(r)} \frac{\partial G_\mu^{(I)}(r)}{\partial \mu} = m_I'(\mu) r \propto \frac{r}{\sqrt{\mu-\mu_c}} \quad (4.35)$$

Because of (4.32) the formula $m_I'(\mu)r$ is actually exact for all r, not only valid for large r, as (3.36). The formula shows that if μ is fixed and different from μ_c the typical BP for large r will just be a linear chain with small outgrowths (see Fig. 4.6, left). However we are interested in a limit where $e^{-m_I(\mu)r}$ survives in the limit $r \to \infty$ and $\mu \to \mu_c$, i.e. $\sqrt{\mu-\mu_c}\, r = \text{const.}$. In this limit we obtain (see Fig. 4.6, right)

$$\langle |L(B)|\rangle_r \sim r^2 \quad \text{i.e.} \quad d_H^{(I)} = 2. \quad (4.36)$$

Finally, we have

$$G_\mu^{(I)}(r) \propto e^{-m_I(\mu)r} \quad \Rightarrow \quad \boxed{\eta_I = 1} \quad (4.37)$$

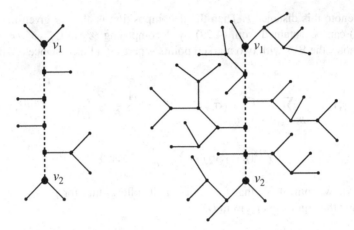

Figure 4.6 Left: typical graph where $\mu > \mu_c$ and r, the link distance between vertices, goes to infinity: we get a linear chain with small outgrowths. Right: typical graph when $m(\mu)r$ is constant when $r \to \infty$. In this case $d_H = 2$. Dashed lines show shortest paths between vertices v_1 and v_2. The outgrowths become critical rooted BPs since $r \to \infty$ forces $\mu \to \mu_c$.

Recall the notation $G_\mu(x) \sim 1/|x|^{D-2+\eta}$ for $m(\mu)|x| \ll 1$. However, our r in (4.37) should be viewed as the the radial distance from 0 to x, i.e. it involves an integration over all point x with $|x| = r$:

$$G_\mu(r) \equiv \int d^D x\, \delta(|x|-r)\, G_\mu(x) \sim \frac{r^{D-1}}{r^{D-2+\eta}} = r^{1-\eta}, \quad rm(\mu) \ll 1. \quad (4.38)$$

Thus (4.37) shows that $\eta_l = 1$ and we observe that Fisher's scaling relation is satisfied:

$$\boxed{\gamma_l = \nu_l(2-\eta_l)} \quad (4.39)$$

4.5 MULTICRITICAL BRANCHED POLYMERS

We have seen that there is a large universality for BPs: a finite number of positive weights w_m, with at least one w_m different from zero for $m > 2$, lead to the scaling limit described above. However, by relaxing the requirement that $w_m \geq 0$ we can obtain different scaling limits. We are thus loosing a strict probabilistic interpretation, but a number of statistical matter systems coupled to BPs will induce such behavior (as we will study in detail in Problem Set 6 for a specific matter model coupled to BPs) and one will encounter similar situations in two-dimensional gravity systems as we will discuss later. Recall that from a technical point of view the universality came because the function $F(Z)$, defined in (4.10) and (4.11) has a simple minimum, Z_c, where $F'(Z_c)=0$, but $F''(Z_c) > 0$. We can clearly obtain that also $F''(Z_c)=0$ by choosing w_m in a suitable way (see Fig. 4.7). As a simple example choose $w_1 = 1$,

Branched Polymers

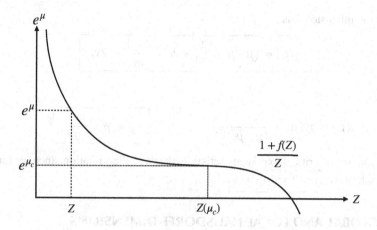

Figure 4.7 An example of $F(Z) = \frac{1+f(Z)}{Z}$ where $F'(Z_c) = F''(Z_c) = 0$ and $F'''(Z_c) < 0$. $F(Z) \to \infty$ for $Z \to 0$ as before, but because w_3 is negative Z_c is no longer a local minumum.

$w_3 = 1$ and $w_4 = -1/12$:

$$F(Z) = \frac{1+Z^2-\frac{1}{12}Z^3}{Z}, \quad Z_c = \sqrt{2}, \quad F'(Z_c) = F''(Z_c) = 0. \tag{4.40}$$

This can be generalized to any order $m > 2$ by appropriate choices of weights w_m:

$$F'(Z_c) = \cdots = F^{(m-1)}(Z_c) = 0, \quad F^{(m)}(Z_c) \neq 0, \tag{4.41}$$

$$\mu - \mu_c \approx c(Z_c - Z)^m, \quad \boxed{Z(\mu) \approx Z(\mu_c) - \tilde{c}(\mu - \mu_c)^{1/m}, \quad \mu \to \mu_c}$$

We denote a model where this situation is realized an mth-*multicritical model* (in Problem Set 5 we will study a number of aspects of multicritical models more closely).

The graphic equation shown in Fig. 4.5 is still valid and we obtain:

$$e^{-\mu} f'(Z(\mu)) = 1 + \frac{Z(\mu)}{Z'(\mu)} \approx 1 - c'(\mu - \mu_c)^{\frac{m-1}{m}} \quad \text{for} \quad \mu \to \mu_c. \tag{4.42}$$

As for ordinary BPs, we have (4.16):

$$G_\mu^{(2)}(x-y) = \frac{[1+f(Z(\mu))]^2}{f'(Z(\mu))} G_{\tilde{\mu}}^{(rw)}(x-y), \tag{4.43}$$

where

$$\tilde{\mu} = -\ln\left(1 + \frac{Z(\mu)}{Z'(\mu)}\right) \approx c'(\mu - \mu_c)^{\frac{m-1}{m}} \quad \text{for} \quad \mu \to \mu_c. \tag{4.44}$$

Thus we obtain

$$\boxed{m_{bp}(\mu) \propto \sqrt{\tilde{\mu}} \propto (\mu - \mu_c)^{\frac{m-1}{2m}}, \quad \nu_{bp} = \frac{m-1}{2m}} \tag{4.45}$$

and for the intrinsic mass

$$m_I(\mu) \propto (\mu-\mu_c)^{\frac{m-1}{m}}, \quad v_I = \frac{m-1}{m} = 2v_{bp} \qquad (4.46)$$

Finally,

$$\chi_{bp}(\mu) = \chi_I(\mu) \propto \frac{c}{(\mu-\mu_c)^{(m-1)/m}} \quad \text{i.e} \quad \gamma_{bp} = \gamma_I = \frac{m-1}{m} \qquad (4.47)$$

Again both set of critical exponents satisfy Fisher's scaling relation since, as for the ordinary BPs, $\eta_{bp} = 0$ and $\eta_I = 1$.

4.6 GLOBAL AND LOCAL HAUSDORFF DIMENSIONS

Until now we have used (4.35)–(4.36) to define the intrinsic Hausdorff dimension by

$$\langle |L(B)| \rangle_r \sim r^{d_H}. \qquad (4.48)$$

We call this the *global Hausdorff dimension* since we can view r as a diameter in the "B" universe, and (4.48) then tell us the volume of a typical such universe. However, a more geometric definition of a Hausdorff dimension is the following: assume our graphs $B \in \mathscr{B}_1$ have a volume $|L(B)| = L \gg 1$. We denote this set of BPs as $\mathscr{B}_1(L)$. Let v be the marked vertex. Let $L_r(B)$ denote the volume of the part of B where the vertices have a link distance less than or equal r to v, i.e. the volume of a "ball" of radius r, centered at v. We then define the *local Hausdorff dimension* d_h by

$$\langle L_r(B) \rangle_L \sim r^{d_h} \quad 1 \ll r \ll L^{1/d_h}. \qquad (4.49)$$

The average is performed in the ensemble $\mathscr{B}_1(L)$. The idea is that L^{1/d_h} is a typical length scale of a graph of volume L and as long as r is much less than this length scale we will have no "finite size" effects. We will now show how to use the two-point function to extract d_h. Let is write

$$G_\mu^{(I)}(r) = \sum_L e^{-\mu L} G_L^{(I)}(r), \qquad (4.50)$$

Here $G_L^{(I)}(r)$ denotes the sum over BPs with two marked vertices and volume L, and the marked points separate a link distance r. The geometric interpretation of $G_L^{(I)}(r)$ is as follows: we perform the weighted sum over BPs B of volume L and weight $\prod_{v \in V(B)} w_{\sigma_v}$. For each B we mark a vertex v_1, then count the number of vertices v_2, located a distance r from v_1, and finally we sum[2] over all v_1. In this way $G_L^{(I)}(r)$

---

[2] In the definition of the two-point function we are fixing marked vertices v_1 and v_2, but effectively, for a graph B without the marked vertices we create different triangulations with marked vertices by moving around and marking the vertices as described. There are a few subtleties related to symmetry factors of the graphs, which we will ignore since they are not important for a generic large graph.

Branched Polymers

estimates (up to normalization) the average "area" $\langle S(r) \rangle_L$ of a "spherical" shell $S(r)$ of radius r. For $1 \ll r \ll L^{1/d_h}$ we expect such shells to behave like r^{d_h-1}, i.e.

$$\boxed{\frac{G_L^{(I)}(r)}{G_L^{(I)}(1)} \propto \langle S(r) \rangle_L \propto r^{d_h-1}, \qquad 1 \ll r \ll L^{1/d_h}} \tag{4.51}$$

Let us use this formula to calculate d_h for (multicritical) BPs. First note that we expect $G_L^{(I)}(r)$ to behave as

$$G_L^{(I)}(r) = e^{\mu_c L} f(r, L), \tag{4.52}$$

where $f(r, L)$ grows slower than exponential for large L. This follows from eq. (4.50) since we know that $G_\mu^{(I)}(r)$ diverges for $\mu < \mu_c$. Close to μ_c we can write

$$G_\mu^{(I)}(r) = \sum_L e^{-(\mu-\mu_c)L} f(r,L) \approx \int dL \, e^{-(\mu-\mu_c)L} f(r,L). \tag{4.53}$$

Using $G_\mu^{(I)}(r) = c e^{-m_I(\mu) r}$, where $m_I(\mu)$ is given by (4.46), we find by inverse Laplace transformation:

$$f(r,L) \propto \int_{-i\infty+c}^{i\infty+c} d\mu \, e^{(\mu-\mu_c)L} e^{-m_I(\mu) r}. \tag{4.54}$$

Expanding $e^{-m_I(\mu) r}$ as $1 + (\mu - \mu_c)^{\frac{m-1}{m}} r + \cdots$ we find for small r:

$$f(r,L) \propto \frac{1}{L^{2-1/m}} r \tag{4.55}$$

and thus from (4.51) and (4.52) that

$$\langle S(r) \rangle_L \propto \frac{f(r,L)}{f(1,L)} = r \quad \Rightarrow \quad \boxed{d_h = 2 \quad \text{for all multicritical models}} \tag{4.56}$$

We conclude that

$$\boxed{\text{for ordinary BPs } d_H^{(I)} = d_h = 2, \text{ but for multicritical BPs } d_H^{(I)} = \frac{m}{m-1} < d_h = 2.}$$

4.7 PROBLEM SETS AND FURTHER READING

We have here discussed only the simplest critical properties of branched polymers or planar trees. A number of more complicated critical aspects are discussed in Problem Sets 5-7, in particular how one couples matter to BPs.

The mathematical study of trees goes back to Galton and Watson in the 19th century but is now an important topic in combinatorics and probability theory. A comprehensive review can be found in the book *Probability on Trees and Networks*, [9]. Here we have mainly been interested in the "continuum limit" of the BPs without actually attempting to address in detail what we mean by "continuum trees". More precise mathematical definitions and results can be found in [10].

5 Random Surfaces and Bosonic Strings

5.1 THE ACTION, GREEN FUNCTIONS AND CRITICAL EXPONENTS

For the relativistic particle we encountered two actions which were geometric and which were classically (and quantum mechanically) equivalent

$$S[P(x,y)] = m_0\, \ell[P(x,y)] = m_0 \int_0^1 d\xi \sqrt{\left(\frac{dX_i}{d\xi}\right)^2}, \quad P: \xi \to X_i(\xi), \quad (5.1)$$

$$S[X,g_{ab}] = \frac{\kappa}{2}\int d\xi \sqrt{g(\xi)}\left[g^{ab}(\xi)\frac{\partial X_i}{\partial \xi^a}\frac{\partial X_i}{\partial \xi^b}+\lambda\right], \quad a,b=1 \quad (5.2)$$

where P denotes a path from y to x in $\mathbb{R}^D$.

We now move from one-dimensional geometric objects (paths) to two-dimensional geometric objects (surfaces). We denote these by F. The two "boundaries" of our paths (y to x) were zero-dimensional (points), and they are naturally replaced by n one-dimensional boundaries of lengths $\ell_1,\ldots,\ell_n$, and we will talk about n-loop functions or n-loop propagators $G(\ell_1,\ldots,\ell_n)$, in the same way as we talked about n-point functions for BPs. This is illustrated in Fig. 5.1. We can contract the loops to points and then we talk about n-point functions $G(x(1),\ldots,x(n))$ and we say that the surface has n punctures. Apart from the boundaries, surfaces also differ from paths by having a non-trivial intrinsic geometry, as we will discuss. In particular they can be topological distinct and differ (apart from the number of boundaries) by the number of *handles*. In Fig. 5.1 we have shown a surface with one handle. For reasons which will be clear later, we will here mainly consider surfaces with no handles, i.e. surfaces which have the topology of the sphere S^2 with a number of boundaries.

Two immediate generalizations of the geometric actions (5.1) and (5.2) for paths to surfaces suggest themselves:

$$S[F] = \kappa A[F] = \kappa \int d^2\xi \sqrt{\det h_{ab}}, \quad h_{ab}(\xi) = \frac{\partial X_i}{\partial \xi^a}\frac{\partial X_i}{\partial \xi^b}, \quad (5.3)$$

$$S[X,g_{ab}] = \frac{\kappa}{2}\int d^2\xi \sqrt{g(\xi)}\left[g^{ab}(\xi)\frac{\partial X_i}{\partial \xi^a}\frac{\partial X_i}{\partial \xi^b}+\lambda\right], \quad a,b=1,2 \quad (5.4)$$

where $A[F]$ denotes the area of the surface $F: \xi \to X_i(\xi)$ in $\mathbb{R}^D$. $S[F]$ defined by $S[X,g_{ab}]$ defined by (5.3) is called the Nambu-Goto action and (5.4) the Polyakov action. One can write down the classical eoms for these actions (treating X_i and g_{ab} as independent variables for $S[X,g_{ab}]$), and they agree for the $X_i(\xi)$ parts, precisely as was the case for the path-actions. From (5.3) it is clear that given some boundaries, the minimum of $S[F]$ will be the surface of minimal area connecting

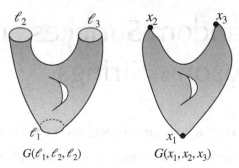

Figure 5.1 A surfaces with three boundary loops and one handle and a surface with three punctures and one handle, contributing to $G(\ell_1,\ell_2,\ell_3)$ and $G(x_1,x_2,x_3)$, respectively.

these boundaries. Such a situation can be quite singular, as is known from the classical variational theory of minimal area surfaces between boundaries, the standard example being the minimal area surface connecting two circles of radius r, the centers separate a distance R. When R is sufficiently large compared to r it is clear that the surface with minimal area will consist of the two disks associated with the circles and an infinitely thin tube connecting these disks. If we for a moment rotate back to spacetimes with Lorentzian signature we can use the actions (5.3) or (5.4) for *relativistic strings*, namely to describe the time-evolution of a spatial boundary, i.e of a *closed* string. This time-evolution will not necessarily lead to the singular configurations mentioned. However, quantizing the theory we are instructed in the path integral to integrate over all possible surfaces, and as we will see, these singular surfaces will come back and haunt us and connect the regularized string theory we consider to BPs.

Using the actions (5.3) or (5.4) for the classical relativistic strings in the path integral we can now define the (Euclidean) n-loop (quantum) functions for such strings (and we call it the *quantum theory of bosonic strings* or *the theory of random surfaces*, this latter notation emphasizing that it is the generalization of RWs to surfaces):

$$G(\ell_1,\ldots,\ell_n) = \int_{\partial F} \mathscr{D}F \, e^{-\kappa A[F]}, \quad \partial F = \{\ell_1,\ldots,\ell_n\}, \tag{5.5}$$

$$G(\ell_1,\ldots,\ell_n) = \int \mathscr{D}[g_{ab}] \int_{\partial F} \mathscr{D}X \, e^{-S[X_i,g_{ab}]} \tag{5.6}$$

As for the particle the notation $\mathscr{D}[g_{ab}]$ means that we should only integrate over intrinsic two-dimensional *geometries*. Different g_{ab} which just correspond to using different coordinate systems should not be counted as independent. In the following we will use the one of the two versions (5.5) and (5.6) that is most convenient for our discussion.

We will now present some general, formal arguments, related just to the fact that we have a path integral over surfaces. For this purpose it is most convenient to use (5.5).

The first thing to note is that interactions between strings seem to be present in the theory without introducing any coupling constants. We can talk about the propagator $G(\ell_1,\ell_2)$ of a string ℓ_1 to a string ℓ_2 by summing over all surface in the path integral (5.5) with boundaries ℓ_1 and ℓ_2. But without introducing any new coupling constant, $G(\ell_1,\ell_2,\ell_3)$ seems to contain the information about a string ℓ_1 propagating and splitting in two strings ℓ_2 and ℓ_3, and ℓ_1 and ℓ_2 joining to ℓ_3. This is a beautiful aspect of string theory and a feature alien to particle physics. The situation is shown in Fig. 5.1

Next, let us consider the two-point function, i.e. the two loops ℓ_1 and ℓ_2 are contracted to points x and y:

$$G(x,y) = \int_{\partial F = \{x,y\}} \mathcal{D}F \, e^{-\kappa A[F]}, \qquad (5.7)$$

We can write

$$G(x,y) = \int_0^\infty dA \, e^{-\kappa A} \int_{\partial F = \{x,y\}} \mathcal{D}F \, \delta(A[F]-A) = \int_0^\infty dA \, e^{-\kappa A} \, \mathcal{N}_2(A(x,y)) \qquad (5.8)$$

where $\mathcal{N}_2(A(x,y))$ denotes the number of surfaces with area A and two marked points fixed at x and y. We thus have the same situation as for the relativistic particle: *the propagator is completely determined if we know the number of surfaces in $\mathbb{R}^D$ with two marked points at x and y and area A.* Of course this number is infinite, and we need (as for the particle) to introduce a regularization in order to perform the counting. At the moment we will just assume we have such a regularization. The same statement is obviously true if we consider the n-point function $G(x(1),\ldots,x(n))$, just with $\mathcal{N}_n(A(x(1),\ldots,x(n)))$, the number of surfaces with n marked points located at $x(1),\ldots,x(n)$ and area A. Note also that these surfaces can self-intersect in $\mathbb{R}^D$. There is nothing in the action which prevents such self-intersection.

We obtain the *susceptibilities* as for RWs and BPs by integrateting the n-point functions over $n-1$ of the points:

$$\chi^{(n)}(\kappa) = \int \prod_{k=1}^{n-1} dx(k) \, G(x(1),\ldots,x(n)) = \int_0^\infty dA \, e^{-\kappa A} \mathcal{N}_n(A). \qquad (5.9)$$

where $\mathcal{N}_n(A)$ denotes the number of surfaces in $\mathbb{R}^D$ with area A and n marked points (and one of them kept fixed in order to eliminate translational invariance of G). Heuristically we have

$$\mathcal{N}_n(A) \approx A \, \mathcal{N}_{n-1}(A), \qquad (5.10)$$

for the same reason as discussed for BPs: it is the same integrals over surfaces, the only difference is that one class of surfaces has one more mark than the other, and this mark can put anywhere on the surface, i.e. the number of ways this can be done is proportional to A. Clearly one needs some kind of regularization to make this into a precise statement, but it should be true for all reasonable regularizations.

The relation between the numbers $\mathcal{N}_n(A)$ and the so-called susceptibility exponents γ_n is the same as we have already encountered for the RWs and BPs. Let us assume we have some regularization of our string theory[1] and that the number of surfaces grows exponentially with A, up to power like subleading corrections:

$$\mathcal{N}_n(A) \propto e^{\kappa_c A} A^{\gamma_n - 1} \left(1 + \mathcal{O}(A^{-1})\right) \tag{5.11}$$

Then

$$\chi^{(n)}(\kappa) \propto \int_0^\infty dA\, e^{-(\kappa - \kappa_c)A} A^{\gamma_n - 1} \left(1 + \mathcal{O}(A^{-1})\right) \xrightarrow[\kappa \to \kappa_c]{} \frac{c_n}{(\kappa - \kappa_c)^{\gamma_n}}. \tag{5.12}$$

κ_c will in general depend on the explicit regularization and in the continuum limit only $\kappa - \kappa_c$ will survive and correspond to a renormalized $\kappa_r = \kappa - \kappa_c$, as we have seen for the particle (and for BPs). Thus the precise exponential growth of the number of surfaces will depend on the regularization, but the subleading power-term relates directly to the continuum limit since this is what determine the divergent power of $\chi^{(n)}(\kappa)$ expressed in terms of the renormalized κ_r. Therefore the subleading power of the number of surfaces should be universal, independent of any (reasonable) regularization, and this turns out to be true.

A trivial consequence of (5.10) is that

$$\gamma_n = \gamma_{n-1} + 1 \qquad \chi^{(n)}(\kappa) \propto -\frac{d\chi^{(n-1)}(\kappa)}{d\kappa}. \tag{5.13}$$

and since $\chi^{(2)}(\kappa) = \chi(\kappa)$, the susceptibility for the two-point function, with an exponent we have called γ, we can write

$$\boxed{\gamma \equiv \gamma_2, \quad \text{and} \quad \chi^{(n)}(\kappa) \xrightarrow[\kappa \to \kappa_c]{} \frac{c_n}{(\kappa - \kappa_c)^{\gamma + n - 2}}} \tag{5.14}$$

It follows from simple geometry that $\boxed{\gamma > 0 \Rightarrow \gamma \leq \frac{1}{2}}$ as we will now argue.

$$\chi^{(n)}(\kappa) = \int \mathcal{D}F\, e^{-\kappa A[F]} \geq \int_{F_1 \cup \cdots \cup F_n} \int \prod_{k=1}^n \mathcal{D}F_k\, e^{-(A[F_1] + \cdots + A[F_n])} = \chi(\kappa)^n \tag{5.15}$$

$$\{F\} \equiv x_n \underset{x_0}{\overset{x_1\; x_2}{\bigodot}}_{x_3} \supseteq x_n \underset{x_0}{\overset{x_1}{\bigodot}}^{x_2} \equiv \{F_1\} \cup \cdots \cup \{F_n\} \tag{5.16}$$

---

[1] A very simple regularization, like the one mentioned for the particle, is to use a hypercubic lattice. For the particle the paths on the hypercubic lattice would follow the links and the geometric action would just be proportional to the number of links. For the string, the surfaces would be made from plaquettes (the sides of a minimal lattice hypercube), and again the action would be proportional to the number of the plaquettes constituting the surface. Everything said about counting can be made precise in this setting.

Random Surfaces and Bosonic Strings

The inequality is satisfied simply because there are more surfaces with n marked points x_i than surfaces with n marked points of the kind shown on the right hand figure. The n separate surfaces are assumed to join in a common "point" or little neighborhood around a point x_0, which is kept fixed while we integrate over $x_1,\ldots,x_n$, in this way producing $\chi(\kappa)^n$. We have used here the property, special for the geometric action, that $A[F_1 \cup \cdots \cup F_n] = A[F_1] + \cdots + A[F_n]$ and also that the decomposition shown to the right in (5.16) essentially is unique. This is the case for $n > 2$, but not for $n = 2$, where one cannot define a unique x_0 as illustrated here:

$$x_1 \quad x_0 \quad x_0' \quad x_2 \qquad (5.17)$$

From (5.15) we conclude

$$\frac{c_n}{(\kappa - \kappa_c)^{n-2+\gamma}} \geq \frac{c^n}{(\kappa - \kappa_c)^{n\gamma}} \quad \Rightarrow \quad \gamma \leq \frac{n-2}{n-1}, \quad n \geq 3, \qquad (5.18)$$

which is the desired result $\gamma \leq \frac{1}{2}$. Below we will study the case $n = 2$ closer and show that under some universality assumptions we obtain $\gamma = \frac{1}{2}$.

We will now apply the same kind of estimate to the two-point function $G(x-y)$ and show that it falls of exponentially.

$$G(x-y) = \int_{\partial F=\{x,y\}} \mathscr{D}F\, e^{-\kappa A[F]} \geq \int_{\partial F_1=\{y,z\}} \mathscr{D}F_1 \int_{\partial F_2=\{z,x\}} \mathscr{D}F_2\, e^{-(A[F_1]+A[F_2])} = G(z-y)G(x-z)$$
$$(5.19)$$

$$\{F\} \equiv y \quad x \;\supseteq\; y \quad z \quad x \equiv \{F_1\} \cup \{F_2\} \qquad (5.20)$$

Again we have the inequality simple because the number of surfaces pinched at z is fewer than the surfaces not pinched. Since $G(x-y) = G(|x-y|)$ we can write

$$-\ln G(|x-y|) \leq -\ln G(|x-z|) - \ln G(|z-y|), \qquad |x-y| = |x-z| + |z-y|. \quad (5.21)$$

Eq. (5.21) states that $-\ln G(|x-y|)$ is a *subadditive* function and from general theory (Fekete's lemma) we know that for such functions

$$\boxed{\lim_{|x-y|\to\infty} \frac{-\ln G(|x-y|)}{|x-y|} = m(\kappa)} \qquad (5.22)$$

If we assume that $G(x-y) \to 0$ for $|x-y| \to \infty$ the mass m has to be non-negative. In addition $G(x-y)$ will be a decreasing function of κ, i.e. $m(\kappa_1) \leq m(\kappa_1)$ for $\kappa_1 < \kappa_2$. We call $m(\kappa)$ the (lowest) mass of the string, and we can write

$$G(x-y) \approx c\, |x-y|^{\alpha} e^{-m(\kappa)|x-y|} \quad \text{for} \quad |x-y|m(\kappa) \gg 1, \qquad (5.23)$$

where the subleading exponent α in not determined by these general arguments. It should be noted that we could have applied precisely the same argument in the case of the free particle to show that the particle propagator, defined by the path integral with the action $S[P] = m_0 \ell[P]$, falls off exponentially.

Arguments similar to the ones leading to the existence of the mass $m(\kappa)$ also lead to the existence of a *string tension*. Let us first define the string tension. Consider the one-loop function $G(\ell_A)$. Thus the surfaces in the path integral have the boundary l_A. In addition *we assume the surfaces have no handles, i.e. that all surfaces have the topology of a disk*. We assume the curve defining ℓ_A is a planar loop in $\mathbb{R}^D$ with area A. We now define the string tension $\sigma(\kappa)$ similarly to the way (5.22) defines the mass $m(\kappa)$:

$$\boxed{\lim_{A \to \infty} \frac{-\ln G(\ell_A)}{A} = \sigma(\kappa)} \tag{5.24}$$

Again we expect from the very nature of the action to have $\sigma(\kappa_1) \leq \sigma(\kappa_2)$ for $\kappa_1 < \kappa_2$. Why do we call $\sigma(\kappa)$ the string tension? We can view $G(\ell_A)$ as the partition function for an ensemble of fluctuating surfaces ("membranes", but with very weird properties since they can self-intersect) where the boundary is kept fixed. The free energy of these membranes will be $F(A) = -\ln G(\ell_A)$ and the tension of the membrane is defined as the change of free energy per unit area when we change the area from A to $A + \Delta A$ by changing the boundary:

$$\Delta F(A) = \sigma(\kappa, A) \Delta A, \quad \text{i.e. for large } A \quad F(A) \approx \sigma(\kappa) A, \tag{5.25}$$

where we have assumed that the free energy is approximately extensive in the variable A for large A. That $\sigma(\kappa, A)$ is independent of A for large A is a remarkable consequence of (5.24) and shows that a relativistic string is very different from an ordinary rubber band, a fact already clear from the classical action (5.3). Consider a thin rectangular, planar surface $F = L_1 \times L_2$, $L_1 \gg L_2$ and increase $L_1 \to L_1 + \Delta L_1$. Then the classical (Euclidean) action will increase by $(\kappa L_2) \Delta L_1$, telling us that the corresponding tension of the rubber band is *independent* of L_1, no matter how much we stretch the band. For an ordinary rubber band the tension will of course increase with the stretching of the band.

From the definition of the one-loop function $G(\ell_A)$ with a planar boundary ℓ_A enclosing a two-dimensional domain of area A in $\mathbb{R}^D$

$$G(\ell_A) = \int_{\partial F = \ell_A} \mathscr{D}F \, e^{-\kappa A[F]}, \tag{5.26}$$

we see that there are more surfaces in the set $F(\ell_A)\}$ of surfaces with boundary ℓ_A than in the two set of surfaces where we have divided A in subset areas A_1 and A_2 along some additional boundary in the interior of the domain defining A. This is illustrate in the case of a rectangle of area A divided into two sub-rectangles of area

Random Surfaces and Bosonic Strings

A_1 and A_2 below:

$$\{F(\ell_A)\} \equiv \quad \supseteq \quad \equiv \{F_1(\ell_{A_1})\} \cup \{F_2(\ell_{A_2})\} \tag{5.27}$$

We thus conclude from (5.26) and (5.27), in the same way as for the two-point function in (5.19), that

$$G(\ell_A) \geq G(\ell_{A_1})\, G(\ell_{A_2}), \quad A = A_1 + A_2. \tag{5.28}$$

This leads (again as for the two-point function) to (5.24) and thus

$$G(\ell_A) \approx A^\alpha \, e^{-\sigma(\kappa) A} \quad \text{for} \quad A\sigma(\kappa) \gg 1. \tag{5.29}$$

Clearly these arguments are very simple and formal, based on counting of surfaces (the number of which is infinite) and the simple geometric form of the action. In order to prove them we need as a starting point to *define* the path integral over surfaces. As for the particle, a regularization is needed in order that we can count. Again there are many ways to introduce such a regularization. If we choose as the action $S[F] = \kappa A[F]$, (5.3), a very simple regularization is to use a hypercubic lattice, as already mentioned in footnote 1, this chapter. Most of the arguments given above can then be made mathematical rigorous. However, we will here use the other geometric action (5.4) and provide a regularization the path integral using that action. One reason for this choice is that we can use part of the regularization when we turn to the study of two-dimensional quantum gravity and so-called "non-critical strings". The first thing we have to deal with in that setting is how to count two-dimensional geometries $[g_{ab}]$.

5.2 REGULARIZING THE INTEGRATION OVER GEOMETRIES

In the case of RWs piecewise linear paths played an important role. In the case of surfaces it will play an equally important role. It will allow us to introduce geometry without having to introduce coordinate systems (and then afterwards have to get rid of this freedom by dividing by *Vol(diff)*). Stepping one dimension up, the natural replacement of a piecewise linear path is a piecewise linear surface, obtained by gluing together triangles. The lengths of links are given and each triangle is considered flat in the interior. In principle we can now calculate the shortest path between two points on the surface (it will be a certain piecewise linear path on the surface) and thus the intrinsic geometry of the surface is given. Note that this can be done without "really" introducing a coordinate system[2]. Consider the sphere S^2 of radius 1 in $\mathbb{R}^3$. We know that this sphere has an intrinsic scalar curvature (Gaussian curvature) 1. Now consider a triangulation of the kind described above, which approximate the sphere well. One would expect that it is also possible to assign a kind of intrinsic curvature to such a triangulation. But where should it be assigned? The interior of the triangles is declared flat, so one cannot in an intuitive way assign curvature to an interior point. One property of the intrinsic curvature is that it is "bending invariant" (Gauss' *Theorema Egregium*). This makes it unnatural to locate the intrinsic curvature on the links, since we can (to some extend) bend the triangulation along the links. We are then left with the vertices of the triangulation as the place to locate the intrinsic curvature, and this can indeed be done in a "natural" way, as will now be described. Geometrically one can "detect" intrinsic curvature by performing a parallel transportation along an infinitesimal curve surrounding a point v on the surface. If the area enclosed by the curve is dA and the so-called deficit angle, the angle between the vector before and after being transported around the curve, is denoted $d\theta$, one has

$$d\theta_v = R_v dA_v + \mathcal{O}(r^3) \tag{5.30}$$

where r is a "typical" diameter in the domain enclosed by the curve. If $R_v = 0$ the surface is locally flat at the point v. This can be understood in a simple way on our piecewise linear surfaces when performing a parallel transportation around a vertex as illustrated in Fig. 5.2. The deficit angle associated with the parallel transportation around a vertex v in a triangulation is

$$\varepsilon_v = 2\pi - \sum_{t \ni v} \alpha_t(v), \tag{5.31}$$

where the summation is over triangles t which have v as a vertex and $\alpha_t(v)$ denote the corresponding angles in the triangles. Of course this relation is not infinitesimal, like the relation (5.30) for a smooth surface and since we have defined the triangulated surface as flat in all other points than the vertices, it is more like assigning a δ-function-like curvature to the vertices. Writing dA_v in (5.30) as $\sqrt{g(\xi)}d^2\xi$ we can integrate the expression (5.30) over the whole surface. Correspondingly we can sum (5.31) over all vertices. Let us for a moment consider closed surfaces. We would then write

$$\text{smooth surfaces:} \int d^2\xi \sqrt{g(\xi)}\, R(\xi) \sim \sum_v \varepsilon_v \ : \text{triangulated surfaces} \tag{5.32}$$

It is now possible to show that $\sum_v \varepsilon_v$ *only depends on the topology of the triangulation*. We will first do that for a particular class of triangulations which will be of

---

[2]Of course we have to label the points in the interior of the triangles in some way, but the geometry in the interior of a triangle is defined by the length of the links and the statment that the interior is "flat".

Random Surfaces and Bosonic Strings

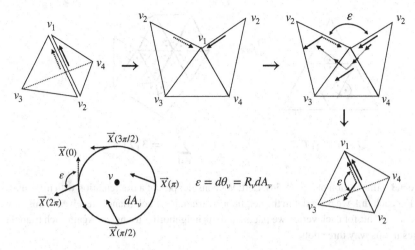

Figure 5.2 The lower left figure illustrates eq. (5.30). If there is a curvature R_v associated with the point v, a vector parallel transported around v will form an angle $R_v dA_v$ with the original vector. The sequence of four figures illustrates parallel transportation around vertex v_1 in a triangulation: We cut open the triangulation along link $\langle v_1 v_2 \rangle$ and unfold the triangles neighboring v_1 in a plane. The two vectors (black and dotted) are parallel since there is no curvature associated with the link $\langle v_1 v_2 \rangle$. In the plane it is trivial to parallel transport the black vector in triangle $\langle v_1 v_2 v_4 \rangle$ to triangle $\langle v_1 v_2 v_3 \rangle$ and compare it to the dotted vector. The angle between them is ε, the deficit angle shown on the figure and defined in (5.31). We can now close the link $\langle v_1 v_2 \rangle$ which was cut open, and we end with the lower right figure, with the black vector parallel transported around v_1.

special interest for us, namely the class of *equilateral triangulations*. For such a triangulation T with no boundaries we have that $\alpha_t(v) = \pi/3$ for all triangles and thus

$$\sum_v \varepsilon_v = 2\pi |V(T)| - \frac{\pi}{3} \sum_v n_v \tag{5.33}$$

where we have introduced the notation: T denotes a triangulation and at the same time the set of triangles in the triangulation. The number of triangles is denoted $|T|$. $V(T)$ denotes the set of vertices in T and $|V(T)|$ the number of vertices. $L(T)$ denotes the set of links in the triangulation and $|L(T)|$ the number of links in the triangulation. Finally, n_v denotes the order of the vertex in the triangulation, which we here define as the number of triangles to which the vertex belongs. From Fig. 5.3 we see that

$$2|L(T)| = 3|T|, \quad \sum_{v \in V(T)} n_v = 3|T| \tag{5.34}$$

and therefore

$$|V(T)| - \frac{1}{6}\sum_v n_v = |V(T)| - \frac{1}{2}|T| = |V(T)| - |L(T)| + |T| \equiv \chi(T) \tag{5.35}$$

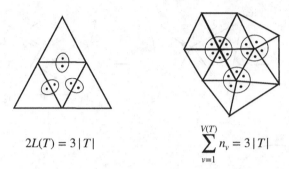

$$2L(T) = 3|T| \qquad \sum_{v=1}^{V(T)} n_v = 3|T|$$

Figure 5.3 Illustration of eqs. (5.34). Left: We consider part of a triangulation with no boundaries. For each link we put a dot in the neighboring triangles. Each triangle receives in this way three dots. Right: for each vertex we put a dot in its neighboring triangles. Again each triangle receives in this way three dots.

where $\chi(T)$ is the so-called *Euler characteristic* of the triangulation. Also, dropping the assumption that the triangles are equilateral, using eq. (5.34) it follows immediately that $\sum_v \varepsilon_v$ is still the same simply by using that the sum of angles in a (flat) triangle is π. Thus

$$\sum_v \varepsilon_v = 2\pi|V(T)| - \pi|T| = 2\pi(|V(T)| - |L(T)| + |T|) = 2\pi\chi(T). \qquad (5.36)$$

In general, if a surface S is covered by a set of polygons then one has

$$|P| - |L| + |V| \equiv \chi(S) = 2 - 2h - n, \qquad (5.37)$$

where $|P|$ is the number of polygons, $|L|$ the number of links and $|V|$ the number of vertices, h the number of handles of the surface and n the number of boundaries. Thus the Euler characteristic depends only on the topology and since the topology of a surface is characterized completely by the number of handles and number of boundaries, the Euler characteristics of a surface determines, for a fixed number of boundaries, the topology of the surface. From the discussion above it is not surprising that we have equality in (5.32), i.e.

$$\int_{\mathcal{M}} d^2\xi \sqrt{g(\xi)}\, R(\xi) = 2\pi\chi(\mathcal{M}) = \sum_{v \in V(T)} \varepsilon_v, \quad \text{(Gauss - Bonnet theorem)} \quad (5.38)$$

where $\mathcal{M}$ is a manifold and T a triangulation with the same topology. The theorem can be extended to surfaces with boundaries:

$$\int_{\mathcal{M}} d^2\xi \sqrt{g(\xi)}\, R(\xi) + \int_{\partial\mathcal{M}} ds\, k_g = \chi(\mathcal{M}) = \sum_{v \in V_I(T)} \varepsilon_v + \sum_{v \in V_B(T)} \varepsilon_v, \qquad (5.39)$$

where k_g denotes the geodesic curvature of the boundary curve and and ds the line element along the curve. $V_I(T)$ denotes the interior vertices while $V_B(T)$ denotes the

Random Surfaces and Bosonic Strings

vertices at the boundary of the triangulation T. Finally, ε_v for a boundary vertex v is changed from (5.31) to

$$\varepsilon_v = \pi - \sum_{t \ni v} \alpha_t, \quad v \in V_B(T). \tag{5.40}$$

It is seen from a figure similar to Fig. 5.2 that for each boundary vertex v, ε_v in (5.40) is just the rotation of the tangent vector moving around the vertex, when the boundary triangles are put down in a plane. This is of course also the interpretation of the infinitesimal term $k_g \, ds$ on $\mathcal{M}$.

According to (5.32) the term in the Einstein-Hilbert action for gravity with contains R does not change in two dimensions as long as we do not change the topology. This is the reason we could leave it out of a two-dimensional theory of gravity as long as we do not consider changing topologies, as already mentioned.

Let us now consider manifolds with h handles and n boundaries. Formally we can write

$$\int_{\mathcal{G}} \mathcal{D}[g_{ab}] \, (\,\cdot\,) = \int_0^\infty dA_{int} \int \mathcal{D}[g_{ab}] \, \delta\!\left(\int d^2\xi \sqrt{g} - A_{int}\right) (\,\cdot\,) \tag{5.41}$$

We denote the space of geometries with the given topology $\mathcal{G}$ and the subspace with a fixed area A_{int} by $\mathcal{G}_{A_{int}}$, and the delta-function in the last integral ensures that this integration is over $\mathcal{G}_{A_{int}}$. We use here the notation A_{int}, where "int" is an abbreviation of "internal", to signify that the area refers to metric g_{ab}, and not to the area of the surface measured by the metric induced from the embedding in $\mathbb{R}^D$. Consider now the space of equilateral triangulations which have the topology of S^2 with n boundaries and where the length of the links is ε. The area of such a triangle is $\frac{\sqrt{3}}{4} \varepsilon^2$. A triangulation of this kind will belong to $\mathcal{G}_{A_{int}}$ if the number of triangles in such a triangulation satisfies

$$|T|\left(\frac{\sqrt{3}}{4} \cdot \varepsilon^2\right) = A_{int}. \tag{5.42}$$

Denote this set of equilateral triangulations $\mathcal{T}(A_{int}, \varepsilon)$. Clearly the number of triangles for a triangulation in this set will go to infinity when $\varepsilon \to 0$. The main conjecture (which can be proven, but we will not do that here) is that this set of triangulation is sufficiently dense in the set $\mathcal{G}_{A_{int}}$ that we can write

$$\sum_{\mathcal{T}(A_{int}, \varepsilon)} (\,\cdot\,) \to \int_{\mathcal{G}_{A_{int}}} \mathcal{D}[g_{ab}] \, (\,\cdot\,) \quad \text{for} \quad \varepsilon \to 0, \tag{5.43}$$

and integrating in addition over the area A_{int}, and denoting the corresponding set of equilateral triangulations $\mathcal{T}(\varepsilon)$, we then formally write

$$\sum_{\mathcal{T}(\varepsilon)} (\,\cdot\,) \to \int_{\mathcal{G}} \mathcal{D}[g_{ab}] \, (\,\cdot\,) \quad \text{for} \quad \varepsilon \to 0. \tag{5.44}$$

Note that the set $\mathscr{T}(\varepsilon)$ is *independent* of ε, viewed as an abstract set of triangulations. ε will only enter in the implementation of $(\cdot)$, which is some function which is defined on a triangulation and may refer explicitly to the length of the links.

One function that has to be included in $(\cdot)$ in (5.44) is the exponential of the action itself. Let us now give a natural definition of this action on a triangulation T. Let us first map this triangulation to a triangulated, piecewise linear surface in $\mathbb{R}^D$ by mapping the vertices $v \in V(T)$ to points $X_i(v) \in \mathbb{R}^D$, and defined the corresponding surface by declaring the straight line in $\mathbb{R}^D$ from $X_i(v_1)$ to $X_i(v_2)$ a link if $\langle v_1 v_2 \rangle \in L(T)$, and similarly $X_i(v_1), X_i(v_2), X_i(v_3)$ for a triangle in $\mathbb{R}^D$ if v_1, v_2, v_2 defines a triangle in T. On this piecewise linear surface in $\mathbb{R}^D$ we can now define a coordinate system ξ such that $X_i(\xi)$ are the coordinates on the surface, and we then find for a closed surface

$$\int d^2\xi \sqrt{g(\xi)} \left[g^{ab}(\xi) \frac{\partial X_i}{\partial \xi^a} \frac{\partial X_i}{\partial \xi^b} \right] = C \sum_{\langle vv' \rangle \in L(T)} (X_i(v) - X_i(v'))^2, \quad C = \frac{1}{\sqrt{3}}. \quad (5.45)$$

Note that if there are no boundaries this equation can also be written

$$\int d^2\xi \sqrt{g(\xi)} X_i(\xi)(-\Delta_g) X_i(\xi) \big] = C \sum_{v,v' \in V(T)} X_i(v)(-\Delta_{vv'}) X_i(v'), \quad (5.46)$$

where Δ_g denotes the *Laplace-Beltrami operator* on S^2 with metric $g_{ab}(\xi)$ and $\Delta_{vv'}$ denotes the *combinatorial Laplacian* on triangulation T. The combinatorial Laplacian is defined as a $|V(T)| \times |V(T)|$ matrix where the entries in the diagonal are $-n_v$, the order of the vertex v, and the vv' entry is 1 if v and v' are neighbors (i.e. belong to the same link), and zero otherwise. The Laplace-Betrami operator is defined as

$$\Delta_g = \frac{1}{\sqrt{g(\xi)}} \frac{\partial}{\partial \xi^a} g^{ab}(\xi) \frac{\partial}{\partial \xi^b}. \quad (5.47)$$

While the rhs of eqs. (5.45) and (5.46) indeed look like reasonable discretizations of the lhs of these equations, we can actually derive the discretized expressions from our piecewise linear surface picture, which is not really a discretization, but rather a special choice of surface. First consider the given abstract triangulation T where each link has length ε as embedded in some higher dimensional flat space $\mathbb{R}^k$ such that distances are preserved. Thus we have a mapping $v \in V(T) \to y(v) \in \mathbb{R}^k$ such that for all links $\langle vv' \rangle \in L(T)$ we have $|y(v) - y(v')| = \varepsilon$. There are theorems which ensure that there exists a sufficient large k such that all Ts can be mapped isometrically to $\mathbb{R}^k$ ($k = 7$, Nash's theorem). Let us introduce a coordinate system for each triangle (the total coordinate system is then the union of these, including transition functions telling us how to go from one to the other coordinate system in regions of overlap (which will be the links)). It is convenient to introduce barycentric coordinates for the triangles. Consider the triangles $t \in T$ defined by the vertices v_1, v_2 and v_3. The coordinates of a point in the triangle will then be

$$y(\xi) = \xi^1 y(v_1) + \xi^2 y(v_2) + (1 - \xi^1 - \xi^2) y(v_3), \quad 0 \le \xi^1 + \xi^2 \le 1, \quad \xi^a \in [0,1]. \quad (5.48)$$

Random Surfaces and Bosonic Strings

This assigns coordinate ξ to a point in the triangle defined by the three vertices v_i and the corresponding values of $X_i(\xi)$ are

$$X_i(\xi) = \xi^1 X_i(v_1) + \xi^2 X_i(v_2) + (1 - \xi^1 - \xi^2) X_i(v_3). \tag{5.49}$$

Since the metric is flat and trivially $\delta_{\alpha\beta}$ in $\mathbb{R}^k$ where y lives we have

$$g_{ab}(\xi) = \delta_{\alpha\beta} \frac{\partial y^\alpha}{\partial \xi^a} \frac{\partial y^\beta}{\partial \xi^b} = \varepsilon^2 \begin{pmatrix} 1 & \frac{1}{2} \\ \frac{1}{2} & 1 \end{pmatrix}, \quad g^{ab} = \frac{4}{3\varepsilon^2} \begin{pmatrix} 1 & -\frac{1}{2} \\ -\frac{1}{2} & 1 \end{pmatrix}, \quad \sqrt{g} = \frac{\sqrt{3}\varepsilon^2}{2}. \tag{5.50}$$

Integration over one triangle thus produces (after a little calculation)

$$\int_t d^2\xi \sqrt{g}\, g^{ab} \frac{\partial X_i}{\partial \xi^a} \frac{\partial X_i}{\partial \xi^b} = \frac{1}{\sqrt{3}} \sum_{\langle vv' \rangle \in L(t)} (X_i(v) - X_i(v'))^2. \tag{5.51}$$

Summing over all triangles then leads to (5.45) (links should only be counted once in neighboring triangles, since they will represent the overlap of the two coordinate systems in the triangles). For a given triangulation, i.e. a given intrinsic geometry g_{ab} of the corresponding piecewise linear surface, the action (5.2) is then

$$S[T,X] = \frac{\kappa}{2\sqrt{3}} \sum_{\langle vv' \rangle \in L(T)} (X_i(v) - X_i(v'))^2 + \frac{\kappa \lambda \sqrt{3}}{8} \varepsilon^2 |T|. \tag{5.52}$$

As for the particle it is convenience in the following to consider the path integral in terms of dimensionless variables, and we thus redefine $(\kappa/\sqrt{3})^{1/2} X \to X$ and $\kappa\lambda\sqrt{3}\varepsilon^2/8 \to \mu$ and our dimensionless action is finally

$$S[X,T] = \frac{1}{2} \sum_{\langle vv' \rangle \in L(T)} (X_i(v) - X_i(v'))^2 + \mu |T| \tag{5.53}$$

We can now define the regularized version of (5.6)

$$G_\mu(\ell_1,\ldots,\ell_n) = \sum_{T \in \mathcal{T}(\ell_1,\ldots,\ell_n)} e^{-\mu|T|} \int \prod_{v \in V(T)/\{\ell_1,\ldots,\ell_n\}} dX(v)\, e^{-S[X,T]} \tag{5.54}$$

Here T denotes an abstract triangulation with n boundaries and $\mathcal{T}(\ell_1,\ldots,\ell_n)$ the set of such triangulations. Each boundary consists of a number of vertices and the associated links, connecting the vertices to a loop. These loops have a double meaning in the notation above. They denote at the same time the boundary-loop in the abstract triangulation *and* its image in $\mathbb{R}^D$ by the map $v \to X(v)$. In (5.54) one does not integrate over the $X(v)$ where v is a boundary vertex in the triangulation: the boundaries are kept fixed in $\mathbb{R}^D$.

Finally, the regularized n-point function for surfaces with h handles and n punctures is defined by

$$G_\mu(x(v_1),\ldots,x((v_n)) = \sum_{T \in \mathcal{T}(v_1,\ldots,v_n)} e^{-\mu|T|} \int \prod_{v \in V(T)/\{v_1,\ldots,v_n\}} dX(v)\, e^{-S[X,T]} \tag{5.55}$$

where $\mathcal{T}(v_1,\ldots,v_n)$ denotes the triangulations with h handles and marked vertices $v_1,\ldots,v_n$, and where the coordinates $X_i(v_k)$, $k=1,\ldots,n$ on the surface are kept fixed, while the rest are integrated over. The regulated susceptibilities are now defined as in (5.9), except that they now, with the use of the action (5.2) instead of (5.1) and the rescaling of X, will be a function of μ:

$$\chi^{(n)}(\mu) = \int \prod_{k=1}^{n-1} dx(v_i)\, G_\mu(x(v_1),\ldots,x(v_n)). \tag{5.56}$$

In particular, let us mention that for the one-point function $G_\mu(x(v_1))$ (which is also equal $\chi^{(1)}(\mu)$ and independent of $x(v_1)$ by translational invariance), one can explicitly perform the Gaussian integrals in (5.55) by using (5.45) and (5.46). Introducing $Y(v) = X(v) - x(v_1)$ we find

$$\int \prod_{v \in V(T)/\{v_1\}} dY(v)\, e^{-S_g[Y,T]} = \left(\frac{(2\pi)^{|V(T)|-1}}{\det(-\Delta'_{vv'}(T))}\right)^{\frac{D}{2}}, \tag{5.57}$$

where $\Delta'_{vv'}(T)$ denotes the $(|V(T)|-1) \times (|V(T)|-1)$ matrix constructed from the combinatoral Laplacian defined for T by deleting the v_1th row and column. Thus we have (using $|V(T)| = |T|/2 + 2 - 2h$)

$$\boxed{G_\mu^{(1)}(x) \equiv \chi^{(1)}(\mu) = \sum_{T \in \mathcal{T}(v_1)} e^{-(\mu-\mu_0)|T|} \left(\frac{(2\pi)^{1-2h}}{\det(-\Delta'_{vv'}(T))}\right)^{\frac{D}{2}}} \qquad e^{\mu_0} = (2\pi)^{D/4}, \tag{5.58}$$

which is a remarkable explicit formula, valid for all D by analytic continuation in D.

Using these regularized functions it is now possible to prove the statements made above for the Green functions. We will not given the proofs here, but let us summarized the statements which can be made.

Theorem 1: There exists a critical value μ_c such that $G_\mu(\ell_1,\ldots,\ell_n)$ is defined by (5.54) is convergent for $\mu > \mu_c$ and divergent for $\mu < \mu_c$. This critical value is independent of the number, positions, and lengths for the boundary loops as well as the number of handles of the surface.

Theorem 2: The two-point function $G_\mu(x,y)$ falls of exponentially with the distance $|x-y|$ between the two points for $\mu > \mu_c$ and the mass

$$m(\mu) = -\lim_{|x-y|\to\infty} \frac{\ln G_\mu(x-y)}{|x-y|} > 0.$$

The mass is independent of the number of handles of the surface. The two-loop function $G_\mu(\ell_1,\ell_2)$ has the same exponential fall off when the distance between the two loops goes to to infinity.

Random Surfaces and Bosonic Strings

Theorem 3: Consider the ensemble of surfaces with no handles ($h=0$). The string tension $\sigma(\mu)$, defined as the exponential fall off of the one-loop function $G_\mu(\ell_A)$ for planar loops ℓ_A enclosing a domain of area A in $\mathbb{R}^D$, exists for any $\mu > \mu_c$ and

$$\sigma(\mu) = -\lim_{A \to \infty} \frac{\ln G_\mu(\ell_A)}{A} > 0.$$

Theorem 1 implies that if we decompose $G_\mu(\ell_1, \ldots, \ell_n)$ in (5.54) in a sum over Green functions constructed from $|T| = N$ triangles

$$G_\mu(\ell_1, \ldots, \ell_n) = \sum_N e^{-\mu N} G_N(\ell_1, \ldots, \ell_l), \tag{5.59}$$

then

$$G_N(\ell_1, \ldots, \ell_n) = e^{\mu_c N} F(\ell_1, \ldots, \ell_m; N), \tag{5.60}$$

where $F(\ell, \ldots, \ell_n; N)$ is subleading in N and

$$G_\mu(\ell_1, \ldots, \ell_n) \sim \sum_N e^{-(\mu - \mu_c) N} F(\ell_1, \ldots, \ell_n; N). \tag{5.61}$$

It is now seen that the only way large N can dominate the sum is when $\mu \to \mu_c$. Recall from (5.42) that $N\varepsilon^2 \propto A_{int}$, the intrinsic area of a surface. As an order of magnitude estimate, (5.61) suggests that $\langle N \rangle \sim \frac{1}{\mu - \mu_c}$ (and we will later prove that this is true for $n \geq 2$, while one has $\langle N \rangle \sim \frac{1}{\sqrt{\mu - \mu_c}}$ for $n = 1$). Thus it natural to take a limit

$$\mu - \mu_c = \Lambda \varepsilon^2 \quad \Rightarrow \quad \langle A_{int} \rangle \propto \langle N \varepsilon^2 \rangle \sim \frac{1}{\Lambda}, \tag{5.62}$$

which is the limit we were aiming for in (5.42), and a limit which is natural from the way we introduced the dimensionless parameter μ in the first place, namely as $\mu \propto \kappa \lambda \varepsilon^2$. The only new thing in (5.62) is that μ undergoes an additive renormalization, but that should not be a surprise since we have already seen this in the case of RWs and BPs, where the constant μ_c was related to the exponential growth of the number of RWs or BPs with length or size, respectively. The origin of μ_c in (5.60) is exactly the same. Note however that we have not yet made any contact with the actual size of the surfaces $X_i(v)$ embedded in $\mathbb{R}^D$, as we did in the case of RWs or BPs. Clearly the behavior of $m(\mu)$ and (and as something new: $\sigma(\mu)$) for $\mu \to \mu_c$ will of utmost importance when trying to do that, precisely as it was the case for RWs and BPs.

Before we study the behavior of $m(\mu)$ and $\sigma(\mu)$ in the limit $\mu \to \mu_c$, we will make a digression and discuss the summation over the number of handles of the surfaces.

5.3 DIGRESSION: SUMMATION OVER TOPOLOGIES

As mentioned above a beautiful aspect of string theory is that viewed from the path integration perspective, as a theory of random surfaces in $\mathbb{R}^D$, a n-loop function is as natural as a two-loop function. This leads to the inclusion of surfaces with handles, since we can now view a two-loop function with one handles as a "time"-evolution

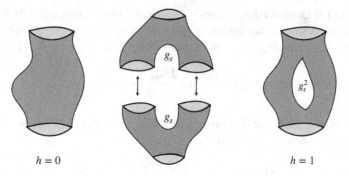

Figure 5.4 Left figure: Propagation of a string with $h=0$. The topology of the surface is that of a cylinder. Middle figure: a string, still with $h=0$, can split in two during propagation, with associated coupling constant g_s, or two strings can join, again with associated g_s. Right figure: Combining the splitting and the joining, this lead to propagation of a string, but now spanning a surface with $h=1$. Associated with $h=1$ is a factor g_s^2 coming from the splitting and joining of the string.

of the two-loop function, where one loop splits in two loops (made possible because we have three-loop functions), which then "later" join again to one loop (again made possible because we have three-loop functions). This is illustrated in Fig. 5.4. Thus from the figure it is clear that if one includes in the two-loop function the propagation of surfaces with one handle, by iteration one has to include surfaces with an arbitrary high number of handles, i..e. we have to sum over all surfaces with two boundaries and all handles:

$$G_\mu(\ell_1,\ell_2) = \sum_{h=0}^{\infty} G_\mu^{(h)}(\ell_1,\ell_2). \tag{5.63}$$

Since we are now discussing the change of topology we have to step back to the Einstein-Hilbert action (3.4). As discussed below eq. (3.7) the reason we dropped the curvature term in the action in (5.2) is that with no topology change of the surface, it would only contribute with a constant, which we calculated in (5.38) and (5.39). However, it is natural to include it when we consider the sum in (5.63):

$$\exp\left\{\frac{1}{2\pi G}\left(\int_M d^2\xi\sqrt{g}R + \int_{\partial M} ds\, k_g\right)\right\} = \exp\left\{\frac{2-n}{G}\right\}\exp\left\{-\frac{2h}{G}\right\}. \tag{5.64}$$

and we then replace (5.63) by

$$G_\mu(\ell_1,\ell_2) = \sum_{h=0}^{\infty} g_s^{2h}\, G_\mu^{(h)}(\ell_1,\ell_2) \qquad \boxed{g_s = e^{-\frac{1}{G}}} \tag{5.65}$$

where we have defined a new coupling constant, the so-called *string coupling constant* in terms of G, the gravitational coupling constant of two-dimensional gravity. Looking at Fig. 5.4 there is a factor g_s associated with a splitting of a string in two (the number of boundaries changes from 1 to 2), and again a factor of g_s associated with the joining of two strings to one, and thus a total factor of g_s^2 associated with string propagation via a surface with one handle compared to string propagation via a surface with no handles.

Using our regularization we have managed to define $G_\mu^{(h)}(\ell_1,\ldots,\ell_n)$ for any $\mu > \mu_c$. Does the regularization also provide us with a definition of $G_\mu(\ell_1,\ldots,\ell_2)$ by a formula like (5.65) where we sum over all topologies? Clearly (5.65) provides us with a perturbation theory, *a string perturbation theory*. We "just" have to calculate the contributions for each h, and then perform the sum. And for any value of h it is clear that the contribution from order $h+1$ will be small if we choose g_s sufficiently small.

First a few general remarks. It should be emphasized that the reason it makes sense to talk about an interesting perturbation theory is Theorem 1, which states that μ_c is the same for all h. This is a remarkable result, in particular because μ_c is not a universal constant. It will depend on the way we have chosen to regularize our theory. However, for any reasonable regularization, the statement in Theorem 1 is true. We need a common μ_c for all h, since our real interest is in the "continuum" limit where the cut-off ε in (5.62) is taken to zero. Next, viewing (5.65) as a perturbation series, can we in principle perform the sum? If the series is convergent, no problem. Below we will show that the series is *not* convergent. However, that should not be so surprising, since most perturbation series that one encounters are only asymptotic series. This implies in the wording used above that although it is true that we for a given order h can choose g_s such that the contribution to order h is small, this choice of g_s cannot be made independently of h. Eventually, for any fixed g_s, the large-h contribution from $G_\mu^{(h)}$ will always be large even if multiplied by g_s^{2h}. This does not necessarily mean that the sum cannot be defined and it does not necessarily mean that there is not a well-defined answer that one can agree upon. A trivial example of this situation is the perturbative series of the anharmonic oscillator in quantum mechanics. The perturbation theory is in this case only an asymptotic series. However, the summation can be performed by several of the standard methods for summing divergent series, e.g. the so-called *Borel summation* (which we define and discuss in Problem Set 8). It provides us with an answer. Is this answer the correct one (clearly one can get any number by stupid summation of a divergent series)? Yes, we know this because we can define the quantum theory of the anharmonic oscillator in a way, which is independent of its perturbation expansion, and we can then prove, using this definition, that *if* one chooses to perform an perturbative expansion, the Borel sum of the perturbation series will give the correct result. In the case of string theory, we need something similar: we need *as a minimum* a regularization which for a non-zero cut-off provides us with well-defined expressions for the n-loop functions summed over all handles h. Since we have finite well-defined expressions for each h and even a perturbation expansion, it is tempting to try to *define* our theory including all h, by simply using (5.54) and declaring that $\mathscr{T}(\ell_1,\ldots,\ell_n)$ is the class of all triangulations with n boundaries, independent of h. However, as we will show below, it does not work. The expression is simply not well-defined except as a formal perturbation series in h.

Nevertheless, we might still be able to sum the divergent perturbation series and obtain results which might be interesting if they point toward new physics, even if the result in this way is not completely well defined. Let us illustrate this, and the nature of the divergent perturbation series by analysing $\chi^{(1)}(\mu)$ given by (5.58). Let us start with the simplest situation, namely choosing $D = 0$. This is then pure two-dimensional quantum gravity (which we will study in some detail in the next chapter). Since we now have a changing topology we incorporate (5.64) and (5.65) and write:

$$\chi^{(1)}(\mu) = \sum_h g_s^{2h-1} \sum_N e^{-\mu N} \sum_{T \in \mathcal{T}_N^{(h)}(v_1)} 1 = \sum_h g_s^{2h-1} \sum_N e^{-\mu N} \mathcal{N}_1^{(h)}(N) \quad (5.66)$$

where the summation over triangulations is a sum over handles h and for given h a sum over the number of triangles, and for given h, N a sum over all such triangulations with one marked vertex v_1. Finally, $\mathcal{N}_1^{(h)}(N)$ denotes the number of triangulations with N triangles, h handles and one marked vertex. One can calculate the asymptotic form of $\mathcal{N}_1^{(h)}(N)$. We will do that in the next chapter in the simplest case of $h = 0$. The result is

$$\mathcal{N}_1^{(h)}(N) = c_h N^{5(h-1)/2} e^{\mu_c N} \left(1 + \mathcal{O}(1/N)\right) \quad (5.67)$$

where c_h is bounded as function of h. For any fixed h the number of triangulations grows exponentially and for a fixed h the critical μ in (5.66) will be the μ_c which appears in this exponential grows. Again, for fixed h, large N will dominate in (5.66) and it makes some sense to use the asymptotic form (5.67) in (5.66) if we are interested in the limit $\mu \to \mu_c$. Doing that we obtain

$$\chi^{(1)}(\mu) \approx \sum_h c_h g_s^{2h-1} \sum_N N^{5(h-1)/2} e^{-(\mu-\mu_c)N}$$

$$\approx \frac{g_s}{\mu-\mu_c} \sum_h c_h \Gamma\left(\frac{5h}{2} - \frac{3}{2}\right) \left(\frac{g_s}{(\mu-\mu_c)^{5/4}}\right)^{2h-2}, \quad (5.68)$$

where we have replaced the summation with an integration, which is allowed for $h > 0$, but not really for $h = 0$. Two aspects are clear: first of all the behavior for $\mu \to \mu_c$ becomes more and more singular for $\mu \to \mu_c$ with increasing h. Thus the sum makes no sense in that limit (which is the one we are interested in!) unless we scale the string coupling constant to zero together with taking $\mu \to \mu_c$. Such a limit is called *the double scaling limit*. We then demand (recalling that $g_s = e^{-1/G}$, G the gravitational coupling constant)

$$\frac{e^{-1/G}}{(\mu-\mu_c)^{5/4}} = e^{-1/G_R}, \quad \text{or} \quad \frac{1}{G} = \frac{5}{4} \ln \frac{1}{\mu-\mu_c} + \frac{1}{G_R}. \quad (5.69)$$

This double scaling limit thus has the intriguing interpretation as a *renormalization* of the gravitational coupling constant G: for $\mu \to \mu_c$ the "bare" $1/G$ goes to infinity, but leaves behind a renormalized G_R. We can now write, ignoring the term in front of the sum in (5.68):

$$\chi^{(1)}(G_R) \propto \sum_h c_h \Gamma\left(\frac{5h}{2} - \frac{3}{2}\right) e^{\frac{\chi(h)}{G_R}}. \tag{5.70}$$

So our "renormalized" $\chi^{(1)}(G_R)$ is a sum of contributions for each h-sector, given by the Einstein action term for that sector, but with a renormalized gravitational coupling constant, and the "action contribution" multiplied by the "number of geometries" with handle h. This is an amazing formula, but unfortunately the series is divergent for any fixed value of G_R, since the factorial factor grows too rapidly. It is not even Borel summable. Nevertheless, it can be summed! We will not discuss here the methods one can use, but they are discussed in Problem Set 8[3]. As already mentioned, unfortunately the result is not unique and the various results have some troublesome features, but it is not ruled out that one might find the correct, physical argument to select the "correct" sum.

If we return to the surfaces embedded in D dimensions one could hope that the integration over X-coordinates could help to "tame" the sum over topologies in (5.58) when $\mathcal{T}(v_1)$ means all triangulations, irrespective of the number of handles. Unfortunately it is not the case as we will now argue. The determinant in (5.58) is the result of the Gaussian integration over $X(v)$s. There is a theorem called *Kirchoff's matrix-tree theorem* which states that the determinant is equal to the number of spanning trees in the triangulation, where a spanning tree is a connected tree-subgraph reaching all vertices. Thus we have the following estimate:

$$1 \leq \det(-\Delta_{vv'}(T)) \leq \prod_{v \in V(T)} n_v, \tag{5.71}$$

since the product of vertex orders is clearly larger than or equal to the number of spanning trees. Let us now be more specific with the class of triangulations we consider (we will also need this in the next subsection). We denote by $\mathcal{T}^{(3)}$ the class which satisfies three conditions. (1): the boundary of the triangles sharing a vertex is a circle (i.e. locally, around the vertex, we have $\mathbb{R}^2$ topology). (2): a link is uniquely defined by its vertices (i.e. we cannot have two links connected to the same two vertices). (3): a triangle is uniquely defined by its three vertices. This implies that $|T| \geq 4$. If a triangulation has h handles it will have h independent non-contractable loops, and for triangulations in class $\mathcal{T}^{(3)}$ this implies that it contains at least h vertices. Recall (5.35): $|V(T)| - |T|/2 = 2 - 2h$. Thus

$$|V(T)| \leq |T| \quad \text{and} \quad \frac{3|T|}{|V(T)|} = 6\frac{|V(T)| + 2h - 2}{|V(T)|} \leq 18. \tag{5.72}$$

---

[3] Also, in Problem Set 9 it is shown have to carry out the calculations hinted above in detail, in the case of BPs, where we enlarge the set of BPs from trees to trees with loops. The number of such polymers will then grow factorially, not exponentially, with the number of links, and we have precisely the problem above. Nevertheless, it is possible to perform the summation over such BPs with loops explicitly.

Now we can estimate that

$$\sum_v \ln n_v \leq |V(T)| \ln \left(\frac{\sum_v n_v}{|V(T)|}\right) = |V(T)| \ln \left(\frac{3|T|}{|V(T)|}\right) \leq |T| \ln 18. \qquad (5.73)$$

Finally, for D positive

$$\left(\det(-\Delta'_{vv'}(T))\right)^{-D/2} \geq e^{-\frac{D\ln 18}{2}|T|}, \qquad (5.74)$$

and using the rhs in (5.58) we are basically getting back to the $D = 0$ situation, just with a shifted μ. For negative D we can use the lower estimate in (5.71) and replace the determinant by 1 and reach the same conclusion. The expression (5.58) is thus infinite unless we invent some fancy summation procedure, as discussed.

5.4 SCALING OF THE MASS

We have already argued that the string susceptibility $\gamma_s \leq \frac{1}{2}$. We will now show that $\gamma_s > 0 \Rightarrow \gamma_s = \frac{1}{2}$. The arguments given here will use triangulations and the action (5.53) and will be heuristic in nature, and the basic assumption is *universality*: the critical behavior should be independent of the detailed class of triangulations used. A rigorous proof can be given using the hypercubic regularization described earlier. We have defined the set of triangulations $\mathscr{T}^{(3)}$. Let us now specific it further, and define $\mathscr{T}_1^{(3)}$ as the set of $\mathscr{T}^{(3)}$-triangulations with one boundary which consists of a double-link, i.e. two links connecting the same two boundary vertices. Similarly, $\mathscr{T}_2^{(3)}$ is defined as the set of $\mathscr{T}^{(3)}$-triangulations with two boundaries, where also the other boundary consists of a double-link. By definition there are no interior double-links. Let us define a larger class of triangulations denoted by $\mathscr{T}^{(2)}$, where we allow double-links (but not triple links etc.), but only if cutting the triangulation along the double-link will separate the triangulation in two disconnected parts. One would not expect our strings defined on this class to exhibit a critical behavior different from the strings defined on $\mathscr{T}^{(3)}$ since also in $\mathscr{T}^{(3)}$ it might be possible to cut a triangulation in two disconnect parts, not a along a "two-loop" but along a "three-loop", as shown in Fig. 5.5. Clearly this difference should not matter when the triangulations are very large, as is the case for those which determine the critical behavior. In class $\mathscr{T}^{(2)}$ one now defines $\mathscr{T}_1^{(2)}$ and $\mathscr{T}_2^{(2)}$ in the same way as for $\mathscr{T}^{(3)}$. Let now $\ell^{(d)}$ denote such a boundary double link. It will depend on coordinates x_1 and x_2 which we do usually not include in the integration. However, if we decided to integrate over x_2, say, in average it will be at a distance of order 1 from x_1, since x_1 and x_2 interact via a Gaussian term. This distance is very small compared to average distances to most vertices in the triangulation if $|T| \gg 1$. We will thus simply approximate the two boundary points by one point x and in this approximation the one-loop function corresponding the $G_\mu(\ell^{(d)})$ simply becomes the one-point function $G_\mu(x)$, *which is independent of x*. In this approximation we see that the difference between between $\mathscr{T}^{(3)}$ and $\mathscr{T}^{(2)}$ becomes local as shown Fig. 5.6, and summarized by the following

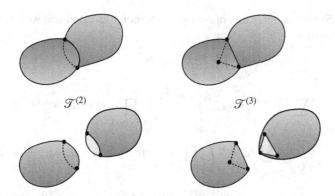

Figure 5.5 A triangulation belonging to $\mathscr{T}^{(2)}$ can be cut in two disconnected parts along any two-loop which is not a boundary. However, "ordinary" triangulations in $\mathscr{T}^{(3)}$ may also be cut in two along a three-loop. Such a three-loop is not necessarily a triangle belonging to the triangulation, as shown in the right part of the figure.

change of assignment to each internal link:

$$e^{-\frac{1}{2}(X(v_1)-X(v_2))^2} \to \left(1+G_\mu^{(1)}\right) e^{-\frac{1}{2}(X(v_1)-X(v_2))^2}, \tag{5.75}$$

since every link in a $\mathscr{T}^{(3)}$ triangulation (which is not a boundary link) can also be a double-link which serves as the boundary for an arbitrary "outgrowth" belonging to $\mathscr{T}_1^{(2)}$.

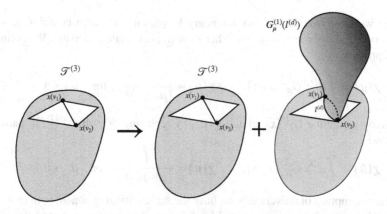

Figure 5.6 An illustration of the relation between the $\mathscr{T}^{(3)}$ and $\mathscr{T}^{(2)}$ triangulations. Any intrinsic link in a $\mathscr{T}^{(3)}$ triangulation can be split open to a double link $l^{(d)}$, to which one can attach an outgrowth $G_\mu^{(1)}(l^{(d)})$. The resulting triangulation now belongs to $\mathscr{T}^{(2)}$.

In (5.75) $G_\mu^{(1)}$ denotes this one-loop or one-point function and we can now make a decomposition:

$$\begin{aligned}
G_\mu^{(1)} &= \sum_{T \in \mathcal{T}_1^{(2)}} e^{-\mu|T|} \int \prod_{v \in T/\{v_1\}} dX(v) \, e^{-S_g[X,T]} \\
&= \sum_{\bar{T} \in \mathcal{T}_1^{(3)}} e^{-\bar{\mu}|\bar{T}|} \left(1+G_\mu^{(1)}\right)^{|L(\bar{T})|-2} \int \prod_{v \in \bar{T}/\{v_1\}} dX(v) \, e^{-S_g[X,\bar{T}]} \\
&= \frac{1}{1+G_\mu^{(1)}} \sum_{\bar{T} \in \mathcal{T}_1^{(3)}} e^{-\bar{\mu}|\bar{T}|} \int \prod_{v \in \bar{T}/\{v_1\}} dX(v) \, e^{-S_g[X,\bar{T}]}, \quad \bar{\mu} = \mu - \frac{3}{2} \ln\left(1+G_\mu^{(1)}\right),
\end{aligned}$$

where $(\bar{\cdot})$ refers to the ensemble $\mathcal{T}^{(3)}$ and non-bar quantities to the ensemble $\mathcal{T}^{(2)}$, and where we have used $|L(\bar{T})| - 2 = \frac{3}{2}|\bar{T}| - 1$ when we have two boundary links. Finally, v_1 is one of the vertices in the boundary loop (we integrate over the other vertex, but it is not important in the scaling limit). The rhs is just $\bar{G}_{\bar{\mu}}^{(1)}$ except for factor $(1+G_\mu^{(1)})^{-1}$ coming from the 1 in $\frac{3}{2}|\bar{T}|-1$, and the decomposition used is shown in Fig. 5.7. Summarizing

$$\boxed{G_\mu^{(1)} = \left(1+G_\mu^{(1)}\right)^{-1} \bar{G}_{\bar{\mu}}^{(1)}, \quad \bar{\mu} = \mu - \frac{3}{2} \ln\left(1+G_\mu^{(1)}\right)}. \tag{5.76}$$

Let us now define the two-loop or two-point function (we will not distinguish when the boundaries are just double-links) in the ensemble $\mathcal{T}_2^{(2)}$ as

$$G_\mu^{(2)}(x-y) = \sum_{T \in \mathcal{T}_2^{(2)}} e^{-\mu|T|} \int \prod_{v \in T/\{v_1,v_2\}} dX(v) \, e^{-S_g[X,T]}, \quad y = x(v_1), \, x = x(v_2), \tag{5.77}$$

where we have two double-link boundary loops, one of which contains vertex v_1 with coordinate $y = x(v_1)$ and the other v_2 with coordinate $x = x(v_2)$. We define the susceptibility as

$$\chi(\mu) = \int d^D y \, G_\mu^{(2)}(x-y), \quad \chi(\mu) \propto \frac{1}{(\mu-\mu_c)^{\gamma_s}} \quad \text{for} \quad \mu \to \mu_c. \tag{5.78}$$

Similarly we define $\bar{G}_{\bar{\mu}}^{(2)}(x-y)$ and $\bar{\chi}(\bar{\mu})$, by replacing $\mathcal{T}^{(2)}$ by $\mathcal{T}^{(3)}$, in particular we have

$$\bar{\chi}(\bar{\mu}) = \int d^D y \, \bar{G}_{\bar{\mu}}^{(2)}(x-y), \quad \bar{\chi}(\bar{\mu}) \propto \frac{1}{(\bar{\mu}-\bar{\mu}_c)^{\gamma_s}} \quad \text{for} \quad \bar{\mu} \to \bar{\mu}_c. \tag{5.79}$$

By the assumption of universality we have the same critical exponent γ_s in (5.78) and (5.79), but μ_c and $\bar{\mu}_c$ will in general be different. The relation between $\bar{\mu}$ and μ is given by (5.76) and in particular we can find the relation for $\mu \to \mu_c$ since we from our general discussion around (5.13) expect

$$G_\mu^{(1)} \to G_{\mu_c}^{(1)} + c(\mu-\mu_c)^{1-\gamma_s} \tag{5.80}$$

$$\bar{\mu}(\mu) - \bar{\mu}(\mu_c) = \tilde{c}(\mu-\mu_c)^{1-\gamma_s} + (\mu-\mu_c) + \cdots, \tag{5.81}$$

Random Surfaces and Bosonic Strings

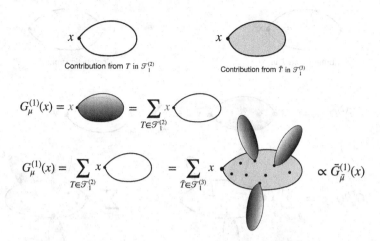

Figure 5.7 The use of the decompostion shown in Fig. 5.6 to write the triangulations in $\mathscr{T}_1^{(2)}$ as triangulations in $\mathscr{T}_1^{(3)}$ with outgrowths. As shown in the lowest figure, a $\mathscr{T}_1^{(2)}$ triangulation can be be decomposed into a $\mathscr{T}_1^{(3)}$ triangulation with $\mathscr{T}^{(2)}$ outgrowths. The dots shown with an outgrowth attached are really links in the $\mathscr{T}_1^{(3)}$ triangulation which are cut open to double links, while the dots without outgrowths symbolize links in $\mathscr{T}_1^{(3)}$ which have not been cut open.

where $G_{\mu_c}^{(1)} > 0$ and *finite*, since $\gamma_s \leq \frac{1}{2}$ and thus the critical part goes to zero for $\mu \to \mu_c$.

We can now make the same graphical decomposition as we did for $G_\mu^{(1)}$ and it is shown in Fig. 5.8. Transferring it to an equation we obtain the analogue of (5.76), only for the two-point function

$$G_\mu^{(2)}(x-y) = \left(1+G_\mu^{(1)}\right)^{-2}\left(\bar{G}_{\bar{\mu}}^{(2)}(x-y) + \int d^D z\, \bar{G}_{\bar{\mu}}^{(2)}(x-z)\, G_\mu^{(2)}(z-y)\right), \quad (5.82)$$

and integrating over y we obtain

$$\chi(\mu) = \left(1+G_\mu^{(1)}\right)^{-2}\left[\bar{\chi}(\bar{\mu}) + \bar{\chi}(\bar{\mu})\chi(\mu)\right] \;\Rightarrow\; \boxed{\chi(\mu) = \frac{\left(1+G_\mu^{(1)}\right)^{-2}\bar{\chi}(\bar{\mu})}{1-\left(1+G_\mu^{(1)}\right)^{-2}\bar{\chi}(\bar{\mu})}}$$
(5.83)

First, the factor $(1+G_\mu^{(1)})^{-2}$ in (5.82) comes for the same reason as the factor $(1+G_\mu^{(1)})^{-1}$ was present in (5.76): A $\mathscr{T}^{(3)}$ graph between two boundary loops has $L(\bar{T}) - 4 = \frac{3}{2}|\bar{T}| - 2$ internal links, the -2 leading to the mentioned factor. As mentioned, (5.76) determines $\bar{\mu}$ as a function $\bar{\mu}(\mu)$, and for $\mu \to \mu_c$ we have (5.81). Now, (5.83) shows that

$$\boxed{\bar{\mu}(\mu_c) > \bar{\mu}_c} \quad (5.84)$$

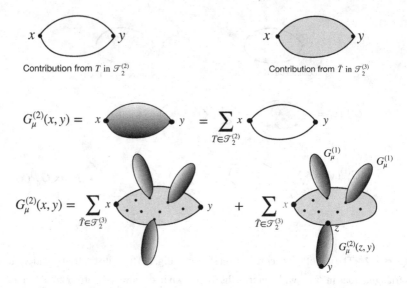

Figure 5.8 The generalization of the one-loop (or one-point) function from Fig. 5.7 to the two-loop (or two-point) function. The lower figure shows how a $\mathcal{T}_2^{(2)}$ triangulation can be represented a $\mathcal{T}_2^{(3)}$ triangulation where some of the links (represented as dots in the figure) can be cut open and a $\mathcal{T}^{(2)}$ outgrowth can be attached. The new aspect compared to Fig. 5.7 is that one of the boundary points can also be on one of the outgrowths in which case this particular outgrowth belongs to $\mathcal{T}_2^{(2)}$ rather than $\mathcal{T}_1^{(2)}$.

simply because the divergence of $\chi(\mu)$ at μ_c occurs for a finite value of $\tilde{\chi}(\bar{\mu})$, namely when the denominator on the rhs of the second equation in (5.83) vanishes. Thus $\tilde{\chi}(\bar{\mu})$ is a perfectly analytic function around the value $\bar{\mu}(\mu_c)$ and we can Taylor expand it around this value:

$$\tilde{\chi}(\bar{\mu}(\mu)) = \left(1+G_{\mu_c}^{(1)}\right)^2 + c\left(\bar{\mu}(\mu)-\bar{\mu}(\mu_c)\right) = \left(1+G_{\mu_c}^{(1)}\right)^2 + c'\left(\mu-\mu_c\right)^{1-\gamma_s} \quad (5.85)$$

where we have used (5.81). Inserting this result in (5.83) we obtain a new value $1-\gamma_s$ for the scaling exponent, which should be compared to the assumed value γ_s:

$$\chi(\mu) \propto \frac{1}{(\mu-\mu_c)^{1-\gamma_s}} \propto \frac{1}{(\mu-\mu_c)^{\gamma_s}} \quad \Rightarrow \quad \boxed{\gamma_s = \frac{1}{2}} \quad (5.86)$$

With this new knowledge let us return to the two-point relation (5.82). A Fourier transformation leads to

$$G_\mu^{(2)}(p) = \left(1+G_{\mu_c}^{(1)}\right)^{-2}\left[\bar{G}_{\bar{\mu}}^{(2)}(p) + \bar{G}_{\bar{\mu}}^{(2)}(p)\,G_\mu^{(2)}(p)\right] \quad (5.87)$$

$$\boxed{G_\mu^{(2)}(p) = \frac{\left(1+G_{\mu_c}^{(1)}\right)^{-2}\bar{G}_{\bar{\mu}}^{(2)}(p)}{1-\left(1+G_{\mu_c}^{(1)}\right)^{-2}\bar{G}_{\bar{\mu}}^{(2)}(p)}} \quad (5.88)$$

and we now expand $\bar{G}^{(2)}_{\bar\mu}(p)$ around $p=0$, remembering that $\bar{G}^{(2)}_{\bar\mu}(p=0) = \int d^D x\, \bar{G}^{(2)}_{\bar\mu}(x) = \bar\chi(\bar\mu)$:

$$\bar{G}^{(2)}_{\bar\mu}(p) = \bar{G}^{(2)}_{\bar\mu}(0) - \bar{c}(\bar\mu)\, p^2 + \cdots = \bar\chi(\bar\mu) - \bar{c}(\bar\mu)\, p^2 + \cdots, \qquad (5.89)$$

Using this and (5.85) we then obtain

$$G^{(2)}_\mu(p) = \frac{\left(1+G^{(1)}_{\mu_c}\right)^{-2}\left(\bar\chi(\bar\mu)-\bar{c}\,p^2+\cdots\right)}{1-\left(1+G^{(1)}_{\mu_c}\right)^{-2}\left(\bar\chi(\bar\mu)-\bar{c}\,p^2+\cdots\right)} \xrightarrow{\mu\to\mu_c} \frac{1+\cdots}{c'\sqrt{\mu-\mu_c}+\bar{c}\,p^2+\cdots} \qquad (5.90)$$

where the $\cdots$ signifies terms of order p^2 or $\sqrt{\mu-\mu_c}$ in the numerator and terms of order p^4, $(\mu-\mu_c)$ and $p^2\sqrt{\mu-\mu_c}$ in the denominator. Thus we can finally write

$$\boxed{G^{(2)}_\mu(p) \approx \frac{1}{\bar c}\frac{1}{m^2(\mu)+p^2}, \quad m(\mu) = \check{c}\sqrt[4]{\mu-\mu_c}\quad \nu=\frac{1}{4}, \quad d_H = 4} \qquad (5.91)$$

We thus have exponents identical to the ones encountered for BPs.

We have been working with dimensionless variables since eq. (5.53). In order to understand the relation of the above results to BPs it is convenient to reintroduce dimensions. We thus write

$$x a(\mu) = x_{ph}, \qquad p = p_{ph}\, a(\mu), \qquad (5.92)$$

where $a(\mu)$ is a length-unit in $\mathbb{R}^D$. We can think of $a(\mu)$ as an average length of a link of the triangles in the triangulations, when they are mapped into $\mathbb{R}^D$. Thus $a(\mu)$ is not necessarily the same as the intrinsic link length $\varepsilon(\mu)$ which appears in (5.62) and which ensures that the intrinsic area of a typical triangulation is finite. The dependence of $a(\mu)$ on μ will be determined by the requirement that the functions $G^{(2)}_\mu(x,y)$ and $G^{(2)}_\mu(p)$ have a non-trivial behavior for $\mu\to\mu_c$. From (5.91) and (5.92) we see that the natural way to obtain a non-trivial limit is to define a "renormalized" physical mass m_{ph} by

$$m(\mu) = m_{ph}\, a(\mu), \qquad a(\mu) = \check{c}\sqrt[4]{\mu-\mu_c}, \qquad (5.93)$$

which implies that

$$G^{(2)}_\mu(p) \xrightarrow{\mu\to\mu_c} \frac{1}{a^2(\mu)}\frac{1}{m^2_{ph}+p^2_{ph}}, \quad \text{and} \quad m(\mu)|x| = m_{ph}|x_{ph}| \qquad (5.94)$$

This way of taking the scaling limit is similar to the way we did it both for the free particle and for BPs, and we obtain the same result! Comparing (5.93) to (5.62) we see that

$$\boxed{\varepsilon(\mu) \propto \frac{m^2_{ph}}{\sqrt{\Lambda}}\, a^2(\mu).} \qquad (5.95)$$

This is a relation similar to (3.24) for the RW, and it reflects the same: since $d_H = 4$, the average area of a surface from the path integral embbeded and measured in $\mathbb{R}^D$ has an area $\langle A_{ext}\rangle \propto 1/(m_{ph}^4 a^2(\mu))$ if the individual triangles in $\mathbb{R}^D$ have an average area proportional to $a^2(\mu)$. If we insist that the average *intrinsic* area of a surface, $\langle A_{int}\rangle$, is finite, like in (5.62), the intrinsic length $\varepsilon(\mu)$ assigned to a link in the triangulation has to be much smaller than $a(\mu)$, as is indeed expressed in relation (5.95). Let us now discuss the BP-picture in more detail.

Consider $G_\mu^{(2)}(x-y)$ given by (5.77) as the partition function for surfaces with two marked vertices separated a distance $|x-y|$ in $\mathbb{R}^D$. From the scaling of the Fourier transformed $G_\mu(p)$ given by (5.94) it is clear that we have

$$G_\mu^{(2)}(x-y) \underset{\mu\to\mu_c}{\to} a^{D-2}(\mu)\, G(x_{ph}-y_{ph}; m_{ph}), \tag{5.96}$$

where $G(x_{ph}-y_{ph}; m_{ph})$ is the continuum propagator (2.12) of the free particle. From (5.77) and (5.96) we have for $\mu \to \mu_c$

$$\langle |T|\rangle_{G_\mu^{(2)}(x)} = -\frac{1}{G_\mu^{(2)}(x-y)}\frac{d}{d\mu}G_\mu^{(2)}(x-y) \propto \frac{1}{\mu-\mu_c} \propto \frac{|x|^4}{m_{ph}^4 x_{ph}^4}, \tag{5.97}$$

explicitly showing that the Hausdorff dimension $d_H = 4$. Similarly, using (5.80) we have

$$\langle |T|\rangle_{G_\mu^{(1)}(x)} = -\frac{1}{G_\mu^{(1)}(x)}\frac{d}{d\mu}G_\mu^{(1)}(x) \underset{\mu\to\mu_c}{\propto} \frac{1}{\sqrt{\mu-\mu_c}} \tag{5.98}$$

$\langle |T|\rangle$ is the average number of triangles and we have a picture of scaling *consistent* with a BP picture where triangles play the role of the links in the BPs and $G_\mu^{(1)}(x)$ plays the role of the rooted BP partition function $Z(\mu)$. For the ensemble of surfaces defined defined by $G_\mu^{(2)}(x-y)$, the smallest number of triangles needed to connect x and y is of order $|x_{ph}-y_{ph}|/\sqrt[4]{\mu-\mu_c}$. But how do we know that the surfaces really look like BPs? The situation is illustrated in Fig. 5.9.

The answer comes from the decomposition of $G_\mu^{(2)}(x-y)$ shown in Fig. 5.8. The iteration of the figure or eq. (5.88) leads to

$$G_\mu^{(2)}(p) = \frac{\bar{G}_{\bar\mu}^{(2)}(p)}{(1+G_\mu^{(1)})^2} + \left(\frac{\bar{G}_{\bar\mu}^{(2)}(p)}{(1+G_\mu^{(1)})^2}\right)^2 + \left(\frac{\bar{G}_{\bar\mu}^{(2)}(p)}{(1+G_\mu^{(1)})^2}\right)^3 + \cdots \tag{5.99}$$

The important point here is that $\bar{G}_{\bar\mu}^{(2)}(p)$ is not critical as $\mu \to \mu_c$ and $G_\mu^{(1)} \to G_{\mu_c}^{(1)}$ for $\mu \to \mu_c$, but $G_\mu^{(2)}(p)$ diverges as shown in (5.94). Note that

$$\sum_{n=1}^\infty x^n = a \Rightarrow \langle n\rangle = a+1, \quad \langle n\rangle := \frac{\sum_n n x^n}{\sum_n x^n}. \tag{5.100}$$

We thus conclude that the number of "blobs" in Fig. 5.9 is $\langle n\rangle \propto 1/\sqrt{\mu-\mu_c}$. For each blob we only have a finite number of triangles associated with a triangulation

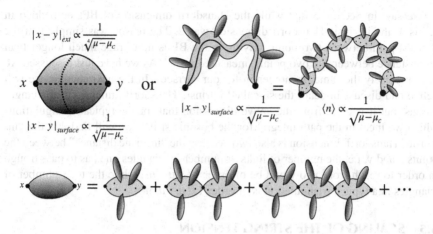

Figure 5.9 $|x-y|_{ext} \propto (\mu-\mu_c)^{-1/4}$ in $\mathbb{R}^D$. How does a typical surface in the path integral connecting x and y look? The upper left figure shows a surface where the distance between x and y within the surface is of the same order as $|x-y|_{ext}$. The middle upper figure shows a surface where the distance between x and y is much larger if one has to stay within the surface. The upper right figure shows how the middle figure is realized via the $\mathscr{T}^{(3)}$ blobs which appear in the decomposition of $G_\mu^{(2)}(x-y)$ shown in Fig. 5.8, and spelt out in detail in the lower figure here. The average number of $\mathscr{T}^{(3)}$ blobs in the figures are $\langle n \rangle \propto (\mu-\mu_c)^{-1/2}$.

$\bar{T} \in \mathscr{T}^{(3)}$ since $\bar{\mu}(\mu_c) > \bar{\mu}_c$ (see (5.84)). Thus a shortest path in a typical triangulation connecting x and y will be of length proportional to $1/\sqrt{\mu-\mu_c}$ since it has to pass through all the blobs. This is much longer than the shortest pass in $\mathbb{R}^D$ between x and y, which, as mentioned, is of order $|x_{ph}-y_{ph}|/\sqrt[4]{\mu-\mu_c}$. It shows that we can consider the $\mathscr{T}^{(3)}$-part of a blob as an effective "BP-link" and and these links then perform a RW from x to y, increasing the length from being proportional to $1/\sqrt[4]{\mu-\mu_c}$ to being proportional to the square of this. The image of the $\mathscr{T}^{(3)}$-part of a $\mathscr{T}^{(2)}$-triangulation in $\mathbb{R}^D$ thus effectively determines the geodesic distance between the marked points x and y if we are forced to stay within the surface. The analogue for "real" BPs is that x and y is connected by a *unique* shortest link-path in a given BP and this link-path is then mapped to a RW path between x and y in $\mathbb{R}^D$. The typical number of links in this RW will be proportional to $1/\sqrt{\mu-\mu_c}$. Then outgrowths in the form of rooted BPs are attached to the vertices of the shortest path between x and y and the average number of links or vertices in such a roooted BP is also proportional to $1/\sqrt{\mu-\mu_c}$. We have the same situation here for the surfaces. The finite number of outgrowths attached to the $\mathscr{T}^{(3)}$-part of a blob are $G_\mu^{(1)}$-outgrowths, which each contain a number of triangles proportional to $1/\sqrt{\mu-\mu_c}$. All together one thus has $1/(\mu-\mu_c)$ triangles in $G_\mu^{(2)}(x)$ as there should be according to (5.97). This shows that the BP-picture indeed is the correct one for the surfaces which dominate the path integral defining $G_\mu^{(2)}(x-y)$.

We saw in Section 3 that while the Hausdorff dimension of BPs embedded in $\mathbb{R}^D$ is 4, the intrinsic Hausdorff dimension was 2. The reason was simply that the shortest path between two point, staying in the BP is in average much longer than the distance between the two points, measured in $\mathbb{R}^D$. As we have just discussed we have precisely the same phenomenon for our surfaces. In the next chapter, we will define and discuss in detail the so-called intrinsic Hausdorff dimension for triangulations, but it is clear from the above discussion that for the typical triangulations which we meet in the path integral for the bosonic string, we will find that the "intrinsic" Hausdorff dimension is also two because the "intrinsic distance" between the points x and y, i.e. the number of links or number of triangles one has to pass trough in order to reach from x to y will be of order $1/\sqrt{\mu-\mu_c}$, while the total number of triangles in a typical triangulation is of order $1/(\mu-\mu_c)$.

5.5 SCALING OF THE STRING TENSION

Theorem 3, below eq. (5.58), tells us that there exists a string tension for $\mu > \mu_c$. However, it does not tell us if $\sigma(\mu)$ scales to zero for $\mu \to \mu_c$. Also, it does not tell us anything about subleading corrections to $G_\mu(\ell_A)$ when $A \to \infty$, where A is the area of the planar loop, embedded in $\mathbb{R}^D$. There can be many subleading correction, but a generic correction comes from the setup, which is such that the length of the boundary ℓ_A has to go to infinity when A goes to infinity. This will create a term in the exponential part of $G_\mu(\ell_A)$, depending on the length of the boundary. Until now we have used the same notation for the set of boundary links ℓ, viewed as a boundary in the triangulation and this boundary mapped into $\mathbb{R}^D$. Let us now denote the number of links in ℓ by $|\ell|$ and the length of the boundary, mapped to $\mathbb{R}^D$ by L_A. In general we expect a behavior

$$G_\mu(\ell_A) = e^{-\sigma(\mu)A - \lambda(\mu)L_A + \cdots} \tag{5.101}$$

As long as $L_A/A \to 0$, this second term will play no role and does not appear in theorem 3.

Again we can write down the relation between $G_\mu(\ell_A)$ and $\bar{G}_{\bar{\mu}}(\ell_A)$, as illustrated in Fig. 5.10:

$$G_\mu(\ell_A) = \bar{G}_{\bar{\mu}}(\ell_A)\left(1+G_\mu^{(1)}\right)^{-|\ell_A|} = \bar{G}_{\bar{\mu}}(\ell_A)\, e^{-\frac{2}{3}(\mu-\bar{\mu})|\ell_A|}. \tag{5.102}$$

As long as $|\ell_A| \leq cL_A$ we conclude from (5.101) that

$$\boxed{\sigma(\mu) = \bar{\sigma}(\bar{\mu})} \tag{5.103}$$

but even if that is not satisfied we have $\sigma(\mu) \geq \bar{\sigma}(\bar{\mu})$. Since we have already seen that for $\mu \to \mu_c$ we have $\bar{\mu}(\mu_c) > \bar{\mu}_c$, we know from theorem 3 that $\bar{\sigma}(\bar{\mu}(\mu_c)) > 0$, and thus that $\sigma(\mu_c) > 0$. Conclusion: *the string tension is not scaling to zero for* $\mu \to \mu_c$.

It is also possible (and relatively easy) to prove directly that the string tension does not scale to zero for $\mu \to \mu_c$. The only assumption used is $|\ell_A| \leq cL_A$. The proof is

Random Surfaces and Bosonic Strings

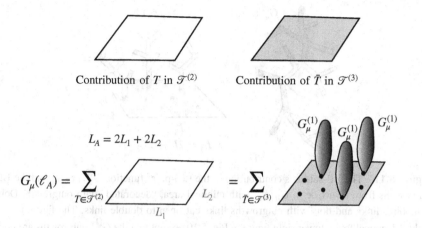

Figure 5.10 The decomposition of triangulations $T \in \mathscr{T}^{(2)}$ contributing to of $G_\mu(\ell_A)$ into triangulations $\bar{T} \in \mathscr{T}^{(3)}$ together with outgrowths from some of the links, cut open to double-links. The grey surface represents a triangulation $\bar{T} \in \mathscr{T}^{(3)}$, the dots on $\bar{T}$ symbolize links, and the dots associated with outgrowths symbolize links cut open to double-links to which there are $G_\mu^{(1)}$ outgrowths attached. These outgrowths are defined in Fig. 5.7. The important point is that even when $\mu \to \mu_c$, the number of dots in the $\mathscr{T}^{(3)}$ triangulation covering the ℓ_A surface is essentially proportional to the minimum number of triangles one can use to cover the surface since $\bar{\mu}$ *is not critical* when $\mu = \mu_c$. Thus almost all triangles in the $\mathscr{T}^{(2)}$ triangulation are located in the outgrowths.

based on the simple estimate that for any triangulation where the $|\ell_A|$ boundary points are distributed along the boundary we have for the action (5.53)

$$S[X,T] \geq \sum_{t \in T} A_t \geq A. \tag{5.104}$$

This follows from the fact that the area A_t of a triangle spanned by points $X(v_1)$, $X(v_2)$ and $X(v_3)$ is less than or equal to one fourth of the squares of the lengths of any two of its sides. Of course the sum of A_t's is larger than or equal to A, the minimal area associated with a surface with planar boundary of length L_A. Using (5.104) one can show that $\sigma(\mu) \geq 1$ for $\mu > \mu_c$ as long as $|\ell_A| \leq c L_A$ (for details consult the book [1]).

Before discussing the physical consequences of the non-scaling of the string tension, note that the first correction to this result can easily be calculated from $\sigma(\mu) = \bar{\sigma}(\bar{\mu})$ since we have:

$$\frac{d\sigma(\mu)}{d\mu} = \frac{d\bar{\mu}}{d\mu} \frac{d\sigma(\mu)}{d\bar{\mu}} = \frac{d\bar{\mu}}{d\mu} \frac{d\bar{\sigma}(\bar{\mu})}{d\bar{\mu}} \propto \frac{d\bar{\mu}}{d\mu} \propto \frac{1}{\sqrt{\mu - \mu_c}}, \tag{5.105}$$

where we first use that $\sigma(\mu) = \bar{\sigma}(\bar{\mu})$, next that $\bar{\sigma}(\bar{\mu})$ is analytical around $\bar{\mu}(\mu_c)$, and finally (5.81) with $\gamma_s = 1/2$. Integrating this relation we obtain

$$\boxed{\sigma(\mu) = \sigma(\mu_c) + c\sqrt{\mu - \mu_c} + \mathscr{O}(\mu - \mu_c), \qquad \sigma(\mu_c) > 0} \tag{5.106}$$

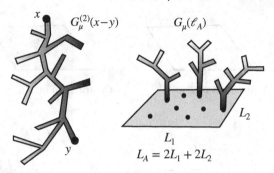

Figure 5.11 The BP-surfaces contributing to the two-point function $G^{(2)}_\mu(x-y)$ and the BP outgrowths from a surface in $G_\mu(\ell_A)$ with minimal area, "decorated" with outgrowth. Dots symbolize links and dots with outgrowths links cut open to double links. The figure to the right is identical to the lower right figure in Fig. 5.10, except that the $G^{(1)}_\mu$ outgrowths are now explicitly replaced by their scaling limit, namely rooted BPs.

Let us now discuss the physical consequence of this non-scaling. The basic scaling, already introduced for the two-point function in (5.93) and (5.94) ensured that $e^{-m(\mu)|x|}$ survived in the limit $\mu \to \mu_c$ as $e^{-m_{ph}|x_{ph}|}$. The natural extension of this is to ensure that $e^{-\sigma(\mu)A}$ survives in the scaling limit as $e^{-\sigma_{ph}A_{ph}}$. Thus we demand

$$\sigma(\mu)A = \sigma_{ph}A_{ph}, \quad A_{ph} := Aa^2(\mu), \quad \text{i.e.} \quad \sigma_{ph} = \frac{\sigma(\mu)}{a^2(\mu)}. \tag{5.107}$$

The only way to obtain a finite σ_{ph} for $\mu \to \mu_c$ is to have a scaling $\sigma(\mu) \propto \sqrt{\mu-\mu_c}$, but from (5.106) we see it is not the case, although (tantalizing!) the correction to the constant term has the right dependence. We conclude from (5.107) and (5.106) that $\boxed{\sigma_{ph} = \infty}$.

What does a typical surface in the path integral contributing to $G_\mu(\ell_A)$ look like? The average number of triangles in such a surface is

$$\langle |T| \rangle_{G_\mu(\ell_A)} = -\frac{1}{G_\mu(\ell_A)} \frac{d}{d\mu} G_\mu(\ell_A) \propto \frac{A}{\sqrt{\mu-\mu_c}} \propto \frac{A_{ph}}{\mu-\mu_c}, \tag{5.108}$$

where we have used (5.101), (5.105) and (5.107). The number of triangles needed to cover a surface with "dimensionless" area A is just A up to a trivial factor, and $A = A_{ph}/a^2(\mu)$. This is basically the number of (grey) $\mathcal{T}^{(3)}$-triangles in a typical surface in the decomposition made in Fig. 5.10. For each link in that minimal $\mathcal{T}^{(3)}$-surface we can potentially have a $G^{(1)}_\mu$ outgrowth with a number of triangles proportional to $1/\sqrt{\mu-\mu_c}$. The total number of triangles for such a surface is thus precisely the number calculated in (5.108) and a typical surface is therefore well represented by a minimal surface with $G^{(1)}_\mu$-outgrowths everywhere. Since these outgrowths also are BPs we have a picture like the one presented in Fig. 5.11.

This theory of minimal surfaces with BP-like excitations seems to have nothing to do with what is usually known as bosonic string theory where one has a finite string tension and where in addition the lowest mass excitation is a *tachyon*, i.e. a particle with a mass where $m^2 < 0$. The tachyon is a sign of a sick theory (all kind of disasters happen when $m^2 < 0$, but we will not go into details here). Our statistical theory does not have this sickness, since we showed by very general arguments that our propagator has to fall off exponentially, i.e. the lowest mass excitation has $m \geq 0$. Is it possible to make contact to the standard bosonic string theory? The answer seems to be yes, as hinted by the correction term in eq. (5.106), and it seems to be linked to the problem of having $|\ell_A|$ fixed boundary points, $|\ell_A| \to \infty$, which at the same time is assigned an intrinsic length $\ell_{intrin} = |\ell_A|\varepsilon(\mu)$ and an extrinsic length $\ell_{extrin} = |\ell_A|a(\mu)$, where ε and a are related by (5.95). However, since there are still subtleties associated with this resolution of the difference between the formally defined continuum bosonic string and our regularized (well-defined) bosonic string, we will not discuss the topic further.

5.6 PROBLEM SETS AND FURTHER READING

The Problem Sets 8 and 9 address the problem of summation of divergent perturbation series, and the double scaling limit studied in Problem 9 for BPs with loops is quite similar to the one encountered in certain string theories when summing over topologies of the string. Here only the very simplest aspects of the summation of divergent perturbation theories have been mentioned. The techniques have developed a lot the last 20 years under the heading *resurgence*. For reviews see [11, 12].

As mentioned a number of the heuristic arguments used in this chapter can be made rigorous for surfaces embedded in a hypercubic lattice. Details can be found in [1]. Finally, the intriguing, yet not fully understood connection between the regularized string theory discussed here (the statistical theory of random surfaces) and ordinary "tachyonic" bosonic string theory can be found in the articles [13, 14].

6 Two-Dimensional Quantum Gravity

6.1 SOLVING 2D QUANTUM GRAVITY BY COUNTING GEOMETRIES

We now consider the case where we have no Gaussian matter fields X_i coupled to two-dimensional quantum gravity. The Einstein-Hilbert action is given by (3.4) with $M = 2$. We have already seen that for two-dimensional gravity the curvature term is topological (eqs. (5.38) and (5.39)) and thus does not contribute to any dynamics unless we consider processes where the topology changes. On the other hand we have already discussed the problems with two-dimensional quantum gravity and topology changes in Chapter 5.3, so in the following we are going to restrict ourselves to two-dimensional manifolds which have the topology of the sphere ($h=0$), but with a number n, $n \geq 0$, of boundaries. It is convenient to associate independent boundary cosmological constants Z_i to each boundary i. In this way our (trivial) action will be (dropping the curvature term in the Einstein-Hilbert action)

$$S[g,\Lambda] = \Lambda \int d^2\xi \sqrt{g(\xi)}, \qquad \text{no boundary cosmological constants} \qquad (6.1)$$

where we denote Λ the cosmological constant (in our old notation of (5.64) it would be $(2\Lambda)/2\pi G$), and including boundary cosmological constants

$$S[g,\Lambda,Z_1,\ldots,Z_b] = S[g,\Lambda] + \sum_{i=1}^{n} Z_i \int ds_i = \Lambda V_g + \sum_{i=1}^{n} Z_i L_{i,g}, \qquad (6.2)$$

where V_g is the volume of spacetime and $L_{i,g}$ is the length of boundary i, calculated using the metric g. We now define the following partition functions, depending on the boundaries:

$$W(\Lambda;Z_1,\ldots,Z_n) = \int \mathcal{D}[g]\, e^{-S[g,\Lambda,Z_1,\ldots,Z_n]}, \qquad (6.3)$$

$$W(\Lambda;L_1,\ldots,L_n) = \int \mathcal{D}[g]\, e^{-S[g,\Lambda]} \prod_{i=1}^{n} \delta(L_i - L_{i,g}), \qquad (6.4)$$

$$W(V;L_1,\ldots,L_n) = \int \mathcal{D}[g]\, \delta(V - V_g) \prod_{i=1}^{n} \delta(L_i - L_{i,g}), \qquad (6.5)$$

$W(\Lambda;Z_1,\ldots,Z_n)$ and $W(\Lambda;L_1,\ldots,L_n)$ are related by a Laplace transformation:

$$W(\Lambda;Z_1,\ldots,Z_n) = \int_0^\infty \prod_{i=1}^{n} dL_i\, e^{-Z_i L_i}\, W(\Lambda;L_1,\ldots,L_n) \qquad (6.6)$$

DOI: 10.1201/9781003320562-6

and likewise, $W(\Lambda; L_1, \ldots, L_n)$ and $W(V; L_1, \ldots, L_n)$ are related by a Laplace transformation:

$$W(\Lambda; L_1, \ldots, L_n) = \int_0^\infty dV\, e^{-\Lambda V}\, W(V; L_1, \ldots, L_n). \tag{6.7}$$

From eq. (6.5) it is seen that

$W(V; L_1, \ldots, L_n) = $ # of geometries with volume V and boundary-lengths L_i (6.8)

It follows that these partition functions of two-dimensional quantum gravity are completely determined if we can *count the number of geometries with volume V and boundary lengths L_i* and that these partition functions in that sense are *entirely entropic*. A main result in this chapter will be that we can perform this counting and find

$$\boxed{W(V; L_1, \ldots, L_n) \propto V^{n-7/2}\, \sqrt{L_1 \cdots L_n}\, \exp\left(-\frac{(L_1+\cdots+L_n)^2}{4V}\right)} \tag{6.9}$$

As usual, in order to perform this counting we first need a regularization of the geometries, and we have it already, namely the one we used when discussing the bosonic string: Dynamical Triangulations (DT), where we consider the subset of geometries defined by equilateral triangles:

$$\int \mathcal{D}[g] \to \sum_{T \in \mathcal{T}} \tag{6.10}$$

where $\mathcal{T}$ denotes a suitable class of equilateral triangulations. As already discussed in the case of the bosonic string, if we have a triangulation T with $|T|$ triangles, and boundaries with l_i links and an assignment of length ε to the links, we relate the continuum quantities to the DT quantities by writing

$$V \sim |T|\varepsilon^2, \qquad L_i \sim l_i \varepsilon \tag{6.11}$$

and we will take a limit where $\varepsilon \to 0$ while $|T|$ and ℓ_i go to infinity in such a way that V and L_i stay fixed. In that limit the number of triangulations will be

$$w(|T|, l_i, \ldots, l_n) = \tag{6.12}$$

$$c\, e^{\mu_c |T|}\, e^{\lambda_c (l_1 + \cdots + l_n)}\, |T|^{n-7/2}\, \sqrt{l_1 \cdots l_n}\, \exp\left(-\frac{(l_1+\cdots+l_n)^2}{\tilde{c}|T|}\right) \left[1 + \cdots\right]$$

where c and $\tilde{c}$, as well as μ_c and λ_c depend on the specific set $\mathcal{T}$ of equilateral triangulations we are using, and the $\cdots$ indicate subleading corrections in $|T|$ and l_i. The exponential growth, depending on μ_c and λ_c will not survive when we convert the counting formula (6.12) to the continuum formula (6.9), but it is important for being able to make this conversion that the number of triangulations only grows exponentially with $|T|$ and that is only the case if we restrict the topology, i.e. the number of handles h of the two-dimensional manifold. In the following $h = 0$.

Two-Dimensional Quantum Gravity

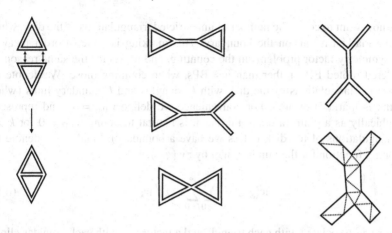

Figure 6.1 To the left: gluing together two triangles in the double link notation. The boundary of the resulting square will still be an uninterrupted line. In the middle: three examples of so-called unrestricted triangulations. The two top triangulations have double links which are not part of triangles and in the bottom triangulation two triangles only share a vertex. However, the boundary is always well-defined in the double link notation. To the right: double-links can be always imitated by regular triangulations, where the links are "thickened" such that they consist of a few triangles..

6.2 COUNTING TRIANGULATIONS OF THE DISK

In order to count the triangulations we have to define the class of triangulations we want to count. In Chapter 5.4, we defined two classes of triangulations, $\mathcal{T}^{(3)}$ and $\mathcal{T}^{(2)}$. In particular $\mathcal{T}^{(3)}$ is a natural class and one can indeed use it (and it has been done). However, we will here choose a somewhat larger class, which at first seems unnatural, but, as we will shortly argue, should be perfectly suitable for extracting a continuum limit when $\varepsilon \to 0$. We will denote the class $\mathcal{T}^{(0)}$ and call it *unrestricted triangulations*. The main reason for choosing this class is that the counting is easier that for $\mathcal{T}^{(3)}$. Let us consider triangulations with one boundary, i.e. triangulations of a disk. Let us use a so-called double-line notation, where we represent the triangles as shown in Fig. 6.1, and where they are glued together to form a larger triangulation, as also shown in the figure.

We now allow for more general boundaries as shown, where the boundary can consist of double links (which should be thought of as having an infinitesimal area between the links if they are not associated with a triangle), and where triangles may only share a single vertex. In all cases the outer boundary lines form a closed curve. Allowing such "degenerate" triangulations should make no difference in the $\varepsilon \to 0$ limit, since one can alway imitate such a degenerate triangulation by a regular one of width ε, again as illustrated in the figure. If it made a difference we should be worried about universality and whether our discretization is a good one. As mentioned one obtains identical results for $\varepsilon \to 0$ for both regular and unrestricted triangulations, although we are not going to prove this here (for a proof see [1]).

We now want to count the number of unrestricted triangulations of the disk, where we have marked a link on the boundary. This marking is done in order to avoid some symmetry factor problems in the counting, and it was for the same reason we considered rooted BPs, rather than just BPs, when counting those. We denote the number of triangulations of the disk with k triangles and l boundary links (where one link is marked) by $w_{k,l}$. For convenience we define $w_{0,0} = 1$ (and represents it graphically as a point (a dot)). Further it is natural to define $w_{k,0} = 0$ for $k > 0$ (no triangulations of the disk unless we have a boundary). Finally, we denote the generating function for the numbers $w_{k,l}$ by $zw(g,z)$:

$$zw(g,z) = \sum_{k=0}^{\infty}\sum_{l=0}^{\infty} g^k z^{-l} w_{k,l}. \qquad (6.13)$$

A factor g is associated with each triangle and a factor z^{-1} with each boundary link[1]. We are using z^{-1} rather than z to enumerate the number of boundary links because the analytic structure of $w(g,z)$ in the complex z-plane will be simpler. For the same reason we write $zw(g,z)$ instead of $w(g,z)$ for the generating function to ensure that $w(g,z) \to 0$ for $|z| \to \infty$. Thus:

$$\boxed{w(g,z) = \sum_{k=0}^{\infty}\sum_{l=0}^{\infty} g^k z^{-(l+1)} w_{k,l} = \sum_{l=0}^{\infty} \frac{w_l(g)}{z^{l+1}}, \qquad w_l(g) = \sum_{k=0}^{\infty} g^k w_{k,l}.} \qquad (6.14)$$

Here $w_l(g)$ is the generating function for triangulations with l boundary links. In particular we have by definition $w_0(g) = 1$ and thus

$$w(g,z) \to \frac{1}{z} \quad \text{for} \quad |z| \to \infty \qquad (6.15)$$

Assume now we have a triangulation of the disk with a marked boundary link. The marked link can belong to a triangle or a double link. This is illustrated in Fig. 6.2. Removing the triangle and marking a link on the boundary of the remaining triangulation again leads to a marked triangulation of the disk, but with one fewer triangle and a boundary where the boundary length is increased by one. Similarly, removing the double-link the triangulation splits into two disconnected disks where we can also mark the boundaries.

Let the original triangulation T consist of k triangles and l boundary links. In the generating function it contributes with a term g^k/z^{l+1}. If it is decomposed to a triangle and new triangulation T', we can write this factor as $g \cdot z \cdot (g^{k-1}/z^{l+2})$ where the factor in the bracket is the weight associated to T' in the generating function. If T is decomposed into the two disconnected triangulations T_1 and T_2 with the number of triangles k_1 and k_2 and the number of boundary links l_1 and l_2 by removing the double-link, we have $k_1+k_2=k$ and $l_1+l_2+2=l$. We can now decompose the factor

---

[1] In combinatorics one uses the word *indeterminate* for the variables like g and z^{-1} in the generating functions. We will only occasionally use that notation.

Two-Dimensional Quantum Gravity

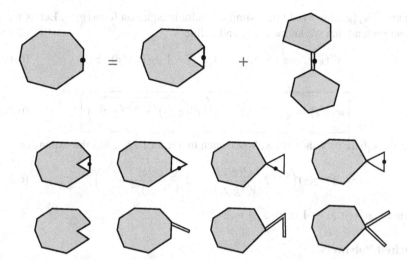

Figure 6.2 Top: Illustrating that the marked boundary link either belongs to a triangle or a double-link. The two lower rows show the various ways the triangle with the marked link can be attached to the rest of the triangulation and some of the ways the mark can be placed on the triangle. When the triangle to be removed has more than one boundary link, removing the triangle results in double-links, such that the new boundary always has one more link than the original boundary. We have not specified where the mark should be placed on the new boundary. In principle one is free to invent ones own procedure for that.

g^k/z^{l+1} associated with T into $\frac{1}{z}(g^{k_1}/z^{l_1+1})(g^{k_2}/z^{l_2+1})$. Summing over all triangulations, as done in the generating function, with weights g^k/z^{l+1}, we then arrive at the following equation:

$$w(g,z) \approx gzw(g,z) + \frac{1}{z}w^2(g,z). \tag{6.16}$$

This equation can easily be solved. However, the devil is in the details and this is the reason we do not use "=" but "≈" in (6.16). The equation is not correct when the triangulations on the lhs have boundary lengths $l = 0, 1$. The $w(g,z)$ on the lhs contains the term $\frac{1}{z}$ coming from $\frac{w_{0,0}}{z}$, but this "point" associated with $l = 0$, cannot be found in the decomposition shown in Fig. 6.2 since the triangulation T' has at least two links, and T_1 and T_2 are connected by a double-link. We thus have to add a term $1/z$ on the rhs of eq. (6.16). Similarly, since T' has at least two links, we should subtract from the $w(g,z)$ on the rhs of eq. (6.16), associated with summing over T' triangulations, the terms $w_0(g)/z$ and $w_1(g)/z^2$ in notation of eq. (6.14). The corrected eq. (6.16) is thus

$$w(g,z) = \frac{1}{z} + gz\left[w(g,z) - \frac{1}{z} - \frac{w_1(g)}{z^2}\right] + \frac{1}{z}w^2(g,z) \tag{6.17}$$

or

$$\boxed{w^2(g,z) = (z - gz^2)w(g,z) - (1 - g(w_1(g) + z))} \tag{6.18}$$

If we knew $w_1(g)$ this would be a simple quadratic equation for $w(g,z)$. Let us for a moment pretend that we know $w_1(g)$ and define

$$V'(g,z) = z - gz^2, \qquad Q(g,z) = 1 - g(w_1(g) + z) \qquad (6.19)$$

Then

$$\boxed{w(g,z) = \frac{1}{2}\left(V'(g,z) - \sqrt{(V'(g,z))^2 - 4Q(g,z)}\right)} \qquad (6.20)$$

where the square root should be chosen such that it for large z has the expansion

$$V'(g,z)\left(1 - 2\frac{Q(g,z)}{(V'(g,z))^2} - 2\frac{Q^2(g,z)}{(V'(g,z))^4} - \cdots\right) \qquad (6.21)$$

ensuring that $w(g,z) \to 1/z$ for $z \to \infty$.

Branched Polymers

Let us chose $g = 0$, i.e. we have no triangles. The "triangulations" are thus boundary graphs consisting double links, as illustrated in Fig. 6.3. These can clearly be viewed as branched polymers, more or less like the rooted branched polymers, only here the "root" is a mark on a double link. As illustrated in Fig. 6.3 this is equivalent to marking the whole double link and in addition one of the vertices of the double link. In this way we have a vertex relative to which we can define "height" (the link distance) of the other vertices, precisely as we can define the height of vertices for rooted BPs as the link distance to the root. From (6.20) we obtain the generating function

$$w(z) = \frac{1}{2}\left(z - \sqrt{z^2 - 4}\right) = \sum_{l=0}^{\infty} \frac{w_{2l}}{z^{2l+1}} \qquad (6.22)$$

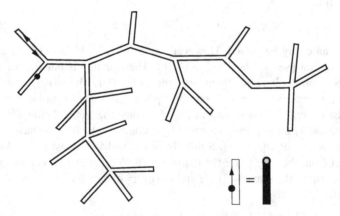

Figure 6.3 Branched polymer constructed from double-links and with one link marked. The boundary has an orientation and the two links constituting the double line points in "opposite" directions. Marking one of the two links in the double link is equivalent to marking the vertex one meets moving forward along the marked link and then in addition marking the whole double link (illustrate in the figure by making it black).

Two-Dimensional Quantum Gravity

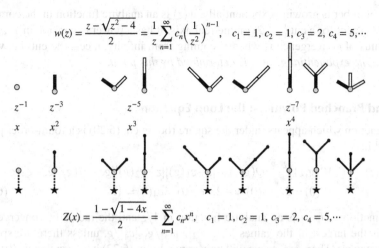

$$w(z) = \frac{z - \sqrt{z^2 - 4}}{2} = \frac{1}{z} \sum_{n=1}^{\infty} c_n \left(\frac{1}{z^2}\right)^{n-1}, \quad c_1 = 1, c_2 = 1, c_3 = 2, c_4 = 5, \cdots$$

$$Z(x) = \frac{1 - \sqrt{1 - 4x}}{2} = \sum_{n=1}^{\infty} c_n x^n, \quad c_1 = 1, c_2 = 1, c_3 = 2, c_4 = 5, \cdots$$

Figure 6.4 Top: BPs of double-links with one marked double link and corresponding marked vertex (as illustrated in Fig. 6.3). The marked link is black and the corresponding marked vertex is grey. They are in bijective correspondence with rooted BPs with one more link. Bottom: the rooted BPs, where the root is shown as a star connected to the marked vertex (grey) as described in the main text.

$$w_{2l} = \frac{(2l)!}{(l+1)! \, l!} \xrightarrow{l \to \infty} \frac{1}{\pi} l^{-3/2} 4^l \left(1 + \mathcal{O}\left(\frac{1}{l}\right)\right) \tag{6.23}$$

and this is exactly the partition function for BPs where arbitrary branching is allowed with equal weight. Let us spell this out in some detail. First, the labeling of the BPs. Labeling a link and one of its vertices of BPs is equivalent to introducing a root vertex of order one and connecting it to the marked vertex such that the marked link is the last link one meets going around the marked vertex counter clockwise, starting with the link connecting the root and the marked vertex. The rooted BP has one more link than our BP with a marked link and vertex, and the relation is illustrated in Fig. 6.4 which also shows that in the standard rooted BP notation with[2] $x \sim 1/z^2$ we have instead of $w(z)$ from (6.22) the partition function $Z(x)$ determined by

$$\frac{1}{x} = \frac{F(Z)}{Z}, \quad F(Z) = \sum_{n=1}^{\infty} Z^n = \frac{Z}{1-Z}, \quad Z(x) = \frac{1 - \sqrt{1-4x}}{2}. \tag{6.24}$$

Thus the BPs related to (6.22) indeed have arbitrary branching with weight 1 and we have (of course) precisely the expected behavior for rooted BPs:

$$w_{2l} \sim l^{\gamma - 2} e^{\lambda_c l} \quad \text{for } l \to \infty, \quad \gamma = \frac{1}{2}, \quad \lambda_c = \ln 4, \tag{6.25}$$

---

[2] Rather than $x \sim 1/z$ we have $x \sim 1/z^2$ since the links are double links, each component contributing a factor $1/z$.

i.e. the number is growing exponentially. $w(z)$ is an analytic function in the complex z-plane except for a cut $[-2,2]$ on the real axis. It has a power expansion in $1/z$, and the radius of convergence is when z, coming from infinity, meets the cut, i.e. when $\frac{1}{z_c} = \frac{1}{2}$. *The exponential growth is determined by this point.*

Beyond Branched Polymers: the Loop Equation

The function which appears under the square root in eq. (6.20) is a fouth-order polynomial in z:

$$f_g(z) = V'(g,z))^2 - 4Q(g,z) = [z-c_1(g)][z-c_2(g)][gz-c_3(g)][gz-c_4(g)],$$
$$c_1(0)=2,\ c_2(0)=-2,\ c_3(0)=c_4(0)=1. \tag{6.26}$$

The function $\sqrt{f_g(z)}$ is an analytic function in $1/z$ and the radius of convergence, $1/z_c$, is the largest of the values $|c_1|$, $|c_2|$, $|c_3|/g$, $|c_4|/g$, unless there are special circumstances. There are, as we will now argue. From (6.22) it is seen that $w_1(0)=0$, and we can to lowest order in g write

$$\begin{aligned} V'(g,z))^2 - 4Q(g,z) &= z^2 - 4 - 2gz^3 + g^2z^4 + 4gz + \mathcal{O}(g^2) \\ &= [z-(2+2g)+\mathcal{O}(g^2)][z+(2-2g)+\mathcal{O}(g^2)][1-gz+\mathcal{O}(g)]^2 \end{aligned} \tag{6.27}$$

This implies that $c_i(g)$ change analytically with g from their values (6.26) and the radius of convergence of $\sqrt{f_g(z)}$ as a function of $1/z$ will be of order $g + \mathcal{O}(1)$ unless $c_3(g)$ is exactly equal to $c_4(g)$, in which case $\sqrt{(gz-c_3(g))(gz-c_4(g))} = \pm(gz-c(g)$, $c(g)=c_3(g)=c_4(g)$ and does not determine the radius of convergence of $\sqrt{f_g(z)}$. This is needed to avoid that the radius of convergence jumps discontinuous from being $1/2$ to g when g changes from being zero to non-zero. There cannot be such a jump, since the radius of convergence determines the exponential growth of the number of boundaries for a fixed number of triangles. Having zero triangles or one triangles and then only boundaries, should not have a dramatic effect on the number of boundaries. Then[3]

$$-\sqrt{V'(g,z))^2 - 4Q(g,z)} = (gz-c(g))\sqrt{(z-c_+(g))(z-c_-(g))}, \tag{6.28}$$

where we have introduced the notation

$$c(g) \equiv c_3(g) = c_4(g), \quad c_1(g) \equiv c_+(g) > c_-(g) \equiv c_2(g), \tag{6.29}$$

where $c(g)$, $c_\pm(g)$ are analytic around $g=0$. We now conclude that (6.20) can be written as

$$\boxed{w(g,z) = \frac{1}{2}\left(z - gz^2 + (gz - c(g))\sqrt{(z-c_+(g))(z-c_-(g))}\right)} \tag{6.30}$$

---

[3] In (6.28) we have chosen the sign of the square root on the rhs to be positive when z is large and positive. This choice is made to ensure that the term $-gz^2$ in $V'(g,z)$ is cancelled by the corresponding term coming from the square root, see (6.21).

Two-Dimensional Quantum Gravity

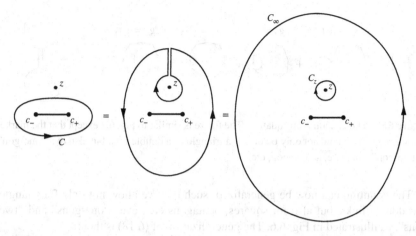

Figure 6.5 Left: contour C encloses the cut $[c_-, c_+]$, but not z. Middle and right: the contour can be deformed to pass beyond z to a contour C_∞ at infinity, provided a little circle C_z around z and winding in opposite direction is left behind.

The requirement that $w(g,z) = 1/z + \mathcal{O}(1/z^2)$ leads, by expanding in $1/z$, to three equations. The cancellation of the term gz^2 is automatic, the term z has to be cancelled by expanding the square root. Also, a constant term is not allowed in the expansion, and finally the term which goes like $1/z$ has to have the coefficient 1. These three equations determine $c(g)$ and $c_\pm(g)$ and lead to a third order equation which can be solved explicitly. However, for our purpose we do not really need the explicit solution. We only need the assumption that $c(g), c_\pm(g)$ are analytic functions in a neighborhood of $g=0$ (which can be checked from the explicit solution).

$w(g,z)$ is now an analytic function in the complex z-plane, except for a cut $[c_-(g), c_+(g)]$ and the radius of convergence of the power series in $1/z$ is determined by $|c_+(g)|$ (we have $|c_+(g)| \geq |c_-(g)|$), which is an increasing function of g, the reason being that with increasing g we have an increased probability of having more triangles and then a larger number of different boundaries. Using eq. (6.14) we can now rewrite equation (6.18) in the following way

$$\oint_C \frac{d\omega}{2\pi i} \frac{V'(g,\omega)}{z-\omega} w(g,\omega) = w^2(g,z), \quad \text{the loop equation} \tag{6.31}$$

The contour C encloses the cut $[c_-(g), c_+(g)]$, but not the point z, as shown in Fig. 6.5. Deforming the contour as also shown in Fig. 6.5 and using the expansions

$$\frac{1}{z-\omega} = -\frac{1}{\omega} \sum_{k=0}^{\infty} \left(\frac{z}{\omega}\right)^k, \quad w(g,\omega) = \sum_{l=0}^{\infty} \frac{w_l(g)}{\omega^{l+1}}, \tag{6.32}$$

to perform the integration along the contour C_∞ at infinity it is seen that the lhs of (6.31) precisely leads to the rhs of eq. (6.18).

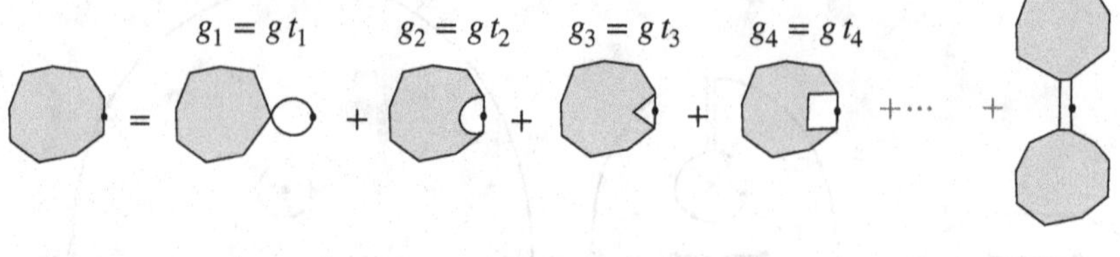

Figure 6.6 The general loop equation. The figure is similar to Fig. 6.2 except that the marked boundary link can now not only belong to a triangle or a double link, but also to a "one-gon", a "two-gon", a square, a pentagon, etc.

The counting can now be generalized, such that we allow not only for triangles and double links, but also for squares, pentagons etc., even "one-gons" and "two-gons", as illustrated in Fig. 6.6. The generalization of (6.18) is then:

$$w(g,z) = g\left(\frac{t_1}{z}+\cdots+t_n z^{n-2}\right) w(g,z) + \frac{1}{z}Q(g,z) + \frac{1}{z}w^2(g,z) \quad (6.33)$$

$$Q(g,z) = 1 - g\sum_{j=1}^{n} t_j \sum_{l=1}^{j-2} z^l w_{j-l-2}(g) \quad (6.34)$$

In these formulas a j-polygon is assigned a weight $g_j = g t_j$, i.e. in the generating function we assign a variable g to each polygon and relative weights t_j to the various kind of polygons. We now use the notation

$$V'(g,z) = z - g(t_1 + t_2 z + \cdots + t_n z^{n-1}), \quad (6.35)$$

and we find again (6.20), but now with $V'(g,z)$ and $Q(g,z)$ generalized to (6.35) and (6.34):

$$w(g,z) = \frac{1}{2}\left(V'(g,z) - \sqrt{(V'(g,z))^2 - 4Q(g,z)}\right) \quad (6.36)$$

As before we can argue that for fixed t_j and g in a neighborhood of 0, we have

$$(V'(g,z))^2 - 4Q(g,z) = M^2(g,z)(z - c_+(g))(z - c_-(g)), \quad (6.37)$$

$$M(g,z) = \sum_{k=1}^{n-1} M_k(g)(z - c_+(g))^{k-1}, \quad (6.38)$$

and (6.36) reads, if we introduce the notation $\vec{g} = (g_1,\ldots,g_n) = g(t_1,\ldots,t_n)$ to emphasize the dependence on multiple g_j:

$$\boxed{w(\vec{g},z) = \frac{1}{2}\left(V'(\vec{g},z) - M(\vec{g},z)\sqrt{(z-c_+(\vec{g}))(z-c_-(\vec{g}))}\right)} \quad (6.39)$$

We can now solve for $M(\vec{g},z)$ (where we suppress the $\vec{g}$ dependence)

$$M(z) = \frac{V'(z)}{\sqrt{(z-c_+)(z-c_-)}} - \frac{2w(z)}{\sqrt{(z-c_+)(z-c_-)}} \quad (6.40)$$

Two-Dimensional Quantum Gravity

Recall from (6.38) that $M(z)$ is a polynomial of order $n-2$. For any polynomial one can write

$$M(z) = \oint_{C_\infty} \frac{d\omega}{2\pi i} \frac{M(\omega)}{\omega - z}, \qquad (6.41)$$

where the contour C_∞ is a contour which can be moved to infinity without crossing z. Using the expression (6.40) for $M(\omega)$ in (6.41), the term with $w(\omega)$ will not contribute since $w(\omega) \to 1/\omega + \mathcal{O}(1/\omega^2)$ for $|\omega| \to \infty$ and we have

$$M(z) = \oint_{C_\infty} \frac{d\omega}{2\pi i} \frac{1}{\omega - z} \frac{V'(\omega)}{\sqrt{(\omega - c_+)(\omega - c_-)}}. \qquad (6.42)$$

By expanding $\dfrac{1}{\omega - z} = \dfrac{1}{\omega - c_+} \sum_{k=0}^{\infty} \left(\dfrac{z - c_+}{\omega - c_+} \right)^k$ we obtain

$$M(z) = \sum_{k=1}^{n-1} M_k (z - c_+)^{k-1}, \quad M_k(\vec{g}) = \oint_{C_\infty} \frac{d\omega}{2\pi i} \frac{V'(\vec{g}, \omega)}{(\omega - c_+(\vec{g}))^{k+\frac{1}{2}} (\omega - c_-(\vec{g}))^{\frac{1}{2}}} \qquad (6.43)$$

The definition of M_k is valid for all integer k, also negative k, even if only $k > 0$ appears in the sum in (6.43). Note also the contour C_∞ in (6.43) can be deformed to any curve enclosing the cut $[c_-, c_+]$ on the real axis. For a polynomial (6.35) of order $n-1$, M_k as defined by (6.43) will be zero for $k > n-1$. We can now write:

$$\begin{aligned} w(z) &= \frac{1}{2} V'(z) - \frac{1}{2} M(z) \sqrt{(z-c_+)(z-c_-)} \\ &= \frac{1}{2} V'(z) - \frac{1}{2} \oint_{C_\infty} \frac{d\omega}{2\pi i} \frac{V'(\omega)}{\omega - z} \frac{\sqrt{(z-c_+)(z-c_-)}}{\sqrt{(\omega - c_+)(\omega - c_-)}}. \end{aligned}$$

Contracting the curve C_∞ back to the curve C shown in Fig. 6.5 we finally obtain

$$\boxed{w(\vec{g}, z) = \frac{1}{2} \oint_C \frac{d\omega}{2\pi i} \frac{V'(\vec{g}, \omega)}{z - \omega} \frac{\sqrt{(z - c_+(\vec{g}))(z - c_-(\vec{g}))}}{\sqrt{(\omega - c_+(\vec{g}))(\omega - c_-(\vec{g}))}}} \qquad (6.44)$$

This a solution to the loop equation (6.31): We have now a closed expression for $w(\vec{g}, z)$, and $c_\pm(\vec{g})$ are uniquely determined by the condition that $w(z) \to 1/z$ for $|z| \to \infty$. Explicitly, expanding the integrand in (6.44) in powers of $1/z$, we obtain:

$$w(z) = \frac{1}{2} \oint_{C_\infty} \frac{d\omega}{2\pi i} \frac{V'(\omega)}{\sqrt{(\omega - c_+)(\omega - c_-)}} + \frac{1}{2z} \oint_{C_\infty} \frac{d\omega}{2\pi i} \frac{[\omega - \frac{1}{2}(c_+ + c_-)] V'(\omega)}{\sqrt{(\omega - c_+)(\omega - c_-)}} + \mathcal{O}\left(\frac{1}{z^2}\right)$$

Thus the condition $w(z) \to 1/z$ for $|z| \to \infty$ can be formulated as

$$\boxed{M_0(\vec{g}) = 0, \quad M_{-1}(\vec{g}) = 2} \qquad (6.45)$$

These are two equations which in principle determine $c_\pm(\vec{g})$.

6.3 MULTILOOPS AND THE LOOP-INSERTION OPERATOR

Let us write the generating function for our generalized triangulations (which contain also squares, pentagons etc.) in detail:

$$w(\vec{g},z) = \sum_{l,k_1,\ldots,k_n} w_{k_1,\ldots,k_n,l} \frac{1}{z^{l+1}} \prod_{j=1}^{n} g_j^{k_j}. \qquad (6.46)$$

In this notation $w_{k_1,\ldots,k_n,l}$ denotes the number of graphs with the topology of the sphere with one boundary, with k_j j-sided polygons, $j = 1,\ldots,n$, and one boundary with l links, where one of the boundary links has a mark.

Let now $n \to \infty$, i.e. we allow polygons with arbitrarily many sides. We introduce the so-called *loop insertion operator* as

$$\boxed{\frac{d}{dV(z)} = \sum_{j=1}^{\infty} \frac{j}{z^{j+1}} \frac{d}{dg_j}.} \qquad (6.47)$$

When $\dfrac{d}{dV(z_2)}$ acts on $w(\vec{g},z_1)$ it changes

$$g_j^{k_j} \to \frac{j}{z_2^{j+1}} k_j g_j^{k_j-1}, \qquad (6.48)$$

which has the interpretation that it removes a j-polygons, i.e. it creates a hole and a corresponding boundary of length j and it associates a new boundary variable z_2 to this boundary. This is the factor $1/z_2^{j+1}$ in (6.48). The factor j is present because we want to mark one of the links on the boundary and it can be done in j ways. The factor k_j is present since there are k_j different j-polygons and we can choose to remove any one of them. Finally, $g_j^{k_j} \to g_j^{k_j-1}$ since we remove one j-polygon. The process is illustrated in Fig. 6.7. Denoting the generating function for graphs with the topologies of the sphere with two marked boundaries by $w(\vec{g},z_1,z_2)$, where z_1 and z_2 are used to enumerate the boundaries, we have

$$\frac{d}{dV(z_2)} w(\vec{g},z_1) = w(\vec{g},z_1,z_2) \qquad (6.49)$$

and by an obvious generalization to n boundaries:

$$\boxed{\frac{d}{dV(z_2)} \cdots \frac{d}{dV(z_n)} w(\vec{g},z_1) = w(\vec{g},z_1,\ldots,z_n)} \qquad (6.50)$$

where $w(\vec{g},z_1,\ldots,z_n)$ is the generating function for the number of spherical graphs with n boundaries, constructed from arbitrary j-polygons. So if we can calculate the generating function for graphs with one boundary, but constructed with arbitrary j-polygons, we in principle have the complete information about the graphs with

Two-Dimensional Quantum Gravity

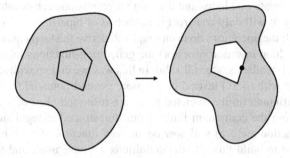

Figure 6.7 The loop insertion operator illustrated: The left part shows a triangulation with a polygon. This polygon is removed, thereby creating a new boundary (a loop) in the triangulation and a link on this boundary is marked. Finally, a new boundary cosmological constant z is assigned to this new boundary. As explained in the main text, the loop insertion operator implements this operation on the generating function for graphs with n loops, turning it into the generating function for graphs with $n+1$ loops.

n boundaries. Note that *after* we have constructed $w(\vec{g}, z_1, \ldots, z_n)$ using the loop-insertion operator, we can put any $g_j = 0$. For instance, if we only want to use triangles (as was our starting point) we simply, after having calculated the general $w(\vec{g}, z_1, \ldots, z_n)$, choose $g_j = 0$ except for g_3 (but we need the general expression in order to apply the loop-insertion operator, since even if we only use triangles (and double links), a boundary can have any length l, and we create this boundary by removing a polygon of length l. In order to remove it, it has to be present in the first place).

6.4 EXPLICIT SOLUTION FOR BIPARTITE GRAPHS

Let us consider a special class of "triangulations" where there are only loops (and in particular only boundary loops) of even lengths. Such triangulations are called *bipartite* since one can show that the vertices can be divided in two groups, yellow and blue, say, such that all links only connect different colored vertices. We again assume that the topology of the graphs are spherical with holes, i.e. boundaries. The graphs have to be constructed from $2j$ sided polygons (any odd sided polygon will create a loop of odd length) and double-links. The function $V'(z)$ thus has the form

$$V'(z) = z - g \sum_{j=2}^{\infty} t_{2j} z^{2j-1}. \tag{6.51}$$

For graphs with the topology of the disk and constructed from even sided polygons, loops will automatically be of even length. However if the topology of the graph is different from the disk, e.g. that of the cylinder, one can still have loops with odd length, as it is seen by gluing together three squares to form a prism with two boundaries of length 3. Thus if we restrict the loop inserting operator to act only with

g_{2k}, it will not create odd loops and thus not all graphs one can construct from even sided polygons. It will only construct the subclass of bipartite graphs. However, in this subclass all the operations done on graphs to derive the loop equation for $w(\vec{g},z)$ and (using the loop insertion operator) the generating functions $w(\vec{g},z_1,\ldots,z_n)$ for spheres with n boundaries, are still valid. In this way we end up with same equations as before, only with (6.51) instead of the more general potential (6.35). We do not expect the continuum limit associated to such a restricted class of triangulations to be different from the continuum limit of our unrestricted triangulations. Since the generating function $zw(\vec{g},z)$ will now be an even function of z we have $c_- = -c_+$. It is convenient to build this into the definitions we have used, and thus we define $c = c_+$ and

$$w(\vec{g},z) = \sum_{j=0}^{\infty} \frac{w_{2j}}{z^{2j+1}} = \frac{1}{2}\left(V'(\vec{g},z) - \tilde{M}(\vec{g},z)\sqrt{z^2 - c^2(\vec{g})}\right), \tag{6.52}$$

$$\tilde{M}(\vec{g},z) = \oint_{C_\infty} \frac{d\omega}{2\pi i} \frac{1}{\omega - z} \frac{V'(\vec{g},\omega)}{\sqrt{\omega^2 - c^2(\vec{g})}} = \oint_{C_\infty} \frac{d\omega}{2\pi i} \frac{1}{\omega^2 - z^2} \frac{\omega V'(\vec{g},\omega)}{\sqrt{\omega^2 - c^2(\vec{g})}}, \tag{6.53}$$

and by expanding $\frac{1}{\omega^2 - z^2} = \frac{1}{\omega^2 - c^2} \sum_{k=0}^{\infty} \left(\frac{z^2 - c^2}{\omega^2 - c^2}\right)^k$ we obtain

$$\tilde{M}(\vec{g},z) = \sum_{k=1}^{\infty} \tilde{M}_k(\vec{g})(z^2 - c^2(\vec{g}))^{k-1}, \quad \tilde{M}_k(\vec{g}) = \oint_C \frac{d\omega}{2\pi i} \frac{\omega V'(\vec{g},\omega)}{(\omega^2 - c^2(\vec{g}))^{k+\frac{1}{2}}}, \tag{6.54}$$

where the curve C can be any curve enclosing the cut $[-c,c]$ on the real axis. The requirement $w(z) \to 1/z$ for $|z| \to \infty$ leads to the condition which replaces (6.45)

$$\tilde{M}_0(\vec{g}) = 2. \tag{6.55}$$

This equation determines $c(\vec{g})$: for given $\vec{g}$ one has to perform the integral (6.54) for $k=0$ and adjust the value of c^2 such that the integral is equal 2.

When applying the loop-insertion operator to functions which can be written as $F(\vec{g},c^2)$, it is convenient to write the loop insertion operator as

$$\frac{d}{dV(z)} = \sum_{j=1}^{\infty} \frac{2j}{z^{2j+1}} \frac{d}{dg_{2j}} \quad \left(\frac{d}{dg_{2j}} = \frac{\partial}{\partial g_{2j}} + \frac{dc^2}{dg_{2j}} \frac{\partial}{\partial c^2}\right)$$

$$= \frac{\partial}{\partial V(z)} + \frac{dc^2}{dV(z)} \frac{\partial}{\partial c^2} \tag{6.56}$$

Let us now calculate how these operators act on the functions which appear in $w(\vec{g},z)$:

1) For $|\omega| < |z|$ we have

$$\frac{\partial V'(\vec{g},\omega)}{\partial V(z)} = \sum_{j=1}^{\infty} \frac{2j}{z^{2j+1}} \frac{\partial}{\partial g_{2j}} \cdot \left(\omega - \sum_{k=1}^{\infty} g_{2k}\omega^{2j-1}\right) = \frac{-2\omega z}{(\omega^2 - z^2)^2} \tag{6.57}$$

and by analytic continuation we extend the result to the entire complex ω-plane and z-plane.

Two-Dimensional Quantum Gravity

2) We now use this result to write

$$\frac{\partial \tilde{M}_k}{\partial V(z)} = \oint_C \frac{d\omega}{2\pi i} \frac{\omega}{(\omega^2-c^2)^{k+\frac{1}{2}}} \frac{\partial V'(\vec{g},\omega)}{\partial V(z)} = \frac{d}{dz} \frac{z}{(z^2-c^2)^{k+\frac{1}{2}}} \quad (6.58)$$

where we first choose C such that it encloses the cut $[-c,c]$, but not z. After that, to evaluate the integral, we are free to deform $C \to C_\infty$, picking up the pole term at z. The final contour integral at C_∞ is zero.

3) It follows from the definition (6.54) of $\tilde{M}_k$ that

$$\frac{\partial \tilde{M}_k}{\partial c^2} = \left(k+\frac{1}{2}\right) \tilde{M}_{k+1} \quad (6.59)$$

4) Acting with $d/dV(z)$ on eq. (6.55) we obtain:

$$0 = \frac{d\tilde{M}_0}{dV(z)} = \frac{\partial \tilde{M}_0}{\partial V(z)} + \frac{dc^2}{dV(z)} \frac{\partial \tilde{M}_0}{\partial c^2} = \frac{-c^2}{(z^2-c^2)^{3/2}} + \frac{1}{2}\tilde{M}_1 \frac{dc^2}{dV(z)}$$

This implies that

$$\frac{dc^2}{dV(z)} = \frac{2}{\tilde{M}_1(\vec{g},c^2)} \frac{c^2}{(z^2-c^2(\vec{g}))^{3/2}} \quad (6.60)$$

From 1)–4) it follows that we finally can write

$$\frac{d}{dV(z)} = \frac{\partial}{\partial V(z)} + \frac{2}{\tilde{M}_1(\vec{g},c^2)} \frac{c^2}{(z^2-c^2(\vec{g}))^{3/2}} \frac{\partial}{\partial c^2} \quad (6.61)$$

We can now apply these expression, when calculating

$$\frac{d}{dV(z_2)} w(\vec{g},z_1) = w(\vec{g},z_1,z_2) \quad (6.62)$$

and after an elementary but lengthy calculation (details are discussed in Problem Set 12), one obtains:

$$\boxed{w(\vec{g},z_1,z_2) = \frac{1}{2(z_1^2-z_2^2)^2}\left[z_2^2\sqrt{\frac{z_1^2-c^2}{z_2^2-c^2}} + z_1^2\sqrt{\frac{z_2^2-c^2}{z_1^2-c^2}} - 2z_1z_2\right]} \quad (6.63)$$

First of all one can check that the formula is not singular for $z_1=z_2$ by rewriting it as

$$w(\vec{g},z_1,z_2) = \frac{F^2(z_1,z_2,c^2)}{\left(\sqrt{z_1^2-c^2}+\sqrt{z_2^2-c^2}\right)^2 \sqrt{z_1^2-c^2}\sqrt{z_2^2-c^2}} \quad (6.64)$$

where

$$F(z_1,z_2,c^2) = \frac{z_2\sqrt{z_1^2-c^2} - z_1\sqrt{z_2^2-c^2}}{\sqrt{z_1^2-c^2} - \sqrt{z_2^2-c^2}} \quad (6.65)$$

Next, it should be emphasized that (6.63) is a remarkable formula: *it does not depend explicitly on $\vec{g}$. The only dependence on $\vec{g}$ comes through c^2, whose dependence on $\vec{g}$ is obtained by solving (6.55). $w(\vec{g}, z_1, z_2)$ is said to be universal.*

This simple functional form of the two-loop function makes it easy to obtain the three-loop function by applying $d/dV(z_3)$ in the form given by (6.61) (see Problem Set 12 for details) and we obtain

$$w(\vec{g}, z_1, z_2, z_3) = \frac{c^4}{2\tilde{M}_1} \frac{1}{(z_1^2 - c^2)^{3/2}} \frac{1}{(z_2^2 - c^2)^{3/2}} \frac{1}{(z_3^2 - c^2)^{3/2}} \qquad (6.66)$$

and by induction one can prove (again, details are provided in Problem Set 12)

$$w(\vec{g}, z_1, \ldots, z_n) = \left(\frac{2}{\tilde{M}_1} \frac{d}{dc^2}\right)^{n-3} \left[\frac{1}{2c^2 \tilde{M}_1} \prod_{k=1}^{n} \frac{c^2}{(z_k^2 - c^2)^{3/2}}\right] \qquad n \geq 3 \qquad (6.67)$$

This is the explicit solution to the counting problem for bipartite graphs constructed from even-sided polygons and double links, with the topology of the sphere with n boundaries. One can find the generating function for the number of graphs with a given number of $2j$-polygons and given boundary lengths l_i, $i = 1, \ldots, n$ (where l_i by construction is even). Again it is remarkable that the generating function can be completely expressed in a condensed form only depending on factors involving $(z_i^2 - c^2)^{-m - \frac{1}{2}}$ where $1 \leq m \leq n-2$, and c^2 and $\tilde{M}_k$, where $1 \leq k \leq n-2$. This follows from rules 1)–4) above, when using (6.67). It turns out that in the scaling limit, to be discussed below, the $\tilde{M}_k$ has a geometric meaning related to so-called intersection indices on Riemann surfaces. However, we have no time to discuss further this interesting topic which bridges between combinatorics and differential topology (some details can be found in the article [17]).

6.5 THE NUMBER OF LARGE TRIANGULATIONS

Let us now write $g_{2k} = g t_{2k}$, where we keep the t_{2k} fixed: we are counting $2k$-polygons with relative weight t_{2k} and g enumerates the polygons, no matter how many sides they have, i.e. we write

$$w(g, z_1, \ldots, z_n) = \sum_{k, l_1, \ldots, l_n} w_{k, l_1, \ldots, l_n} g^k \prod_{i=1}^{n} \frac{1}{z_i^{l_i + 1}}. \qquad (6.68)$$

In this formula $w_{k,l_1,\ldots,l_n}$ is the number of graphs with k polygons, the $2j$-gon counted with relative weight t_{2j}, and with n boundaries of lengths l_i, $i = 1, \ldots, n$, where l_i is even for the bipartite graphs we are considering.

Let us first concentrate on the situation where $k \to \infty$ while $l_1, \ldots, l_n$ are kept fixed. We have seen for RWs, BPs, and the bosonic string that the large k limit is associated with non-analytic behavior of the generationg function. We expect the same here. When we look at the expression for $w(g, z_1, \ldots, z_n)$ there are only two

Two-Dimensional Quantum Gravity

potential sources of such non-analytic behavior: (a) $M_1(g,c^2) \to 0$ for $g \to g_c$, or (b) $c^2(g)$ becomes non-analytic for $g \to g_c$ (where g_c is a function $g_c(t_{2j})$ of the weights t_{2j}. But we keep these fixed, as mentioned above). We will see that (a) and (b) happen at the same point g_c.

Assume that $\tilde{M}_1(g,c^2(g)) \to 0$ for $g \to g_c$. Introduce the following notation

$$\Delta g \equiv g_c - g, \qquad \Delta(c^2) \equiv c^2(g_c) - c^2(g), \qquad \tilde{M}_2^c \equiv \tilde{M}_2(g_c, c^2(g_c)). \tag{6.69}$$

For a given g, solving (6.55) leads to to a value $c^2(g)$ such that $\tilde{M}_0(g, c^2(g)) = 2$, i.e.

$$\begin{aligned} 2 = \tilde{M}_0(g_c, c^2(g_c)) &= \tilde{M}_0(g+\Delta g, c, c^2(g)) + \Delta(c^2) \\ &= \tilde{M}_0(g, c^2(g)) + \frac{\partial \tilde{M}_0}{\partial g}\Delta g + \frac{\partial \tilde{M}_0}{\partial c^2}\Delta(c^2) + \cdots \end{aligned} \tag{6.70}$$

Now use that $\tilde{M}_0(g, c^2(g)) = 2$ and $\frac{\partial \tilde{M}_0}{\partial c^2} = \frac{1}{2}\tilde{M}_1(g, c^2(g))$ (rule 3) above) and expand again

$$\begin{aligned} 0 = \tilde{M}_1(g_c, c^2(g_c)) &= \tilde{M}_1(g+\Delta g, c, c^2(g) + \Delta(c^2)) \\ &= \tilde{M}_1(g, c^2(g)) + \frac{\partial \tilde{M}_1}{\partial g}\Delta g + \frac{\partial \tilde{M}_1}{\partial c^2}\Delta(c^2) + \cdots \end{aligned} \tag{6.71}$$

Using again $\frac{\partial \tilde{M}_1}{\partial c^2} = \frac{3}{2}\tilde{M}_2(g, c^2(g)) = \frac{3}{2}\tilde{M}_2(g_c, c^2(g_c)) + \cdots$ we can finally write

$$0 = \frac{\partial \tilde{M}_0}{\partial g}\Delta g - \frac{3}{4}\tilde{M}_2^c (\Delta(c^2))^2 + \mathcal{O}((\Delta g)^2, \Delta(c^2)\Delta g) \tag{6.72}$$

In a Appendix to this section we will show that provided all $t_{2j} \geq 0$ and at least one t_{2j} is not zero, then both $\partial \tilde{M}_0/\partial g$ and $\tilde{M}_2^c$ are negative. Thus we have reached our conclusion

$$\Delta(c^2(g)) \sim \sqrt{\Delta g} \quad \text{or} \quad c^2(g) = c^2(g_c) - \text{const.} \sqrt{g_c - g} + \mathcal{O}(g_c - g). \tag{6.73}$$

The point g_c where $\tilde{M}_1(g) = 0$ is also the point where $c^2(g)$ ceases to be an analytic function of g and the singular behavior is a square root singularity.

From (6.71) and (6.73) we have for $g \to g_c$.

$$\tilde{M}_1(g, c^2(g)) = -\frac{3}{2}\tilde{M}_2^c \Delta(c^2) \propto \sqrt{\Delta g} \Rightarrow \tag{6.74}$$

$$\frac{\partial}{\partial c^2} \frac{1}{\tilde{M}_1(g, c^2)} = -\frac{3\tilde{M}_2(g, c^2)}{2\tilde{M}_1^2(g, c^2)} \propto \frac{1}{\Delta g} \tag{6.75}$$

It now follows that the most singular behavior of $w(g, z_1, \ldots, z_n)$ is obtained by differentiating $\tilde{M}_1(g, c^2)$ a maximal number of times in formula (6.67), and we obtain a singular behavior

$$\left(\frac{1}{\tilde{M}_1}\frac{d}{dc^2}\right)^{n-3}\left[\frac{1}{\tilde{M}_1}\cdots\right] \propto \left[\frac{1}{\tilde{M}_1^{2n-5}} + lst\right] \propto \left[\frac{1}{(\Delta g)^{n-5/2}} \cdots + lst\right], \tag{6.76}$$

where *lst* means "less singular terms". Thus we obtain the following behavior for the generating function

$$w(g, z_1, \ldots, z_n) \propto \frac{1}{(\Delta g)^{n-5/2}} \prod_{i=1}^{n} \frac{1}{(z_i^2 - c^2(g_c))^{3/2}} + lst \qquad (6.77)$$

Using the expansion

$$\frac{1}{(\Delta g)^{n-5/2}} = \frac{1}{g_c^{n-5/2}} \frac{1}{(1 - g/g_c)^{n-5/2}} = \frac{1}{g_c^{n-5/2}} \sum_{k=0}^{\infty} \binom{n-7/2+k}{k} \left(\frac{g}{g_c}\right)^k \qquad (6.78)$$

as well as

$$\frac{1}{(z^2 - c^2(g_c))^{3/2}} = \frac{1}{z^3} \sum_{j=0}^{\infty} \binom{j+1/2}{j} \left(\frac{c^2(g_c)}{z^2}\right)^j \qquad (6.79)$$

we obtain from (6.68) and (6.77), using (as discussed in Problem Set 5)

$$\binom{m+k}{k} = \frac{(m+k)!}{m!\, k!} \to \frac{k^m}{m!}\left(1 + \mathcal{O}\!\left(\frac{1}{k}\right)\right) \quad \text{for} \quad k \to \infty \qquad (6.80)$$

that

$$w_{k,l_1,\ldots,l_n} \propto k^{n-7/2} \sqrt{l_1 l_2 \cdots l_n} \left(\frac{1}{g_c}\right)^k \prod_{i=1}^{n} (c(g_c))^{l_i} \left(1 + \mathcal{O}\!\left(\frac{1}{k}, \frac{1}{l_i}\right)\right) \qquad (6.81)$$

This result is *universal*. The only dependence which refers to the choice of t_{2j}, the relative weights of the polygons, is $g_c(t_{2k})$ and $c(g_c(t_{2k}))$. These two constants determine the exponential growth of the number of triangulations, both wrt number of polygons and wrt lengths of the boundaries. The actual exponential rate of growth is thus not universal, but the fact that the growth *is* exponential is universal, but more importantly: *the leading power corrections to the exponential growth are universal and independent of the choice of t_{2j}.*

When deriving formula (6.12) we have actually assumed that $l_i \ll \sqrt{k}$. If $l_i \sim \sqrt{k}$ (which is not unnatural, since it is a graph where the boundary lengths are of the order we would expect for macroscopic boundaries), eq. (6.77) is not quite correct. In (6.77) we should really have used

$$\frac{1}{(z^2 - c^2(g))^{3/2}} = \frac{1}{z^3} \sum_{l=0}^{\infty} \binom{l+1/2}{l} \left(\frac{c^2(g_c)}{z^2}\right)^l \left(\frac{c^2(g)}{c^2(g_c)}\right)^l, \qquad (6.82)$$

and the last factor is no longer close to 1 when $l \sim 1/\ln \frac{c^2(g_c)}{c^2(g)} \propto 1/\sqrt{\Delta g}$. In the sum (6.78) a typical value $\langle k \rangle$ of k (where the function summed over has a maximum) is likewise of the order $1/\ln(g_c/g) \propto 1/\Delta g$. Therefore, precisely when l_i becomes of the order of $\sqrt{\langle k \rangle}$ (6.81) needs to be modified and a more careful treatment leads to (6.12). Instead of doing that we will derived (6.12) in the "continuum limit", where the relation between the l_i's and k becomes well-defined and given as in (6.11). We now turns to this.

6.6 THE CONTINUUM LIMIT

Recall the continuum formulas (6.1)–(6.8) for two-dimensional quantum gravity, which we for convenience repeat here

$$S[g,\Lambda,Z_1,\ldots,Z_n] = S[g,\Lambda] + \sum_{i=1}^n Z_i \int ds_i = \Lambda V_g + \sum_{i=1}^n Z_i L_{i,g}, \tag{6.83}$$

$$W(\Lambda;Z_1,\ldots,Z_n) = \int \mathcal{D}[g] \, e^{-S[g,\Lambda,Z_1,\ldots,Z_n]}, \tag{6.84}$$

$$= \int_0^\infty dV \, e^{-\Lambda V} \int_0^\infty \prod_{i=1}^n dL_i \, e^{-Z_i L_i} W(V;L_1,\ldots,L_n) \tag{6.85}$$

The corresponding discretized expressions are

$$S_T(\mu,\lambda_1,\ldots,\lambda_n) = \mu k + \sum_{i=1}^n \lambda_i l_i \tag{6.86}$$

$$w(\mu,\lambda_1,\ldots,\lambda_n) = \sum_{T \in \mathcal{T}(n)} e^{-S_T(\mu,\lambda_1\ldots,\lambda_n)} \tag{6.87}$$

$$= \sum_k e^{-\mu k} \sum_{l_1,\ldots,l_n} e^{-\sum_i \lambda_i l_i} w_{k,l_1,\ldots,l_n} \tag{6.88}$$

Here $W(V;L_1,\ldots,L_n)$ denotes the formal number of geometries with volume V and boundary lengths L_i (see (6.8)), and similarly $w_{k,l_1,\ldots,l_n}$ is the number of "triangulations" made of k polygons and with boundary lengths l_i. Thus we can make the following identification with our generation function:

$$z_1 \cdots z_n w(g,z_1,\ldots,z_n) \equiv w(\mu,\lambda_1,\ldots,\lambda_n), \quad g = e^{-\mu}, \quad \frac{1}{z_i} = e^{-\lambda_i}, \tag{6.89}$$

where the factors $z_1 \cdots z_n$ just follows the convention (6.13) and where we identify

$$g = e^{-\mu}, \quad \frac{1}{z_i} = e^{-\lambda_i}, \quad g_c = e^{-\mu_c}, \quad \frac{1}{c(g_c)} = e^{-\lambda_c}. \tag{6.90}$$

With these definitions we can write

$$\left(\frac{g}{g_c}\right)^k = e^{-(\mu-\mu_c)k} = e^{-\Lambda V} \quad \text{if} \quad \boxed{V = k\varepsilon^2, \quad \mu - \mu_c = \Lambda \varepsilon^2} \tag{6.91}$$

$$\left(\frac{c(g_c)}{z}\right)^l = e^{-(\lambda-\lambda_c)l} = e^{-ZL} \quad \text{if} \quad \boxed{L = l\varepsilon, \quad \lambda - \lambda_c = Z\varepsilon} \tag{6.92}$$

Note that for fixed Λ and $\varepsilon \to 0$ we have $\mu \to \mu_c$ and thus $g \to g_c$ and we can write

$$\frac{\Delta g}{g_c} = \mu - \mu_c = \Lambda \varepsilon^2. \tag{6.93}$$

In agreement with (6.11) ε can be given the interpretation of the link length in the "triangulation" and V and L then represent the continuum volume (area) of the triangulation and continuum length of a boundary. Of course the boxed relations in (6.91) and (6.92) only make sense as continuum relations in the limit where $k, l \gg 1$. If we insist on a limit where V and L are finite while $\varepsilon \to 0$, we have in this limit $l \sim \sqrt{k}$, exactly the situation discussed above in connection with (6.82). Recalling the exponential growths of $w_{k,l_1,\ldots,l_n}$ in (6.81), we see that what appear in the sum over k and l_i in (6.87) are precisely terms $e^{-(\mu-\mu_c)k}$ and $e^{-(\lambda_i-\lambda_c)l_i}$, i.e. with the boxed indentifications $e^{-\Lambda V}$ and $e^{-Z_i L_i}$, and these allow us to make a very direct translation from (6.88) to (6.85) *proved we introduce the concept of renormalized cosmological and boundary cosmological constants* Λ *and* Z_i, as done in (6.91) and (6.92). We call it a renormalization of the cosmological constant for the following reason: μ is dimensionless, and the cosmological term in the action (6.86) is μk which we according the identifications above would write as $\mu k = \Lambda_0 V$, where then $\Lambda_0 = \mu / \varepsilon^2$. We call Λ_0 the *bare* cosmological constant. Now the relation with Λ in (6.91) can be written as

$$\boxed{\Lambda_0 = \frac{\mu_c}{\varepsilon^2} + \Lambda}, \tag{6.94}$$

which is a so-called additive renormalization of the bare cosmological constant, needed in order to obtain finite answers from the path integral. The constant to be subtracted from Λ_0 can be identified as coming from the entropy of configurations since μ_c determines the exponential growth of the number of configurations with the same discrete volume (the same number of polygons). The situation is entirely identical to what happened in the case of a free relativistic particle (see (2.31) and (2.41)). A similar interpretation can now be given to the renormalization of the boundary cosmological constant, represented as boxed equations in (6.92).

With the above indentifications we can now easily take the continuum limit of our multi-loop functions (6.63)–(6.67). In the limit $\varepsilon \to 0$, i.e. $g \to g_c$ we have

$$c^2(g) = c^2(g_c) - \Delta(c^2) = c^2(g_c) - \text{cnst.} \sqrt{\Delta g} = c^2(g_c) - \text{cnst.} \sqrt{\Lambda} \varepsilon. \tag{6.95}$$

$$z_i^2 - c^2(g) = \text{cnst.} \left(Z_i + \sqrt{\Lambda}\right) \varepsilon \tag{6.96}$$

after a suitable rescaling of Λ, and from (6.74)

$$\tilde{M}_1(g, c^2(g)) = \text{cnst.} \sqrt{\Lambda} \varepsilon \tag{6.97}$$

Thus

$$\begin{aligned} w(\vec{g}, z_1, \ldots, z_n) &= \left(\frac{2}{\tilde{M}_1} \frac{d}{dc^2}\right)^{n-3} \left[\frac{1}{2c^2 \tilde{M}_1} \prod_{k=1}^{n} \frac{c^2}{(z_k^2 - c^2)^{3/2}}\right] \\ &\to \frac{\text{cnst.}}{\varepsilon^{7n/2-5}} \left(-\frac{d}{d\Lambda}\right)^{n-3} \left[\frac{1}{\sqrt{\Lambda}} \prod_{k=1}^{n} \frac{1}{(Z_k + \sqrt{\Lambda})^{3/2}}\right] \end{aligned} \tag{6.98}$$

and we can write for $n \geq 3$ and in the limit $\varepsilon \to 0$:

$$w(\vec{g}, z_1, \ldots, z_n) \to \frac{\text{cnst.}}{\varepsilon^{7n/2-5}} W(\Lambda, Z_1, \ldots, L_n) \qquad (6.99)$$

$$W(\Lambda; Z_1, \ldots, Z_n) = \left(-\frac{d}{d\Lambda}\right)^{n-3} \left[\frac{1}{\sqrt{\Lambda}} \prod_{k=1}^{n} \frac{1}{(Z_k + \sqrt{\Lambda})^{3/2}}\right] \qquad (6.100)$$

Finally, we obtain for the two-loop function:

$$W(\Lambda; Z_1, Z_2) = \frac{1}{2(Z_1 - Z_2)^2} \left(\frac{Z_1 + Z_2 + 2\sqrt{\Lambda}}{\sqrt{Z_1 + \sqrt{\Lambda}}\sqrt{Z_2 + \sqrt{\Lambda}}} - 2\right) \qquad (6.101)$$

$$= \frac{1}{2\left(\sqrt{Z_1 + \sqrt{\Lambda}} + \sqrt{Z_2 + \sqrt{\Lambda}}\right)^2 \sqrt{Z_1 + \sqrt{\Lambda}}\sqrt{Z_2 + \sqrt{\Lambda}}} \qquad (6.102)$$

From (6.6) and (6.7) we can calculate $W(\Lambda, L_1, \ldots, L_n)$ and $W(V, L_1, \ldots, L_n)$ by inverse Laplace transformations. The result is for $n \geq 3$

$$W(\Lambda; L_1, \ldots, L_n) \propto \left(-\frac{d}{d\Lambda}\right)^{n-3} \left[\frac{1}{\sqrt{\Lambda}} \prod_{k=1}^{n} \sqrt{L_i} e^{-\sqrt{\Lambda} L_i}\right] \qquad (6.103)$$

$$W(V; L_1, \ldots, L_n) \propto V^{n-7/2} \sqrt{L_1 \cdots L_n} \exp\left(-\frac{(L_1 + \cdots + L_n)^2}{4V}\right) \qquad (6.104)$$

Please recall that $W(V, L_1, \ldots, L_n)$ is formally the number of geometries of the sphere with volume V and n boundaries of lengths L_i. We have managed to "count" the number of these geometries.

One can show that (6.104) is actually valid also for $n = 0, 1$ and 2. For $n = 0$ we have, using (6.7)

$$W(V) \propto V^{-7/2} \quad \text{i.e. formally} \quad W(\Lambda) \propto \Lambda^{5/2}, \qquad (6.105)$$

where we write "formally" since the Laplace transform (6.7) is singular for $W(V) \propto V^{-7/2}$. Again, using (6.7), we find

$$W(\Lambda, L_1, L_2) \propto \frac{\sqrt{L_1 L_2}}{L_1 + L_2} e^{-\sqrt{\Lambda}(L_1 + L_2)} \qquad (6.106)$$

$$W(\Lambda, L) \propto L^{-5/2} \left(1 + \sqrt{\Lambda} L\right) e^{-\sqrt{\Lambda} L} \qquad (6.107)$$

We can now use (6.6) to calculate $W(\Lambda, Z)$ and obtain

$$W(\Lambda, Z) \propto \left(Z - \frac{1}{2}\sqrt{\Lambda}\right) \sqrt{Z + \sqrt{\Lambda}} \qquad (6.108)$$

Actually, due to the factor $L^{-5/2}$, the Laplace transform leading to (6.108) is singular for $L \to 0$, so we have thrown away an infinite constant in (6.108). We would have obtained the same problem had we tried directly to take the scaling limit starting from $w(g,z)$: some constant terms survive (see eq. (7.6) below for an explicit formula), which we formally see by looking at the scaling factor $\varepsilon^{5-7n/2}$ in (6.98). It diverges for $n > 1$ and that is why the continuum part dominates any non-universal part. However, for the disk function it is opposite. The continuum part is subleading (scales as $\varepsilon^{3/2}$) compared to constant terms. However, these constant terms will go away if we differentiate $w(g,z)$ wrt g or z sufficiently many times and will not play a role for large triangulations. Thus we dismiss the finite part in the limit $\varepsilon \to 0$ as irrelevant for continuum physics. The same statements made for $n=1$ is even more true in the case of $n=0$, given by (6.105), since in this case we do not have the exponential function in (6.104) to provide a regularization at $V=0$, and the Laplace transform in V becomes singular at $V=0$.

The disk function $W(\Lambda,L)$ is called the *Hartle-Hawking wave function* of the (two-dimensional) universe. An interpretation of this wave function is that it is the amplitude for a universe to evolve from nothing to size L. Unfortunate, this evolution is in spacetimes with Euclidean signature, and it has never been clear precisely how one should rotate the result back to spacetimes with Lorentzian signature.

Finally, considering the limit where $L_i \ll \sqrt{V}$, such that we can ignore the exponential function in (6.104), we see that $W(V,L_1,\ldots,L_n)$ represents what we in the case of bosonic strings called the susceptibility of a string with n boundaries (only are they here entirely intrinsic) and we have (like for bosonic strings)

$$W(V,L_1,\ldots,L_n) \propto V^{n-2+(\gamma-1)}, \qquad \gamma = -\frac{1}{2}, \qquad (6.109)$$

valid for $n \geq 3$. Similarly, in the limit where $L_i \ll 1/\sqrt{\Lambda}$ we have from (6.103)

$$W(\Lambda,L_1,\ldots,L_n) \propto \frac{1}{\Lambda^{n-2+\gamma}}, \qquad \gamma = -\frac{1}{2}, \qquad (6.110)$$

We see that a difference between the bosonic strings (which can be viewed as two-dimensional gravity coupled to D scalar fields X_i) and pure two-dimensional gravity is that γ changes from $1/2$ to $-1/2$. This brings up the interesting question of how two-dimensional gravity behaves when coupled to other matter fields than scalar fields. We have no space to this discussion, except for the few words said in the next section.

6.7 OTHER UNIVERSALITY CLASSES

Let us again emphasize the *universality* of the continuum limit. The class of graphs used and the weights associated with the different polygons are not important as long as $t_{2k} \geq 0$, $k \geq 2$, and at least one t_{2k}, is positive. In this sense the situation is quite similar to the one for RWs and BPs. In the case of BPs we saw that by allowing some of the weights to become negative, one could reach different universality classes of BPs. The same is true in the case of our two-dimensional gravity models. Recall that the critical point g_c was determined by

$$\tilde{M}_1(g_c, c^2(g_c)) = 0, \qquad \tilde{M}_2(g_c, c^2(g_c)) \neq 0. \qquad (6.111)$$

As shown in the Appendix following this section, $\tilde{M}_2(g_c, c^2(g_c)) < 0$ follows from $t_{2k} \geq 0$ and one $t_{2k} > 0$, $k \geq 2$. If we relax the condition that $t_{2k} \geq 0$ we can obtain a more general scaling at a g_c characterized by

$$\tilde{M}_1(g_c) = \cdots = \tilde{M}_{m-1}(g_c) = 0, \qquad \tilde{M}_m(g_c) \neq 0, \quad m > 2. \tag{6.112}$$

Approaching such a point one can show, using rule (1)–(4) (eqs. (6.57)–(6.60)), that for $\Delta g = g_c - g$ going to zero one has

$$\Delta(c^2) = c^2(g_c) - c^2(g) \propto (\Delta g)^{1/m}. \tag{6.113}$$

The situation is thus very similar to the one encountered for the multicritical BPs and we call the continuum gravity model obtained in this limit the mth *multicritical gravity model* (and we will study (6.111) and (6.112) in detail in Problem Set 10). It is possible to show that this continuum model corresponds to pure gravity coupled to a so-called (p,q) rational conformal field theory where $(p,q) = (2, 2m-1)$. One can obtain more general conformal field theories coupled to two-dimensional quantum gravity if we in addition to negative t_{2k} also allow for infinitely many t_{2k} being different from zero. Again, the situation, from a technical point of view of taking the scaling limit, is very similar to the what happens for BPs (as we will discuss in Problem Set 10).

As hinted in eq. (6.113) we can obtain different critical exponents when matter is coupled to two-dimensional gravity. The situation is most beautifully illustrated in the case of the so-called Ising (spin) model coupled to two-dimensional gravity. The Ising model is the simplest possible spin model, where spins are located at the vertices of a lattice, can take values ± 1 and only couple to neighboring spins. If the lattice is a two-dimensional regular lattice the model can be solved analytically (the so-called Onsager solution), and we have the following picture: there exists a critical temperature, T_c, where the spin system undergoes a second-order phase transition from a high temperature unmagnetized phase to a low-temperature magnetized phase. At the phase transition the spin-spin correlation length will diverge, and the spin-spin correlation functions can be described by a conformal field theory with so-called central charge $c = 1/2$ (in the notation mentioned above it is a $(p,q) = (3,4)$ theory. Thus it is not one of the $(p,q) = (2, 2m-1)$ conformal field theories associated the multicritical models). The Ising model can also be defined on the triangulations used to define two-dimensional gravity, and also the combined model of gravity and Ising spin can be solved analytically. Again there is critical temperature T_c', separating a magnetized and unmagnetized phase. However, the critical exponents α, β and γ_m for the spin system are different from the Onsager exponents for the Ising model on a regular lattice. The interaction with fluctuating geometries changes the exponents. But even more remarkable, the susceptibility exponent γ of two-dimensional gravity is also changed from $\gamma = -1/2$ to $\gamma = -1/3$. The change is only at the critical temperature T_c', where the spin-spin correlation length diverges. For $T \neq T_c'$ we have $\gamma = -1/2$, as for two-dimensional gravity without Ising spins. There is thus an intricate interaction between matter and geometry precisely when the matter interaction becomes long range. Unfortunately there is no space for covering this in these notes, but as a compensation Problem Set 11 discusses a mean-field version the Ising model coupled to DT, which captures well this interaction between geometry and matter.

6.8 APPENDIX

If $t_{2j} \geq 0$ and there exists a $j > 1$ such that $t_{2j} > 0$ then $\tilde{M}_k(g, c^2) < 0$ for $k > 1$. First, we note that

$$\frac{1}{(\omega^2 - c^2)^{k+\frac{1}{2}}} = \frac{1}{\omega^{2k+1}} \sum_{i=0}^{\infty} d_i(k) \frac{c^{2i}}{\omega^{2i}}, \quad d_i(k) > 0. \tag{6.114}$$

Next, we have for $k > 1$

$$\tilde{M}_k = \oint_{C_\infty} \frac{d\omega}{2\pi i} \frac{\omega^2 - g\sum_{j=2}^{\infty} t_{2j}\omega^{2j}}{(\omega^2 - c^2)^{k+\frac{1}{2}}} = -g \sum_{j=0}^{\infty} t_{2(k+j)} d_j(k) c^{2j} < 0. \tag{6.115}$$

For $k = 1$ we get

$$\tilde{M}_1 = 1 - g \sum_{j=0}^{\infty} t_{2(1+j)} d_j(1) c^{2j}, \tag{6.116}$$

It is thus seen that $\tilde{M}_1(0, c^2(0)) = 1$ (it is the case of BPs and $c^2(0) = 4$). The same kind of calculation shows that

$$\frac{\partial \tilde{M}_0}{\partial g} = \oint_{C_\infty} \frac{d\omega}{2\pi i} \frac{\frac{\partial}{\partial g}(\omega^2 - g\sum_{j=2}^{\infty} t_{2j}\omega^{2j})}{(\omega^2 - c^2)^{\frac{1}{2}}} = -g \sum_{j=2}^{\infty} t_{2j} d_j(0) c^{2j} < 0 \tag{6.117}$$

We can use this information to show that $c^2(g)$ is an increasing function of $g \in [0, g_c]$ ($c^2(0) = 4$ (BPs)). We have already argued for that intuitively, since increasing g implies an increasing number of polygons in an average "triangulation" which consists of double-lines and polygons. Thus one can have more boundaries of different types and the same length than if we have fewer polygons, and $c^2(g)$ determines the exponential growth of the number of boundaries as a function of the length of the boundaries. However using (6.117) we have directly

$$2 = \tilde{M}_0(g, c^2(g)) \Rightarrow 0 = \frac{\partial \tilde{M}_0}{\partial g} + \frac{\partial \tilde{M}_0}{\partial c^2} \frac{dc^2}{dg} = \frac{\partial \tilde{M}_0}{\partial g} + \frac{1}{2} M_1(g, c^2(g)) \frac{dc^2}{dg} \tag{6.118}$$

Since $M_1(0, c^2(0)) = 1$ and the first zero of $M_1(g, c^2(g))$ is at $g = g_c$, $M_1(g, c^2(g)) > 0$ for $g \in [0, g_c[$ and consequently $dc^2/dg > 0$ in the same interval.

Finally note that one has explicitly

$$\frac{\partial \tilde{M}_0}{\partial g} = \frac{1}{g} \oint_{C_\infty} \frac{d\omega}{2\pi i} \frac{(\omega V'(\omega) - \omega^2)}{(\omega^2 - c^2)^{\frac{1}{2}}} = \frac{1}{g}\left(\tilde{M}_0 - \frac{1}{2} c^2(g)\right) = -\frac{1}{2g}(c^2(g) - 4) \tag{6.119}$$

Thus it is possible to write the important relation (6.72) explicitly as

$$\boxed{\Delta g = \frac{3g(-\tilde{M}_2^c)}{2(c^2(g) - 4)} (\Delta(c^2))^2} \tag{6.120}$$

6.9 PROBLEM SETS AND FURTHER READING

Problem Set 10 shows how it is possible to derive general scaling for the two-dimensional models (including the above mentioned multicritical models) in a quite simple way, which is very similar to the way it was done for BPs. More details about the critical exponents related to this general scaling can be found in the original article on which Problem Set 10 is based [15]. Problem Set 11 provides the simplest possible model for the interaction between geometry and matter where both set of critical exponents are changed. Finally, Problem Set 12 is a guide to how in detail to derive the n-loop functions using the loop insertion operator and the so-called "moments M_k. They were introduced in [16] to analyze the n-loops functions for surfaces with no handles and generalized in [17] to surfaces of arbitrary topology, triggering the development of what is now known as *topological recursion*, a topic which has become an important research area in mathematics (see the review [18]).

Some of the expressions for the loop correlators have also been derived working entirely within continuum quantum fields, in the context of what is called *quantum Liouville theory*, as mentioned. We will not try to define quantum Liouville theory here, but one way to view it is to consider a conformal field theory coupled to two-dimensional quantum gravity. It is then possible, using the so-called *conformal anomaly*, to integrate out the conformal fields and one is left with a modified gravity action, where one has to integrate over geometries. In this theory one can now calculate certain observables exactly when the topology of the two-dimensional manifold is simple, again using conformal invariance arguments. Among the observables one can calculate is the one-loop function and one finds agreement with the calculations using the regularized one-loop functions provided here and then taking the scaling limit. It is important since it provides a very strong argument in favor of the correctness of the regularization used in these notes. Also it should be noted that there are many observable, like multiloop functions for surfaces with complicated topology, which can be calculated in the discretized approach, but not yet using quantum Liouville theory.

During the last 20 years there has been very significant progress in mathematics in understanding rigorously what is meant by integrating over two-dimensional geometries. The references [19, 20, 21] are lecture notes where this progress is described (somewhat pedagogically).

7 The Fractal Structure of 2D Gravity

7.1 UNIVERSALITY AND THE MISSING CORRELATION LENGTH

In the last section we saw how universal scaling limits describing aspects of two-dimensional quantum gravity could be obtained. While we found critical points, critical surfaces and approached them in various ways, which provided a wonderful realization of the Wilsonian point of view, where the continuum quantum theory is related to the approach to critical surfaces, somehow the most important and intuitive part of this picture was missing. The primary intuitive reason for the Wilsonian universality is the existence of a correlation length which diverges when we approach the critical surface. It is this divergence of a correlation length which makes the underlying lattice structure irrelevant and allows us to define a continuum theory with no reference to the lattice. But where is this correlation length when we consider two-dimensional quantum gravity? A priori it is not so clear how to define a correlation length in a theory of quantum gravity. In the path integral we have to integrate over all geometries, but at the same time a correlation length, being a "length", has to refer to a geometry. Still, we will show in this section that one can define a two-point function on the triangulations with a correlation length $\xi(\mu)$ which diverges as $|\mu - \mu_c|^{-\nu}$ when we approach the critical point μ_c, and where the scaling exponent $\nu = 1/4$ determines the Hausdorff dimension ($d_H = 4$) and where the susceptibility exponent γ calculated from this two-point function precisely is the $\gamma = -1/2$ already determined in the former chapter.

7.2 THE TWO-LOOP PROPAGATOR

Let us return to the set up where we only allow for triangles and double-links. The disk function was given by (6.30), which we repeat here for convenience:

$$w(g,z) = \frac{1}{2}\left(z - gz^2 + (gz - c(g))\sqrt{(z - c_+(g))(z - c_-(g))}\right) \qquad (7.1)$$

Taking the continuum limit for this model is slightly different from the situation for the bipartite triangulations. since we have that $c_+(g) \neq -c_-(g)$. The critical behavior is still obtained when $M_1(g) \to 0$ and we have that $c_+(g)$ at this point becomes a non-analytic function of g. Let us here list the behavior for $\Delta g = g_c - g \to 0$, the critical value of the boundary cosmological constant being $z_c = c_+(g_c)$:

$$c(g) = g_c z_c \left(1 + \frac{1}{2}\alpha\sqrt{\Delta g}\right) + \mathcal{O}(\Delta g) \tag{7.2}$$

$$c_+(g) = z_c\left(1 - \alpha\sqrt{\Delta g}\right) + \mathcal{O}(\Delta g) \tag{7.3}$$

$$c_-(g) = c_-(g_c) + \mathcal{O}(\Delta g) \tag{7.4}$$

Here α is a constant which can be calculated like we did in (6.120) for the bipartite graphs. We now define the continuum cosmological constant Λ and boundary cosmological constant Z as above:

$$z = z_c(1 + \varepsilon Z), \quad \Delta g = \alpha^{-2}\varepsilon^2\Lambda, \tag{7.5}$$

These relations are the equivalent to (6.95) and (6.96), and the factor α^{-2} in (7.5) is a rescaling of the cosmological constant, done in order to ensure that $z - c_+(g) \propto Z + \sqrt{\Lambda}$. A similar rescaling was performed in (6.96) and it is just to obtain nice-looking formulas. Taking the limit $\varepsilon \to 0$ we write $w(g,z)$ as

$$w(g,z) = \frac{1}{2}\left(z - gz^2 + \varepsilon^{3/2} g_c z_c^{3/2} W(\Lambda, Z) + \mathcal{O}(\varepsilon^2)\right), \tag{7.6}$$

where

$$W(\Lambda, Z) = \left(Z - \frac{1}{2}\sqrt{\Lambda}\right)\sqrt{Z + \sqrt{\Lambda}} \tag{7.7}$$

This is precisely (6.108), which we derived by a Laplace transformation of $W(V,L)$. As discussed below eq. (6.108) the part $z - gz^2$ has no continuum limit, but will play no role in the following[1]. The continuum multiloop functions $W(\Lambda, Z_1, \ldots, Z_n)$ are given by (6.100) and (6.102). What one observes is that the following formula is valid

$$\lim_{Z_1 \to \infty} Z_1^{3/2} W(\Lambda, Z_1, \ldots, Z_n) \propto \left(-\frac{d}{d\Lambda}\right) W(\Lambda, Z_2, \ldots, Z_n) \tag{7.8}$$

It is even true for $n=2$ and for $n=1$, although in particular the $n=1$ case requires some additional arguments (which we will not present here). The interpretation is as follows: from the relation (6.6) between $W(\Lambda, Z_1, \ldots, Z_n)$ and $W(\Lambda, L_1, \ldots, L_n)$ it is seen that $Z_1 \to \infty$ corresponds to $L_1 \to 0$. Thus we are contracting the marked loop with boundary cosmological constant Z_1 to a "marked point", and we multiply by the factor $Z_1^{3/2}$ to get rid of a remaining allover Z_1 factor associated with the marked point. But this marked point can be anywhere on the surface. Thus the number of surfaces with a marked point is related to the number of surfaces without a marked point by multiplying with the volume (area) of the surface. This is precisely implemented by $\left(-\frac{d}{d\Lambda}\right)$. Recall that Λ only appears in the continuum action (6.2) as ΛV and thus differentiating $W(\Lambda, Z_1, \ldots, Z_n)$ defined as the path integral (6.3) brings down a factor V. Eq. (7.8) is illustrated in Fig. 7.1. We will later use this procedure to contract marked loops to marked points.

---

[1] In certain situations, not discussed here it *can* play a role, see footnote 3, Chapter 8.

The Fractal Structure of 2D Gravity

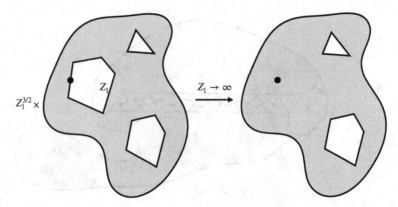

Figure 7.1 Illustration of eq. (7.8). By sending the boundary cosmological constant Z_1 to infinity one will contract the corresponding boundary loop with a marked link to a marked point in the triangulation, leaving a factor $Z_1^{-3/2}$.

Let us now consider the two-loop function, but with the additional constraint that every point on the "exit" loop has a fixed distance to the "entrance" loop. In order to formulate this in a precise way, we return to the discrete formulation in terms of triangles and double-links. We mark a link on the entrance loop, but not on the exit loop. These choices are made for convenience, as we will explain later. We denote the two-loop function by

$$G(g, l_1, l_2; r), \qquad g = e^{-\mu}. \tag{7.9}$$

The relation between g and μ is the standard one we have been using, given by (6.90). l_1 denotes at the same time the entrance loop and the number of links in the entrance loop (one link marked), and similarly l_2 denotes the exit loop and the number of links in the exit loop. r denotes the *graph distance* between l_2 and l_1 in a given triangulation with the two boundary loops. Given a link in l_2 and a link in l_1 we define the graph distance between these links as shortest path through neighboring triangles which connects the two links, the length of the path counted as as number of triangles present in the path. One can imagine the path as a piecewise linear path passing through the centers of the neighboring triangles. The situation is illustrated in Fig. 7.2. The distance between a given link in l_2 and the boundary l_1 is defined as the minimum of the distances between the given link in l_2 and the links in l_1. We now require that *each link in l_2 has the same distance r to l_1* (note that this does not ensure that each link in l_1 has the same distance to l_2, but there is of course at least one link in l_1 which has the distance r to l_2). This way of defining the distance between the boundaries ensures that we have the composition law:

$$G(g, l_1, l_2; r_1 + r_2) = \sum_{l=2}^{\infty} G(g, l_1, l; r_1)\, G(g, l, l_2; r_2). \tag{7.10}$$

This is where the choice of marking and non-marking of the boundary loops comes into play. With our choice there is no additional l weight factor in the sum, related

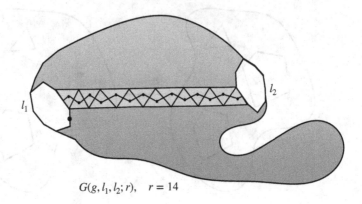

$G(g, l_1, l_2; r), \quad r = 14$

Figure 7.2 Illustration of a configuration (a surface with cylinder topology) with boundary loops of l_1 and l_2, separated a graph distance $r = 14$.

to the way one can "turn" the two cylinders relative to each other, when gluing them together to one cylinder.

We will think of the graph distance defined this way as the *geodesic distance* between the two loops for a given triangulation in the piecewise linear geometry defined by the equilateral triangulation. Of course this is not strictly speaking correct, but we expect for very large generic triangulations and very large distances that the real geodesic distance will be proportional to the graph distance and we will in the following not distinguish between the two.

We now use the same "moves" as shown in Fig. 6.2, except that we apply them to the cylinder surfaces used in the calculation of $G(g, l_1, l_2; r)$. We "peel" away a triangle from the entrance loop, moving "closer" to the exit loop, or, when we meet a double-link, we chop away a "baby" universe, as shown in Fig. 7.3. Algebraically, we can write the operation as follows:

$$G(g, l, l'; r) = g\tilde{G}(g, l+1, l'; r) + 2 \sum_{l''=0}^{l-2} w_{l''}(g) G(g, l-l''-2, l'; r), \quad (7.11)$$

where $w_l(g)$ as usual denotes the disk function with a boundary consisting of l links. $g\tilde{G}(g, l+1, l'; r)$ (the left graph in Fig. 7.3) is not really of the form $gG(g, l+1, l'; r)$ since removing a triangle in general will spoil the property that exit links have a distance r to the entrance loop. However, applying the removal of triangles l times will "in average" get us one step closer to the exit loop, i.e. to $G(g, l, l', r-1)$. Thus we write

$$g\tilde{G}(g, l+1, l'; r) = gG(g, l+1, l'; r) - \frac{1}{l} \frac{\partial G(g, l, l'; r)}{\partial r}, \quad (7.12)$$

where the factor $1/l$ in front of the derivative term refers to $1/l^{th}$ of the l times we in average have to remove a triangle to get from r to $r-1$. Clearly this is not a rigorous result the way it is presented here. The operation $-\frac{1}{l}\frac{\partial}{\partial r}$ is rather intuitive, at best.

The Fractal Structure of 2D Gravity

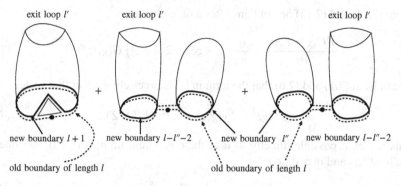

Figure 7.3 Graphically illustration of equation (7.11). The entrance loop of the cylinder triangulation has length l, and the marked link can belong to a triangle or to a double link. After removing the double link and the corresponding baby universe the entrance loop of the resulting cylinder can have any length less than or equal $l-2$.

However, it can be made rigorous, but it is rather tedious[2], and we will simply accept (7.12). Inserted in (7.11) we obtain:

$$\frac{\partial G(g,l,l';r)}{\partial r} = -l G(g,l,l';r) + g l\, G(g,l+1,l';r) + \qquad (7.13)$$

$$2l \sum_{l''=0}^{l-2} G(g,l-l''-2,l';r)\, w_{l''}(g).$$

The last term is a kind of convolution. We thus introduce the (discrete) Laplace transformation, which turns convolutions into products:

$$\hat{G}(z) := \sum_{l=0}^{\infty} \frac{G(l)}{z^{l+1}}, \qquad G(l) = \oint \frac{dz}{2\pi i} z^l\, \hat{G}(z). \qquad (7.14)$$

$$H(l) = \sum_{l'=0}^{l} G(l')F(l-l') \implies z\hat{H}(z) = z\hat{G}(z)\cdot z\hat{F}(z) \qquad (7.15)$$

The inversion formula in (7.14) assumes the contour is in the region where $\hat{G}(z)$ is analytic, and the convolution formula uses the rearrangement

$$\sum_{l=0}^{\infty}\sum_{l'=0}^{l} \frac{G(l-l')F(l')}{z^l} = \sum_{l'=0}^{\infty}\sum_{l=l'}^{\infty} \frac{G(l-l')}{z^{l-l'}}\frac{F(l')}{z^{l'}} \qquad (7.16)$$

We now introduce the discrete Laplace transform for variable l in $G(g,l,l';r)$ and by an abuse of notation we still denote it $G(g,z,l';r)$:

$$G(g,z,l';r) = \sum_{l=0}^{\infty}\frac{G(g,l,l';r)}{z^{l+1}}. \qquad (7.17)$$

---

[2]For the really dedicated readers we can refer to the article [22].

From (7.14) and (7.15) one obtains after a little algebra

$$\frac{\partial G(g,z,l';r)}{\partial r} = \frac{\partial}{\partial z}\left[(z - gz^2 - 2w(g,z))\, G(g,z,l';r)\right] \quad (7.18)$$

Recall from (7.1) and (7.6) that the term in $(\cdot)$ is precisely the part that scales:

$$z - gz^2 - 2w(g,z) \propto -\varepsilon^{3/2} W(\Lambda, Z) \quad (7.19)$$

This makes it possible directly to take the continuum limit of (7.18). Assume the scaling (7.5) and in addition

$$\varepsilon l = L, \quad \varepsilon l' = L', \quad \varepsilon^\delta r \propto R. \quad (7.20)$$

Naively, we would expect that $\delta = 1$. However, that will not be the case. First note that

$$\sum_l = \frac{1}{\varepsilon}\sum_l \varepsilon \to \frac{1}{\varepsilon}\int dL \quad (7.21)$$

Next, it then follows from the composition rule (7.10) that the continuum limit of $G(g,l,l';r)$ has to scale as

$$G(g,l,l';r) \propto \varepsilon\, G(\Lambda, L, L'; R) \quad (7.22)$$

Then (7.14) leads to, for $\varepsilon \to 0$,

$$G(g,z,l';r) \propto \int_0^\infty dL\, e^{-LZ} G(\Lambda, L, L'; R) = G(\Lambda, Z, L'; R) \quad (7.23)$$

From (7.18) we now obtain

$$\varepsilon^\delta \frac{\partial}{\partial R} G(\Lambda, Z, L'; R) = -\frac{1}{\varepsilon}\frac{\partial}{\partial Z}\left(\varepsilon^{3/2} W(\Lambda, Z) G(\Lambda, Z, L'; R)\right) \quad (7.24)$$

Thus $\delta = \frac{1}{2}$: the "geodesic" distance R scales anomalously:

$$\boxed{\dim[R] = \frac{1}{2}\dim[L] = \frac{1}{4}\dim[V]} \quad (7.25)$$

and we have the equation

$$\boxed{\frac{\partial}{\partial R} G(\Lambda, Z, L'; R) = -\frac{\partial}{\partial Z}\left(W(\Lambda, Z) G(\Lambda, Z, L'; R)\right).} \quad (7.26)$$

The function $G(\Lambda, Z, L'; R)$ describes the "propagation" of a spatial universe, where the length distribution is dictated by the boundary cosmological constant Z (and the boundary has a mark), a distance R to a spatial universe where the boundary has length L'. The distance R is an intrinsic geodesic distance for each geometry which

The Fractal Structure of 2D Gravity

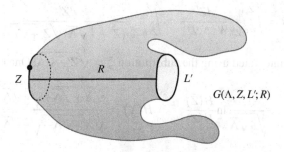

Figure 7.4 Graphically illustration of a surface with the topology of a cylinder contributing to $G(\Lambda, Z, L'; R)$. The entrance loop, which has a mark and a boundary cosmological constant Z, is separate a geodesic distance R from the exit loop which has length L'.

contributes to path integral defining $G(\Lambda, Z, L'; R)$. The situation is illustrated in Fig. 7.4.

Eq. (7.26) is an ordinary 1st order partial differential equation of the form

$$\frac{\partial f(x,y)}{\partial y} = -\frac{\partial (w(x) f(x,y))}{\partial x}, \qquad f(x,y_0) = h(x), \tag{7.27}$$

where the last equation serves as a boundary condition. The solution to this equation is

$$f(x,y) = h(\bar{x}(y;x)) \frac{w(\bar{x}(y;x))}{w(x)}, \tag{7.28}$$

where $\bar{x}(y)$ is a solution to the ordinary differential equation (the so-called characteristic equation for the partial differential equation):

$$\frac{d\bar{x}}{dy} = -w(\bar{x}), \qquad \bar{x}(y_0) = x, \tag{7.29}$$

In eq. (7.28) we have written $\bar{x}(y_0) \equiv \bar{x}(y_0; x)$ to emphasize the dependence on x via the boundary condition. The solution to the characteristic equation is

$$y - y_0 = \int_{\bar{x}(y)}^{x} \frac{dx'}{w(x')}. \tag{7.30}$$

We can directly apply this to (7.26) if we impose the natural boundary condition

$$G(\Lambda, L, L'; R=0) = \delta(L - L'). \tag{7.31}$$

This implies that

$$G(\Lambda, Z, L'; R=0) = \int_0^\infty dL\, e^{-ZL} G(\Lambda, L, L'; R=0) = e^{-ZL'} \tag{7.32}$$

which will serve as our boundary condition for (7.26). Corresponding to (7.29) and (7.30) we have

$$\frac{d\bar{Z}}{dR} = -W(\Lambda, \bar{Z}), \qquad \bar{Z}(0) = Z, \tag{7.33}$$

$$R = \int_{\bar{Z}(R;Z)}^{Z} \frac{dZ'}{W(\Lambda,Z')} = \int_{\bar{Z}(R;Z)}^{Z} \frac{dZ'}{(Z' - \frac{1}{2}\sqrt{\Lambda})\sqrt{Z'+\sqrt{\Lambda}}}, \quad (7.34)$$

which is easily integrated using the substitution $\xi = \sqrt{Z'+\sqrt{\Lambda}}$. One obtains:

$$R = \frac{1}{\sqrt{\frac{3}{2}\sqrt{\Lambda}}} \ln \frac{H(\bar{Z})}{H(Z)}, \quad H(X) = \frac{\sqrt{X+\sqrt{\Lambda}}+\sqrt{\frac{3}{2}\sqrt{\Lambda}}}{\sqrt{X+\sqrt{\Lambda}}-\sqrt{\frac{3}{2}\sqrt{\Lambda}}}, \quad (7.35)$$

From this we can find $\bar{Z}(R;Z)$ and $W(\Lambda,\bar{Z})$

$$\bar{Z}(R;Z) = \frac{1}{2}\sqrt{\Lambda} + \frac{3}{2}\sqrt{\Lambda}\left[\frac{\left(H(Z)e^{\sqrt{\frac{3}{2}\sqrt{\Lambda}}R}+1\right)^2}{\left(H(Z)e^{\sqrt{\frac{3}{2}\sqrt{\Lambda}}R}-1\right)^2} - 1\right] \quad (7.36)$$

$$W(\Lambda,\bar{Z}(R;Z)) = \left(\bar{Z}(R;Z) - \frac{1}{2}\sqrt{\Lambda}\right)\sqrt{\frac{3}{2}\sqrt{\Lambda}}\left[\frac{H(Z)e^{\sqrt{\frac{3}{2}\sqrt{\Lambda}}R}+1}{H(Z)e^{\sqrt{\frac{3}{2}\sqrt{\Lambda}}R}-1}\right] \quad (7.37)$$

Our final solution is thus

$$\boxed{G(\Lambda,Z,L';R) = \frac{W(\Lambda,\bar{Z}(R;Z))}{W(\Lambda,Z)} e^{-\bar{Z}(R;Z)L'}} \quad (7.38)$$

and by a Laplace transformation in L':

$$G(\Lambda,Z,Y;R) = \frac{W(\Lambda,\bar{Z}(R;Z))}{W(\Lambda,Z)} \frac{1}{Y+\bar{Z}(R;Z)} \quad (7.39)$$

7.3 THE TWO-POINT FUNCTION

However, we are more interested in the limit where the exit and entrance loops are contracted to points. For the exit loop this is easy, we just take $L' \to 0$ in (7.38). From (7.8) we also know how to contract the entrance loop to a marked point, namely by multiplying with $Z^{3/2}$ and taking $Z \to \infty$. Since $Z^{3/2}/W(\Lambda,Z) \to 1$ in this limit we obtain

$$G(\Lambda;R) = W(\Lambda;\bar{Z}(R;Z=\infty)) \quad (7.40)$$

Since $H(Z) \to 1$ for $Z \to \infty$ the expression for $W(\Lambda;\bar{Z}(R;Z=\infty))$ becomes quite simple:

$$\boxed{G(\Lambda;R) = c\Lambda^{3/4} \frac{\cosh\sqrt[4]{\Lambda}\tilde{R}}{\sinh^3\sqrt[4]{\Lambda}\tilde{R}}} \quad \tilde{R} = \frac{1}{2}\left(\frac{3}{2}\right)^{1/2}R, \quad c = \left(\frac{3}{2}\right)^{3/4} \quad (7.41)$$

The Fractal Structure of 2D Gravity

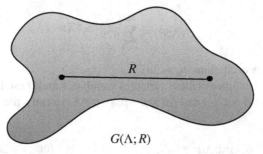

Figure 7.5 Graphically illustration of $G(\Lambda;R)$. We sum over all geometries where two marked points are separated a geodesic distance R.

We denote $G(\Lambda,R)$ the two-point function and it has the interpretation as the partition function for universes where two marked points are separated a geodesic distance R, as shown in Fig. 7.5. It has the following definition in terms of a path integral over geometries:

$$G(\Lambda;R) = \int \mathscr{D}[g]\, e^{-\Lambda \int d^2\xi \sqrt{g}} \iint d^2\xi_1 d^2\xi_2\, \sqrt{g(\xi_1)}\sqrt{g(\xi_2)}\, \delta(D_g(\xi_1,\xi_2)-R) \qquad (7.42)$$

where $D_g(\xi_1,\xi_2)$ is the geodesic distance between ξ_1 and ξ_2, measured in a metric $g_{ab}(\xi)$ defining a given geometry in the path integral.

Formula (7.41) is quite amazing. It is simple to derive from the dynamical triangulation formalism by counting triangulations, simple to define in the scaling limit (the continuum limit), but impossible to calculate directly from the continuum definition (7.42) because the geodesic distance $D_g(\xi_1,\xi_2)$ is an immensely complicated function of the metric g for a general geometry. We also see that the "quantum average" of $D_g(\xi_1,\xi_2)$ indeed is very "quantum" since the dimension of R is different from the dimension of $D_g(\xi_1,\xi_2)$ one would expect for the geodesic distance of a nice smooth geometry. This is of course only possible if a typical geometry appearing in the path integral is not at all nice and smooth at the scale set by R. But since R was arbitrary, this has to be true for geometries at all scales. We will return to discuss this further below.

It is seen that $G(\Lambda;R)$ behaves very much like an ordinary 2-point function:

$$G(\Lambda;R) \propto \frac{1}{\tilde{R}^3} \qquad \text{for } \tilde{R} \ll \frac{1}{\sqrt[4]{\Lambda}} \qquad (7.43)$$

$$G(\Lambda;R) \propto \Lambda^{3/4}\, e^{-2\sqrt[4]{\Lambda}\tilde{R}} \qquad \text{for } \tilde{R} \gg \frac{1}{\sqrt[4]{\Lambda}} \qquad (7.44)$$

In addition the 2-point function has a "stringy" feature: it has an infinity set of equidistance mass excitations (for a real string it is actually the not the mass excitations m_n, but m_n^2 which are equidistancely separated). If we use

$$\frac{\cosh x}{\sinh^3 x} = 4e^{-2x}\frac{1+e^{-2x}}{(1-e^{-2x})^3}, \quad \frac{1+z}{(1-z)^3} = \frac{d}{dz}z\frac{d}{dz}\frac{1}{1-z} = \sum_{n=1}^{\infty} n^2 z^{n-1}, \qquad (7.45)$$

we can write

$$G(\Lambda; R) \propto \sum_{n=1}^{\infty} n^2 e^{-2n\sqrt[4]{\Lambda}\tilde{R}} \tag{7.46}$$

and we have mass excitations $m_n \propto 2n\sqrt[4]{\Lambda}$.

Returning to the dimensionless variables variables r and μ used before taking the continuum limit ($R \propto \varepsilon^{1/2} r$ and $\Delta\mu = \mu - \mu_c \propto \varepsilon^2 \Lambda$), we can write (7.42), (7.43) and (7.44) as

$$G_\mu(r) \propto \Delta\mu^{\frac{3}{4}} \frac{\cosh \Delta\mu^{\frac{1}{4}} r}{\sinh^3 \Delta\mu^{\frac{1}{4}} r} \approx \begin{cases} r^{-3} & \text{for } r \ll \Delta\mu^{-\frac{1}{4}} \\ \Delta\mu^{\frac{3}{4}} e^{-2\Delta\mu^{\frac{1}{4}} r} & \text{for } r \gg \Delta\mu^{-\frac{1}{4}} \end{cases} \tag{7.47}$$

Following our discussion of intrinsic critical exponents for BPs (see (4.38)), we see from the short distance behavior of the two-point function, $G_\mu(r) \propto r^{1-\eta}$, that the exponent $\eta = 4$. This is a quite unusual exponent! [3]. Also we read off from the exponential decay of $G_\mu(r)$ that the exponent $\nu = 1/4$. We also know from (6.109) that $\gamma = -1/2$ and thus the unusual value of η ensures that Fisher's scaling relation is satisfied for $G_\mu(r)$:

$$\boxed{\gamma = \nu(2-\eta), \quad \gamma = -\frac{1}{2}, \quad \eta = 4, \quad \nu = \frac{1}{4}} \tag{7.48}$$

While we appealed to general considerations when using eq. (6.109) to argue that $\gamma = -1/2$, it can also be shown directly from (7.47) using the elementary definition of susceptibility in terms of the two-point function:

$$\chi(\Delta\mu) = \sum_{r=1}^{\infty} G_\mu(r) = \text{cnst.} - \frac{1}{6} \Delta\mu^{\frac{1}{2}} + \cdots \tag{7.49}$$

In the discussions related to spin systems, RWs and BPs the critical exponent γ of the susceptibility was defined by the divergence of $\chi(\Delta\mu)$ for $\Delta\mu \to 0$, namely $\chi(\Delta\mu) \propto (\Delta\mu)^{-\gamma}$. That of course assumes that $\chi(\Delta\mu)$ *is* divergent for $\Delta\mu \to 0$, which was the case. Here we have $\gamma = -1/2$ and we will define the susceptibility by the leading non-analytic term, which in this case is $\sqrt{\Delta\mu}$. Also, one should not be surprised that the coefficient multiplying $\sqrt{\Delta\mu}$ is negative (despite $\chi(\Delta\mu)$ of course being a positive function). If $\gamma = -1/2$, the susceptibility exponent of the three-point function will be $\gamma + 1 > 0$, i.e. the three-point function will diverge for $\Delta\mu \to 0$, and it is of course positive. But we essentially get the three-point function by $\chi^{(3)}(\Delta\mu) \propto -\frac{d}{d\mu} \chi(\Delta\mu)$,

---

[3] In ordinary quantum field theory in flat spacetime one considers $\eta = 2$ to be an upper bound on the anomalus scaling dimension, the reason being that the propagator then behaves like $1/|p|^{2-\eta}$ for large momentum. If $\eta > 2$ the propagator is growing with large momentum, making any probabilistic interpretation of scattering processes in quantum field theory problematic. This is also the reason we in the discussion of bosonic string theory, keeping an eye on Fisher's scaling relation $\gamma = \nu(2-\eta)$, said that having a $\nu > 0$ and a $\gamma > 0$ goes hand in hand. However, surprisingly, for the two-point function of instrinsic 2d gravity the situation is different.

as discussed in the case of the bosonic string. Thus the coefficient multiplying $\sqrt{\Delta\mu}$ in (7.49) has to be negative[4].

Maybe the most important consequence of the exponential behavior shown for the two-point function $G_\mu(r)$ is, using the now standard arguments from RWs and BPs, that *the global Hausdorff dimension, d_H, of the set of spherical triangulations is 4*. Let us now show that also the *local* Hausdorff dimension $d_h = 4$, by studying the geometric meaning of $G(\Lambda; R)$.

7.4 THE LOCAL HAUSDORFF DIMENSION IN 2D GRAVITY

The two-point function $G(\Lambda; R)$ defined for a given cosmological constant Λ is related to the two-point function $G(V; R)$ defined for a given volume V by a Laplace transformation

$$G(\Lambda; R) = \int_0^\infty dV\, e^{-\Lambda V} G(V; R). \quad (7.50)$$

The continuum definition of $G(V; R)$ is then (from (7.42) and (7.50))

$$G(V; R) = \int \mathcal{D}[g]\, \delta\left(\int d^2\xi \sqrt{g} - V\right) \iint d^2\xi_1 d^2\xi_2\, \sqrt{g(\xi_1)}\sqrt{g(\xi_2)}\, \delta(D_g(\xi_1, \xi_2) - R) \quad (7.51)$$

Thus $G(V; R)$ is proportional to the number of geometries with volume V and where in addition two marked points are separated a geodesic distance R. We will now provide a more precise picture of this, which relates $G(V; R)$ to the *local Hausdorff dimension*, exactly as we did in the case of BPs.

For a given point with coordinates ξ_1 we define the "area" (in 2d, like here, the length) of a spherical shell located a geodesic distance R from ξ_1 as

$$S_V(\xi_1, R; g) = \int d^2\xi \sqrt{g(\xi)}\, \delta\left(D_g(\xi, \xi_1) - R\right) \quad (7.52)$$

The average of $S_V(\xi_1)$ over the whole manifold is

$$S_V(R; g) = \frac{1}{V} \int d^2\xi \sqrt{g(\xi)}\, S_V(\xi, R; g) \quad (7.53)$$

The quantum average of $S_V(R; g)$ over all geometries with volume V is now

$$\langle S_V(R)\rangle = \frac{1}{W(V)} \int \mathcal{D}[g]\, \delta\left(\int d^2\xi \sqrt{g} - V\right) S_V(R; g) \quad (7.54)$$

where the partition function $W(V)$ according to (6.105) is given by

$$W(V) = \int \mathcal{D}[g]\, \delta\left(\int d^2\xi \sqrt{g} - V\right) \propto V^{-7/2} \quad (7.55)$$

---

[4] Note that we still have an equation like (5.83), if we considered different classes of triangulations $\mathcal{T}^{(2)}$ and $\mathcal{T}^{(3)}$ as was the case for the bosonic string. However, if $\gamma < 0$ the susceptibility does not go to infinity when $\mu \to \mu_c$. Thus we cannot conclude that (5.84) is valid, i.e. that $\bar{\mu}(\mu_c) > \bar{\mu}_c$, which was the main reason we could argue for the BP picture of bosonic strings.

From (7.52) - (7.55) we obtain

$$\langle S_V(R) \rangle = \frac{G(V;R)}{VW(V)} \propto V^{5/2} G(V;R) \tag{7.56}$$

So $G(V;R)$ has a simple geometric interpretation: for a fixed V it is proportional to the quantum average area of a spherical shell of radius R.

For a smooth two-dimensional geometry g we have for R sufficiently small

$$S_V(R;g) \propto R \quad \text{for} \quad R \ll \frac{1}{\sqrt{V}} \tag{7.57}$$

For a smooth d-dimensional geometry g we have

$$S_V(R;g) \propto R^{d-1} \quad \text{for} \quad R \ll \frac{1}{V^{1/d}} \tag{7.58}$$

If the space is fractal with Hausdorff dimension d_h we have (this is the *definition* of d_h)

$$\langle S_V(R) \rangle \propto R^{d_h-1} \quad \text{for} \quad R \ll \frac{1}{V^{1/d_h}} \tag{7.59}$$

Let us now calculate $\langle S_V(R) \rangle$ using (7.56). From (7.50) we have by an inverse Laplace transformation:

$$G(V;R) = \int_{-i\infty}^{i\infty} \frac{d\Lambda}{2\pi i} e^{V\Lambda} G(\Lambda;R) \tag{7.60}$$

We expand $G(\Lambda;R)$ in powers of Λ:

$$G(\Lambda;R) \propto \frac{1}{\tilde{R}^3} - \frac{\Lambda \tilde{R}}{15} + \frac{4}{189} \Lambda^{3/2} \tilde{R}^3 + c_5 \Lambda^2 \tilde{R}^5 + c_7 \Lambda^{5/2} \tilde{R}^7 + \cdots \tag{7.61}$$

Now use

$$\int_{-i\infty}^{i\infty} \frac{d\Lambda}{2\pi i} e^{V\Lambda} \Lambda^n = \frac{d^n}{dV^n} \delta(V), \quad \int_{-i\infty}^{i\infty} \frac{d\Lambda}{2\pi i} e^{V\Lambda} \Lambda^{n-1/2} = \frac{1}{\Gamma(-n+\frac{1}{2}) V^{n+\frac{1}{2}}}. \tag{7.62}$$

We discard the contributions from Λ^n terms since they corresponds to zero volume V and obtain

$$\langle S_V(R) \rangle \propto R^3 \left(1 + \mathcal{O}\left(\frac{R^4}{V}\right)\right), \tag{7.63}$$

and comparing with (7.59) we conclude that $d_h = 4$.

How is it possible that $\langle S_V(R) \rangle$ is not proportional to R for small R? Let us assume that the smooth geometries constitutes a dense set in the set of all continuous geometries entering in the path integral. In order to talk about this in a meaningful way one has to have a measure defined on the set of continuous geometries, much

The Fractal Structure of 2D Gravity

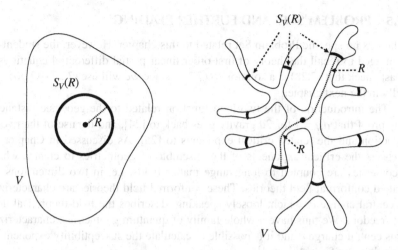

Figure 7.6 Left: when $R \leq R_g$ for a given point, $S_V(R)$ is a nice connected curve. Right: when $R \gg R_g$ this is no longer the case. In the scaling limit, starting out with triangulations, there will be infinitely many disconnected parts of $S_V(R)$ with probability one, no matter how small R is. This is the reason we can have $S_V(R) \propto R^3$.

like we have the Wiener measure in the case of RWs. Important progress has been made in this direction in mathematics in the recent years, but we have no space to discuss it here [5]. Let us just assume that we have such a measure. Now for each smooth geometry g we use in the path integral (7.51) we can find an R_g such that

$$S_V(R;g) \propto R \quad \text{for} \quad R \leq R_g. \tag{7.64}$$

This situation is illustrated on the left part of Fig. 7.6. However, in (7.51) the R is chosen independent of g and is a parameter outside the integration and the result $d_h = 4$ shows that for any R, no matter how small, there will be many more smooth geometries g for with $R \gg R_g$ than there will be smooth geometries g where $R \leq R_g$. For the geometries where $R \gg R_g$ the geometry looks more like the one shown in the right part of Fig. 7.6, and in such a situation there is no reason why (7.64) should be valid. In fact, with probability 1, if we pick randomly a smooth geometry g, we will obtain $R \gg R_g$. The same statement would be "even more true" if applied to the full set of continuous geometries which enter in the path integral (7.51). It is a beautiful result that this set of rather unwieldy geometries has a well-defined Hausdorff dimension, namely $d_h = d_H = 4$, and in a sense it is the generalization of the RW result, where the set of random walks has $d_H = 2$, the double of the dimension expected for a smooth path, only are we in the case of geometries talking about entirely intrinsic properties, while we in the case of RWs talked about properties of the RWs embedded in $\mathbb{R}^D$.

---

[5]The interested reader can consult [23] for a review as well as [19, 20, 21] for related details.

7.5 PROBLEM SETS AND FURTHER READING

There is no specific Problem Set related to this chapter. However, the student is encouraged to recall the theory of first order linear partial differential equations or at least check that (7.28) *is* a solution to (7.27), since we will use (7.27)–(7.30) repeatedly in the next chapter.

The introduction of the two-loop function related to the geodesic distance and the proof that $d_H = 4$ for 2d gravity goes back to [24], and the use of the two-point function and the related critical exponents to [25]. As discussed in Chapter 6 one expects the critical exponents of the ensemble of geometries to chance when the geometries are coupled to long range matter fields, i.e. in two dimensions to so-called conformal field theories. These conformal field theories are characterized by a central charge c, which, loosely speaking, describes the field-theoretical degrees of freedom. We thus have a whole family of quantum geometries, characterized by this central charge c and it is possible to calculate the susceptibility exponent γ as a function of c for $c \in]-\infty, 1]$

$$\gamma(c) = \frac{c - 1 - \sqrt{(c-1)(c-25)}}{12}, \tag{7.65}$$

a formula first proved [26], and in a framework closer to the one used here in [27, 28]. From Fisher's scaling relation we thus expect the critical exponents ν and η of the two-point function to be functions of c and similarly the Hausdorff dimension $d_H = 1/\nu$. *It is a major unsolved problem to find $d_H(c)$ for the geometries characterized by the central charge c*. Presently we only know $d_H = 4$ for $c = 0$ (no matter coupled to gravity). Using the discretized formulations of matter coupled to triangulations it is actually possible to calculate some two-point functions, much in the spirit reported in this chapter for some of the multicritical models mentioned in Chapter 6 (see e.g. [29]), but unfortunately the distance r used in these calculations is not the graph distance since some of the polygons appear with negative weight.

It is seen that formula (7.65) breaks down when $c > 1$ since $\gamma(c)$ becomes complex. We have actually already considered the case $c > 1$, since the bosonic string theory discussed in Chapter 5 corresponds to $c = D$, where D is the number of scalar fields $X_i(\xi)$, $i = 1, \ldots, D$. As we saw there, seemingly the surfaces degenerated to BPs, indicating that we have a kind of phase transition as a function of c at $c = 1$. Again, *it is an unsolved problem to understand this transition at $c = 1$*.

8 The Causal Dynamical Triangulation model

8.1 LORENTZIAN VERSUS EUCLIDEAN SET UP

The two-dimensional Euclidean gravity model we have studied satisfies the Wilsonian criterium for universality: to a large extent it is independent of the details of the short distance regularization. We were not restricted to use triangulations as building blocks, but could use any (finite) combination of polygons as building blocks, as long as the weights of polygons were all positive, and we would obtain the same continuum multi-loop functions when the dimensionless cosmological coupling constant $\mu \to \mu_c$ in such a way that $\mu = \mu_c + \varepsilon^2 \Lambda$, where the link length in the graphs went to zero while the *continuum* cosmological constant Λ survived. In that limit the average number N of polygons in the graphs also diverged for multi-loop functions with three or more loops and we could talk about a finite continuum limit of the volume $V \propto N\varepsilon^2$, where N denoted the number of polygons. We had $\langle V \rangle \propto \Lambda^{-1}$ if we did not fix the volume of spacetime, but considered the model with a fixed cosmological constant Λ. Further, by studying the two-point function as a function of the so-called geodesic distance, we identified the correlation length which diverged when we approached the critical point μ_c. In this sense the two-point function acted precisely as the two-point function of a spin system and the universality of the results could be understood as a result of the divergent correlation length, in the same way as universality of phase transitions of spin systems can be understood as the result of a divergent correlation length between the spins, which makes many details of the short distance lattice structure and interactions irrelevant for scaling limit.

We also stated that when we coupled matter to 2d gravity we could change the critical behavior of the ensemble of polygons when the matter system itself had long range interactions (and in addition the long range correlation in geometry would change the critical properties of the matter system). We mentioned that this change of critical behavior could many times be obtained by assigning negative weights to some of the polygon building blocks. This whole complex of systems provides a lattice regularization of two-dimensional Euclidean quantum gravity coupled to conformal field theories, and is denoted *Euclidean Dynamical triangulations* (EDT) or *Quantum Liouville Theory*.

We will now introduce a new, and different universality class of 2d models, denoted *Causal Dynamical Triangulations* (CDT). Historically, the motivation was that time and space might be more different than it appears in the truly Euclidean approach we have pursued so far. One could emphasize this by insisting that the starting point was to consider Lorentzian geometries with a global proper time (and in particular thus a local causal structure, which gave rise to the name CDT, when one implemented this via triangulations) , and then perform the rotation to Euclidean signature by rotating this global time to imaginary global "time". In this way we arrive with a set of Euclidean geometries which are more restricted than the ones we have studied so far. We triangulate these geometries as before, using equilateral triangles, and using the corresponding Regge action we arrive at a new statistical system of two-dimensional geometries which we still denote CDT, despite the rotation to Euclidean signature. By our choice of geometries we have broken the symmetry between space and (Euclidean) time and from a Wilsonian point of view it is then a distinct possibility that our statistical system of geometries (CDT) will be in a different universality class[1] than the statistical systems of geometries denoted EDT where this symmetry is manifest. We will see that it is indeed the case.

8.2 DEFINING AND SOLVING THE CDT MODEL

We label "time" by an integer coordinate t. For each time coordinate t "space" will be assumed to have the topology of S^1. Space at time t will consists of l_t links glued together via l_t vertices such that the topology of space is S^1. Given space at t and space at $t+1$ we now fill out the "slab" in between by equilateral triangles, such that a triangle has two vertices with time coordinate t and one with time coordinate $t+1$ or oppositely has one vertex with time coordinate t and two vertices with time coordinate $t+1$. The triangles are glued together such that they form a triangulation with the topology of a cylinder where one boundary consists of l_t links and the other boundary consists of l_{t+1} links. The total number of triangles is l_t+l_{t+1}, but for given l_t and l_{t+1} there are of course many ways we can glue together the triangles to form a cylinder with the given lengths of boundaries. Continuing this way we construct triangulations which have slices of constant time labelled by $i=0,1,2,\ldots,t$ and link-lengths $l_i > 0$ (we do not allow slices of constant time without at least one link). This triangulation $\mathcal{T}$ has the topology of the cylinder with boundaries of lengths l_0 and l_t. The left panel in Fig. 8.1 shows such a triangulation with the cylinder presented as an annulus where the circles represent the spatial slices at times 0,1 and 2. For the purpose of combinatorics it is convenient to mark one of the links (and its "first" vertex if the loop is oriented counter clockwise) on the spatial boundary loop with l_0 links (the "entrance" loop), but have no marked link on the boundary loop with l_t links (the "exit" loop). This choice of labeling is similar to the one we used in the last chapter when we considered the two-loop function $G(g,l_1,l_2;r)$, and we choose it for the same reason: from a combinatorial point of view it makes the gluing of two cylinders easier and we will have a composition law similar to (7.10), only with r replaced by t. The total number of triangles and the total number of vertices in $\mathcal{T}$

---

[1]The way we defined EDT above, it was not really a single universality class, since the universality classes were labeled by a continuum parameter, the central charge c of the conformal field theory coupled to the two-dimensional geometry. When comparing EDT to CDT, we will from now on have in mind the specific model where there is no conformal matter coupled to two-dimensional geometries, i.e. in the labeling mentioned, the case $c=0$.

The Causal Dynamical Triangulation model

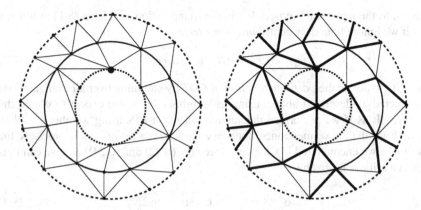

Figure 8.1 Left figure: A CDT triangulation of the cylinder (represented as an annulus). Constant time slices corresponding to $t = 0, 1, 2$ are circles. The boundary loops are dashed (exit loop) and dotted (entrance loop). A vertex on the entrance loop $t=0$ is marked. Right figure: the corresponding branched polymer (thick links on the figure) on the CDT triangulation. An artificial vertex at $t=-1$ connected to each vertex at the $t=0$ loop ensures a bijection between the CDT triangulations with boundaries at times 0 and t and rooted branched polymers of *height t* (the root connects the vertex at $t=-1$ to the marked vertex at $t=0$).

will be

$$N_{\mathcal{T}} = 2\sum_{i=1}^{t-1} l_i + l_0 + l_t, \quad V_{\mathcal{T}} = \sum_{i=0}^{t} l_i, \quad \text{i.e.} \quad 2V_{\mathcal{T}} = N_{\mathcal{T}} + l_0 + l_t \qquad (8.1)$$

where the first equation simply reflects that each internal spatial link is the spatial link of two neighboring triangles, while a boundary link is a spatial link for just one triangle. The action associated with such a triangulation will as usual just be the cosmological term:

$$S[\mathcal{T}] = \mu N_{\mathcal{T}}, \qquad (8.2)$$

where μ is the dimensionless cosmological constant.

It is possible to make a bijective map from this class of triangulations to branched polymers as shown in the right panel of Fig. 8.1 (see [30] for details). Let a vertex at a loop at time $i < t$ be connected with links to k vertices at the loop at time $i+1$. Moving counter clockwise around the vertex, we declare that all links except the last one will belong to the BP. In addition we have added a marked vertex which we connect to all vertices at the entrance loop corresponding to $i=0$. The BP defined in this way has thus a marked vertex and a corresponding marked link, which is the link connected to the marked vertex on the entrance loop. If we define the *height* of a vertex in the BP as the link distance from the marked vertex, it follows by construction that the vertices at height i are precisely the vertices at the loop at time $i-1$ in the triangulation. Given a BP with a marked vertex and a corresponding marked link, one can reconstruct the triangulation and in this way prove the bijection. The number of links in the BP, L_{BP}, is equal to the number of vertices $V_{\mathcal{T}}$ in the triangulation, and from (8.1) in follows that if we ignore boundary contributions we have

$$S[\mathcal{T}] = \mu N_{\mathcal{T}} \approx 2\mu L_{BP}. \tag{8.3}$$

Thus, if we define the *partition function of CDT* by summing over all triangulations constructed as described above, using as weights $e^{-\mu N_{\mathcal{T}}}$, we expect to obtain the same result as if we performed the summation of all BPs using weights $e^{-(2\mu)L_{BP}}$ (apart from the above mentioned boundary terms). The set of BPs is precisely the set of BPs we encountered in Chapter 5 (see eqs. (6.22) and (6.24)) and we can thus expect a critical μ_c given by

$$e^{-2\mu_c} = \frac{1}{4}, \quad \text{i.e.} \quad \mu_c = \ln 2. \tag{8.4}$$

Below we will verify, using very simple arguments, that this is indeed true. Because of the strong link to BPs we also expect that the CDT theory of geometry belongs to a different universality class than the EDT theory, and as we will see this is the case.

We will be interested in a continuum limit of the above lattice construction, where, somewhat similar to what we did in EDT, we take the number of boundary links, the number of triangles and the number of time steps to infinity in such a way that one one can take the link length ε to zero while keeping the continuum boundary lengths, the continuum area and the continuum time finite. In order to implement this we start by keeping l_0 and l_t fixed and sum over all "cylindrical" triangulations $\mathcal{T}(l_0, l_t; t)$ of the kind described above, with fixed l_0, l_t and t, using the action (8.2). Let us denote this amplitude

$$G_\mu(l_0, l_t; t) \equiv G(g, l_0, l_t; t) = \sum_{\{\mathcal{T}(l_0, l_t; t)\}} g^{N_{\mathcal{T}(l_0, l_t; t)}}, \quad g = e^{-\mu} \tag{8.5}$$

Having two cylindrical triangulations $\mathcal{T}(l_0, l_{t_1}; t_1)$ and $\mathcal{T}(l_{t_1}, l_{t_2}; t_2 - t_1)$, they can be glued together to a single cylindrical triangulation $\mathcal{T}(l_0, l_{t_2}; t_2)$ along the boundaries of lengths l_{t_1}. No additional symmetry factor is related to this gluing because we have chosen to mark the entrance loop and not the exit loop, as already mentioned. Thus we can write

$$G(g, l, l'; t_1 + t_2) = \sum_{l''} G(g, l, l''; t_1) G(g, l'', l'; t_2). \tag{8.6}$$

As a special case of (8.6) we can write

$$G(g, l, l'; t+1) = \sum_{l''} G(g, l, l''; 1) G(g, l'', l'; t). \tag{8.7}$$

Thus it is clear that we can find $G(g, l, l')$ by iteration if we only know $G(g, l, l'; 1)$[2]. In order to find $G(g, l, l'; 1)$ we introduce (as usual) the generating function for

---

[2] One could have used the same argument in the case of EDT, eq. (7.10), and one can indeed find $G(g, l.l'; 1)$ in the EDT case and in this way find the two-loop EDT function. We refer to [1] for details.

$G(g,l,l';t)$ and define

$$G(g,x,y;t) = \sum_{l,l'} x^l y^{l'} G(g,l,l';t), \quad x = e^{-\lambda_{en}}, \ y = e^{-\lambda_{ex}} \qquad (8.8)$$

where we have indicated that the indeterminate x,y, if positive and real, also can be given an interpretation as boundary cosmological constants for the entrance and exit loop (as was also the case in EDT where we just used $z_1 = 1/x$ and $z_2 = 1/y$ as the indeterminate in the generating function instead of x and y). In this way we can invert (8.8) if needed:

$$G(g,l,l';t) = \oint \frac{dx}{2\pi i x^{l+1}} \oint \frac{dy}{2\pi i y^{l'+1}} G(g,x,y;t), \qquad (8.9)$$

where the integration contours enclose $x = y = 0$ and lie within the convergence radii of the power series in x and y. It follows from

$$\oint \frac{dz}{2\pi i\, z^{n+1}} = \delta_{0,n} \quad n \in \mathbb{Z}. \qquad (8.10)$$

The relation (8.6) can now be written as

$$G(g,x,y;t_1+t_2) = \oint \frac{dz}{2\pi i z} G(g,x,z^{-1};t_1) G(g,z,y;t_2), \qquad (8.11)$$

where the integration contour encloses $z=0$ and for fixed g,x,y lies inside the radius of convergence $r(g,y)$ for $G(g,z,y;t_2)$ and inside the radius of convergence $r(g,x)$ for $G(g,x,z^{-1};t_2)$ as a power series in $1/z$, i.e. in the region $z \geq 1/r(g,x)$. This is possible when we consider g,x,y less than their critical values, given by eq. (8.15) below, since then $r(g,x) > 1$.

It is now easy to find $G(g,x,y;1)$, just looking at Fig. 8.1:

$$G(g,x,y;1) = \sum_{k=0}^{\infty} \left(gx \sum_{l=0}^{\infty} (gy)^l \right)^k - \sum_{k=0}^{\infty} (gx)^k = \frac{g^2 xy}{(1-gx)(1-gx-gy)}. \qquad (8.12)$$

Formula (8.12) is simply a book-keeping device for all possible ways of evolving from an entrance loop of any length in one step to an exit loop of any length. The subtraction of the term $1/(1-gx)$ has been performed to exclude the degenerate cases where either the entrance or the exit loop is of length zero.

We now use (8.12) in (8.11) with $t_1 = 1$ and $t_2 = t$.

$$G(g,x,z^{-1};1) = \frac{g^2 x}{(1-gx)^2} \frac{1}{z - \frac{g}{1-gx}}, \qquad (8.13)$$

and the integration contour in (8.11) should include $z = g/(1-gx)$ and $z = 0$, but $z=0$ does not contribute since $G(g,z,y;t)/z$ is finite for $z \to 0$ (the entrance loop has length $l \geq 1$). We thus obtain

$$\boxed{G(g,x,y;t+1) = \frac{gx}{1-gx} G\left(\frac{g}{1-gx}, y; g; t\right).} \qquad (8.14)$$

This equation can be solved by iteration (see [31] for details). However, rather than doing that and then deriving the continuum limit of $G(g,x,y;t)$, we will use it to directly "guess" the continuum limit. Let us assume that there are critical points g_c, x_c and y_c like in 2d EDT, such that we can write

$$g = g_c e^{-\Lambda \varepsilon^2/2}, \quad x = x_c e^{-X\varepsilon}, \quad y = y_c e^{-Y\varepsilon}, \quad T = \varepsilon t, \qquad (8.15)$$

where ε denotes the link length, while Λ, X and Y are the continuum cosmological constant and the continuum boundary cosmological constants. The last relation, $T = \varepsilon t$, was absent for 2d EDT since we had no time slicing like here in CDT. In 2d EDT we could have used the geodesic distance from a point or an entrance loop as "time", but the corresponding links at a distance r from the entrance loop did not form a connected loop, but branched out in many loops, reflecting that the Hausdorff dimension of 2d EDT is 4 and that the geodesic distance has an anomalous dimension (recall Fig. 7.6). Here in CDT the situation is seemingly different and successive time slices labeled by the integer t stay connected by construction. Thus the relation $T \propto \varepsilon t$ between the continuum time T and the dimensionless integer lattice time t is reasonable. We will adjust T such that the constant of proportionality is 1. We will now assume that $G(g,x,y;t)$ has a limit $\varepsilon^\eta G_\Lambda(X,Y;T)$ when $\varepsilon \to 0$. This is similar to the situation we encountered in 2d EDT for $G(g,x,y;r)$ and from (8.11) it folows that $\eta = -1$, like in the EDT case. An assigment

$$G(g,x,y;t) \to \varepsilon^{-1} G_\Lambda(X,Y;T) \quad \text{for} \quad \varepsilon \to 0 \qquad (8.16)$$

is only meaningful if we in eq. (8.14) have

$$\frac{g_c x_c}{1 - g_c x_c} = 1, \quad \frac{g_c}{1 - g_c x_c} = 1, \quad \text{i.e.} \quad x_c = 1, \ g_c = \frac{1}{2}. \qquad (8.17)$$

It is seen that we indeed have confirmed the prediction (8.4). Inserting (8.17) back in (8.14) and using (8.15) we obtain to lowest order in ε

$$G(g, 1-\varepsilon X, y; T+\varepsilon) = (1-2\varepsilon X) G(g, 1-\varepsilon[X+\varepsilon(\Lambda-X^2)], y; T), \qquad (8.18)$$

i.e. (suppressing the arguments y and g in $G(g,x,y;t)$)

$$\varepsilon \frac{\partial G(1-\varepsilon X; T)}{\partial T} = -2\varepsilon X\, G(1-\varepsilon X; T) + \varepsilon(\Lambda - X^2) \frac{\partial G(1-\varepsilon X; T)}{\partial X} \qquad (8.19)$$

or

$$\boxed{\frac{\partial G_\Lambda(X,Y;T)}{\partial T} = -\frac{\partial}{\partial X}\left((X^2 - \Lambda) G_\Lambda(X,Y;T)\right)} \qquad (8.20)$$

This partial differential equation has the same structure as the one we already meet in 2d EDT and we can solve it precisely in the same way (see (7.39)):

$$G_\Lambda(X,Y;T) = \frac{\bar{X}^2(T;X) - \Lambda}{X^2 - \Lambda} \frac{1}{\bar{X}(T;X) + Y}, \qquad (8.21)$$

where $\bar{X}(T;X)$ is the solution to the characteristic equation

$$\frac{d\bar{X}}{dT} = -(\bar{X}^2 - \Lambda), \quad \bar{X}(T=0;X) = X. \tag{8.22}$$

i.e.

$$\bar{X}(T;X) = \sqrt{\Lambda} \, \frac{(\sqrt{\Lambda}+X) - e^{-2\sqrt{\Lambda}T}(\sqrt{\Lambda}-X)}{(\sqrt{\Lambda}+X) + e^{-2\sqrt{\Lambda}T}(\sqrt{\Lambda}-X)}. \tag{8.23}$$

We can now introduce the continuum boundary length L as we did for EDT

$$L \equiv \varepsilon l, \quad x^l = x_c^l \, e^{-\varepsilon lX} = e^{-LX}, \tag{8.24}$$

and provided we, like in (8.16), make the identification

$$G(g, l_1, l_2; t) \to \varepsilon \, G_\Lambda(L_1, L_2; T) \quad \text{for} \quad \varepsilon \to 0. \tag{8.25}$$

it is seen that with a change of variables from x, y to X, Y the integration contours in (8.9) change from circles to integration along the imaginary X, Y axes in the limit when $\varepsilon \to 0$ and the continuum limit of (8.9) reads:

$$G_\Lambda(L_1, L_2; T) = \int_{-i\infty}^{i\infty} \frac{dX}{2\pi i} \int_{-i\infty}^{i\infty} \frac{dY}{2\pi i} \, e^{L_1 X + L_2 Y} G_\Lambda(X, Y; T) \tag{8.26}$$

This is just the inverse Laplace transformation of the continuum limit of (8.8):

$$G_\Lambda(X, Y; T) = \int_0^\infty dL_1 \int_0^\infty dL_2 \, e^{-XL_1 - YL_2} G_\Lambda(L_1, L_2; T), \tag{8.27}$$

From the solution (8.21) we can easily perform the inverse Laplace transformation wrt Y as in (8.26) and we obtain (corresponding to (7.38))

$$G_\Lambda(X, L; T) = \frac{\bar{X}^2(T;X) - \Lambda}{X^2 - \Lambda} \, e^{-\bar{X}(T;X)L}. \tag{8.28}$$

Using (8.23) we see that we have the following large T behavior

$$G_\Lambda(X, L; T) \xrightarrow{T \to \infty} \frac{4\Lambda e^{-\sqrt{\Lambda}L}}{(X + \sqrt{\Lambda})^2} \, e^{-2\sqrt{\Lambda}T}, \tag{8.29}$$

$$G_\Lambda(L_1, L_2; T) \xrightarrow{T \to \infty} 4\Lambda L_1 e^{-\sqrt{\Lambda}(L_1 + L_2)} \, e^{-2\sqrt{\Lambda}T} \tag{8.30}$$

i.e. the two-loop functions fall off exponentially. The factor L_1 in (8.30) is present only because we have chosen to mark a point on the entrance loop.

The disk amplitude played an important role in our EDT theory. It had the interpretation of the Hartle-Hawking wave function of the universe and it was the building block for all higher loop functions. Looking at Fig. 8.1, the right panel defines a triangulation with the topology of the disk. It is a special configuration in the sense that it has a special point in the center and at least we should in addition sum over all times t. In the continuum we formally achieve this by starting with $G_\Lambda(X,L;T)$ and contracting the loop $L \to 0$ and then integrating wrt T. We thus *define* the CDT disk function as

$$W_\Lambda(X) \equiv \int_0^\infty dT\, G_\Lambda(X,L=0;T) = \int_{\sqrt{\Lambda}}^X \frac{d\bar{X}}{\bar{X}^2-\Lambda} = \frac{1}{X+\sqrt{\Lambda}}. \qquad (8.31)$$

The integral can be performed by using (8.22) to change integration variable from T to $\bar{X}(T)$. The limit $T \to \infty$ corresponds according to (8.23) to $\bar{X} = \sqrt{\Lambda}$. Taking the Laplace transform we obtain

$$W_\Lambda(L) = \int_{-i\infty}^{+i\infty} \frac{dX}{2\pi i} e^{LX} W(\Lambda,X) = e^{-\sqrt{\Lambda}L}, \qquad (8.32)$$

i.e. the CDT disk amplitude falls off exponentially with L, the decay determined by the square root of the cosmological constant.

We now *define* the following two-point function in CDT: The starting point is the two-loop function $G_\Lambda(L_1,L_2;T)$. We contract the length L_2 of the exit loop to zero as we did for $W_\Lambda(X)$. We also contract the entrance loop to zero, but to compensate for the factor L_1 which we "artificially" introduced by marking a point on the entrance loop, we divide by L_1 before taking the limit $L_1 \to 0$. For functions $F(L)$ where $F'(0)$ exists it can be done by writing:

$$G(L) = LF(L), \quad G'(0) = F(0), \quad \text{i.e.} \quad F(0) = \int_{-i\infty}^{i\infty} \frac{dX}{2\pi i} e^{LX} X G(X)\bigg|_{L=0}. \qquad (8.33)$$

We now define the two-point function $G_\Lambda(T)$ as the sum over all CDT cylinder surfaces where the entrance and exit loops are contracted to points as described above, and where *a marked point has a distance T to the entrance loop (point)*. A typical such surface is shown in Fig. 8.2. From the figure it follows that

$$\begin{aligned}G_\Lambda(T) &= \lim_{L_1 \to 0} \frac{1}{L_1} \int dL\, G_\Lambda(L_1,L;T) L W(L) = -\int_{-i\infty}^{i\infty} \frac{dX}{2\pi i} \frac{X(\bar{X}^2-\Lambda)}{X^2-\Lambda} \frac{dW_\Lambda(\bar{X})}{d\bar{X}} \\ &= -(\bar{X}^2-\Lambda) \frac{dW_\Lambda(\bar{X})}{d\bar{X}}\bigg|_{|X|=\infty} = \frac{dW_\Lambda(\bar{X}(T;\infty))}{dT} = e^{-2\sqrt{\Lambda}T}.\end{aligned} \qquad (8.34)$$

The first equality follows from the figure: the loop at distance T has a length L and the marked point can be anywhere. The surface can now continue in all possible ways compatible with CDT surfaces until a spatial loop contracts to a point, i.e. precisely as $W_\Lambda(L)$. The next equality uses (8.33) and (8.28). The third equality follows from deforming the integration contour to a circle at infinity, where $\bar{X}(T;X)$ is independent of X, as seen from (8.23) which also leads to the final result [3].

---

[3] The result differs from (7.40) where there is no differentiation wrt T (or better R in eq. (7.40)). The difference can be traced back to the non-scaling part of the disk amplitude $w(g,z)$ given by (7.6), although it seemingly cancels out in the differential equation (7.18) which leads to (7.26). It would nevertheless

The Causal Dynamical Triangulation model

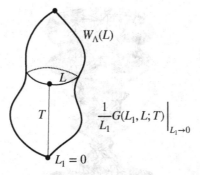

Figure 8.2 First, the propagation from a contracted entrance loop (length $L_1 = 0$) to a point at distance T. This point belongs to a loop of length L, where all points have distance T from the entrance loop. After that the universe continues to evolve until it eventually disappears (indicated by the upper dot). That continued evolution is described by $W_\Lambda(L)$.

If we return to discrete variables we have

$$G_\mu(t) \propto e^{-2\sqrt{\mu-\mu_c}t}, \quad \mu_c = -\ln g_c = \ln 2, \tag{8.35}$$

i.e. the two-point function behaves precisely as the two-point function for intrinsic BPs, and it has the same critical exponents:

$$\boxed{\nu_{cdt} = \frac{1}{2} \quad (\text{i.e. } d_H = 2), \quad \gamma_{cdt} = \frac{1}{2}, \quad \eta_{cdt} = 1.} \tag{8.36}$$

Of course it is not surprising, given the bijective mapping between BPs and CDT configurations, but we have now shown it by explicit calculations. Also, the result is manifest different from the EDT result where we had a propagator behavior $G_\mu(t) \propto e^{-\sqrt[4]{\mu-\mu_c}t}$ for large t and $G_\mu(t) \propto t^{-3}$ for small t (where t denoted the link distance between two marked points), and where the corresponding critical exponents were

$$\nu_{edt} = \frac{1}{4}, \quad \gamma_{edt} = -\frac{1}{2}, \quad \eta_{edt} = 4. \tag{8.37}$$

Our conclusion is that the CDT ensemble of 2d geometries belongs to a different universality class.

By taking the Laplace transform of eq. (8.20) we obtain

$$\frac{\partial}{\partial T}G_\Lambda(L_1,L_2;T) = -\hat{H}(L_1)G_\Lambda(L_1,L_2;T), \quad \hat{H}(L) = -L\frac{d^2}{dL^2} + \Lambda L \tag{8.38}$$

---

enter if one tried to define the two-point function as in Fig. 8.2, starting out at a discretized level. In EDT the non-scaling part of $w(g,z)$ cannot be ignored because of the fractal nature of the geometries. The chance that the loop where the black dot in the figure is located has a macroscopic length is simply zero and for a microscopic loop, the corresponding dominating contribution from $w(g,z)$ will be the non-scaling part of (7.6). A detailed discussion can be found in [31].

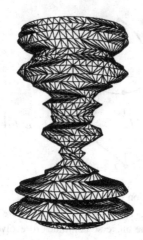

Figure 8.3 A "typical" configuration contributing to the path integral defining the amplitude (8.39). Time is in the vertical direction and the configuration is a triangulation which at time t_n consists of a number of links $l(t_n)$ which are drawn as a circle of length $l(t_n)$. The triangulation is generated by a so-called Monte Carlo simulation of the CDT system. In such simulation one computer-generates the CDT triangulations with the relative probabilities with which they are represented in the path integral.

Thus we can write

$$G_\Lambda(L_1, L_2; T) = \langle L_2 | e^{-\hat{H}T} | L_1 \rangle, \tag{8.39}$$

where $\hat{H}$ is the Hamiltonian for the evolution of our spatial universe of length L. It is an Hermitian operator on the positive real axis (L has to be non-negative) with the scalar product

$$\langle \Psi_2 | \Psi_1 \rangle = \int_0^\infty \frac{dL}{L} \Psi_2^*(L) \Psi_1(L). \tag{8.40}$$

One can find the eigenfunctions and eigenvalues of $\hat{H}$:

$$\hat{H}\Psi_n = E_n \Psi_n, \quad E_n = 2n\sqrt{\Lambda}, \quad \Psi_n(L) \propto L_n^{(-1)}(2\sqrt{\Lambda}L) e^{-\sqrt{\Lambda}L}, \quad n = 1, 2, \ldots \tag{8.41}$$

where $L_n^{(\alpha)}(x)$ is a polynomial of order n (more precisely a generalized (or associated) Laguerre polynomial). Formally "the wave function of the universe" $W_\Lambda(L) = \Psi_0(L)$ is also an eigenfunction of $\hat{H}$, corresponding to $E_0 = 0$. However, it is not a normalizable eigenfunction when using the scalar product (8.40). In Fig. 8.3 we have shown a typical configuration contributing to the path integral defining the propagator (8.39). If T is sufficiently large the ground state of $\hat{H}$, Ψ_1, will dominate the expression (8.39) and in that approximation we find that

$$\langle L(t) \rangle \propto \frac{1}{\sqrt{\Lambda}}, \quad P(L(t)) = \frac{\Psi_1^2(L)}{L} = 4\Lambda L\, e^{-2\sqrt{\Lambda}L}, \quad 0 \ll t \ll T, \tag{8.42}$$

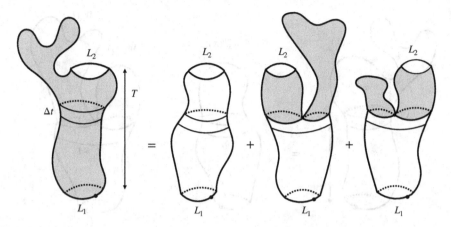

Figure 8.4 The GCDT cylinder amplitude $G_\Lambda(L_1, L_1; T)$, expressed in terms of the CDT cylinder amplitude and the GCDT disk function. As time progresses either no split occurs (the CDT cylinder amplitude) or no split happens before time t and then a split happens between t and $t + dt$ with "probability" $g_s dt$. After that one part develops as a baby universe, the other part develops as the GCDT cylinder amplitude. To count all such configurations we have to integrate over t. Note that baby universes can develop in time which extends beyond T before they eventually disappear in the vacuum!

where $P(L(t))$ denotes the probability distribution for the length $L(t)$ of the spatial universe at time t. Thus looking at Fig. 8.3 and making a normalized histogram for the lengths of the spatial circumferences shown for the discrete times t_n should reproduce the $P(L)$ in eq. (8.42), assuming that the time T is large enough for the ground state of $\hat{H}$ to dominate in the region $0 \ll t \ll T$.

8.3 GCDT: SHOWCASING QUANTUM GEOMETRY

Above we defined the CDT model. One can ask if it is possible to generalize the model without leaving the universality class. If we consider the cylinder amplitude, a natural generalization is still to have a time foliation but allow outgrows like shown in Fig. 8.4. In this way the topology is still that of a cylinder, but we allow the creation of baby universes, which have the topology of a disk. We denote this theory *Generalized* CDT (GCDT). At this point we have actually not defined the GCDT disk amplitude starting from any triangulation. Thus the figure involves both an unknown cylinder amplitude and an unknown disk amplitude. However we will show that consistency of quantum geometry allows us to determine the amplitudes before actually providing a definition via triangulations and taking a scaling limit!

We already allowed the creation of baby universes in EDT, so are we not just getting back to EDT? The difference is that in the case of EDT the creation of baby universes were allowed already at the discretized level without any constraint, and when we took the continuum limit the number of such baby universes became quite

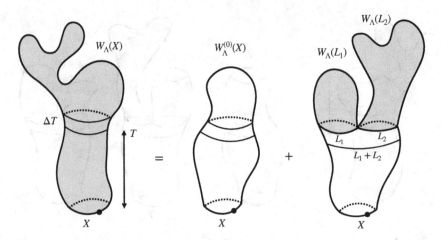

Figure 8.5 The disk amplitude $W_\Lambda(X)$ for GCDT: either no split takes place (which results in the CDT disk amplitude $W_\Lambda^{(0)}(X)$), or at a time interval between T and $T+\Delta T$ it splits with "probability" $g_s \Delta T$. Before T the universe propagates with the CDT cylinder amplitude and after the split anything can happen at future times, except that the two universes have to have the topologies of disks. They will thus contribute $W_\Lambda(L_1)$ and $W_\Lambda(L_2)$ to the total GCDT disk amplitude if the loop of length L_1+L_2 splits in two loops of lengths L_1 and L_2. We have to integrate wrt T, L_1 and L_2 to count all possible configurations.

dominant, and strictly infinite in the continuum limit, as illustrated in Fig. 7.6. Here we are already in the continuum and for Fig. 8.4 to make sense in the continuum there should only be a finite number of baby universes for a finite continuum time T.

In the same way as Fig. 8.4 is a kind of consistence relation if we allow for the creation of baby universe, we can find a consistence relation for the (undefined) disk amplitude itself. It is shown in Fig. 8.5. It involves the cylinder amplitude and disk amplitude from CDT and we know these. We can thus write down an actual equation corresponding to Fig. 8.5

$$W_\Lambda(X) = W_\Lambda^{(0)}(X) + g_s \int_0^\infty dT \int_0^\infty dL_1 \int_0^\infty dL_2 \, (L_1+L_2) G_\Lambda^{(0)}(X, L_1+L_2; T) W_\Lambda(L_1) W_\Lambda(L_2)$$

$$= W_\Lambda^{(0)}(X) + g_s \int_0^\infty dT \, \frac{\bar{X}^2(T) - \Lambda}{X^2 - \Lambda} \frac{dW_\Lambda^2(\bar{X})}{d\bar{X}} \bigg|_{\bar{X} = \bar{X}(T)}$$

$$= W_\Lambda^{(0)}(X) + g_s \frac{W_\Lambda^2(\sqrt{\Lambda}) - W_\Lambda^2(X)}{X^2 - \Lambda}. \tag{8.43}$$

The meaning of the coupling constant g_s is explained in the figure captions of Figs. 8.4 and 8.5. The superscript $^{(0)}$ refers to the CDT functions, which are explicitly given by (8.28) and (8.31). Further, the factor $L = L_1 + L_2$ is present because the loop at time T is pinched at a point and that can be at L different points, morally speaking (in a discretized version the loop would have l links and l vertices and could be pinched in l ways). The second line follows from inserting (8.28) and performing

The Causal Dynamical Triangulation model

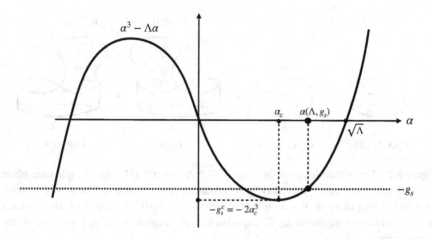

Figure 8.6 The graphic solution $\alpha(\Lambda, g_s)$ to eq. (8.45). For given Λ there is no solution when $g_s > g_s^c = 2\alpha_c^3 = \frac{2}{3\sqrt{3}}\Lambda^{3/2}$. For $g_s = g_s^c$ the solution is $\alpha(\Lambda, g_s^c) = \alpha_c = \sqrt{\Lambda/3}$. This implies that $\alpha(\Lambda, g_s)$ will decrease from $\sqrt{\Lambda}$ for $g_s = 0$ to its minimum value α_c when $g_s = g_s^c$.

the L integrals, which just lead from $W(L)$ to the Laplace transform $W(\bar{X})$. Finally, the third line follows from (8.22) which allows us to replace the T integration by an integration over $\bar{X}$, as was also done in (8.31). We can now solve for $W_\Lambda(X)$:

$$2g_s W_\Lambda(X) = \Lambda - X^2 + \hat{W}_\Lambda(X), \quad \hat{W}_\Lambda(X) = \sqrt{(X^2 - \Lambda)^2 + 4g_s\left(g_s W^2(\sqrt{\Lambda}) + X - \sqrt{\Lambda}\right)}. \tag{8.44}$$

We want $W_\Lambda(L)$ to be a continuous deformation of $W_\Lambda^{(0)}(L) = e^{-\sqrt{\Lambda}L}$ when g_s is close to zero. In particular we want it to fall off exponentially with L. For $g_s = 0$ the 4th-order polynomial under the square root in (8.44) has double zeros at $X = \pm\sqrt{\Lambda}$. For small g_s these double zeros will split, unless $W_\Lambda(\sqrt{\Lambda})$ is fine tuned, and this will result in two cuts in $\hat{W}_\Lambda(X)$, one cut close to $-\sqrt{\Lambda}$ and the other cut close to $\sqrt{\Lambda}$. We cannot allow the cut close to $\sqrt{\Lambda}$, since by inverse Laplace transformation it will result in an exponential growing $W_\Lambda(L)$. Thus we have to insist that $W_\Lambda(\sqrt{\Lambda})$ is fine tuned such that the fourth-order polynomial has a double zero on the positive real axis. As simple calculation then leads to

$$\hat{W}_\Lambda(X) = (X - \alpha)\sqrt{(X + \alpha)^2 - \frac{2g_s}{\alpha}}, \quad \alpha^3 - \Lambda\alpha + g_s = 0. \tag{8.45}$$

One has to choose the solution $\alpha(\Lambda, g_s)$ to the third order equation which is closest to $\sqrt{\Lambda}$, as illustrated in Fig. 8.6 and we then have a power expansion of α in powers of $g_s/\Lambda^{3/2}$:

$$\alpha(\Lambda, g_s) = \sqrt{\Lambda}\left(1 - \frac{g_s}{2\Lambda^{3/2}} + \cdots\right) = \sqrt{\Lambda}\, F\!\left(\frac{g_s}{\Lambda^{3/2}}\right), \quad F(0) = 1, \tag{8.46}$$

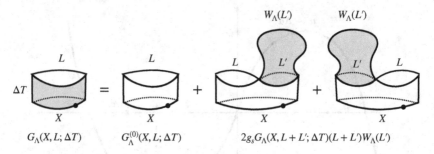

Figure 8.7 The infinitesimal propagation $G_\Lambda(X,L;\Delta T)$ for GCDT: either no split takes place or a split will occur with "probability" $g_s\Delta T$. After the split one part will be a baby universe with the topology of a disk. For the split to occur, the exit loop (with length $L+L'$) has to pinch at a point. This can happen in $L+L'$ ways. We have to integrate wrt L' to count all possible configurations.

and a corresponding expansion of $W_\Lambda(X)$.

Knowing $W_\Lambda(X)$ we can now return to Fig. 8.4 and find an equation for $G_\Lambda(X,L;T)$. Rather than using Fig. 8.4, which involves a time integration from 0 to T, it is more convenient to use the infinitesimal version of it, where time only changes by ΔT as indicated on the figure and shown in detail in Fig. 8.7. This figure leads to the following equation for $G_\Lambda(X,L;\Delta T)$:

$$\begin{aligned}G_\Lambda(X,L,\Delta T) &= G_\Lambda^{(0)}(X,L,\Delta T) + 2g_s\Delta T\int dL'(L+L')G_\Lambda^{(0)}(X,L+L',\Delta T)W_\Lambda(L')\\ &= e^{-XL} - \Delta T\frac{\partial}{\partial X}\left([(X^2-\Lambda)+2g_sW(X)]e^{-XL}\right) + \mathcal{O}(\Delta T^2) \quad (8.47)\end{aligned}$$

Here we have used (8.28) and (8.22) to write

$$G_\Lambda^{(0)}(X,L;\Delta T) = e^{-XL} - \Delta T\frac{\partial}{\partial X}\left((X^2-\Lambda)e^{-XL}\right) + \mathcal{O}(\Delta T^2), \quad (8.48)$$

Thus we see that the only change going from CDT to GCDT is the replacement

$$X^2-\Lambda \;\to\; X^2-\Lambda+2g_sW_\Lambda(X) = \hat{W}_\Lambda(X). \quad (8.49)$$

We can finally write

$$\begin{aligned}G_\Lambda(X,Y;T+\Delta T) &= \int dL\, G_\Lambda(X,L,\Delta T)\, G_\Lambda(L,Y;T)\\ &= G_\Lambda(X,Y;T) - \Delta T\frac{\partial}{\partial X}\left(\hat{W}_\Lambda(X)G_\Lambda(X,Y;T)\right) \quad (8.50)\end{aligned}$$

which leads to the generalization of (8.20)

$$\boxed{\frac{\partial G_\Lambda(X,Y;T)}{\partial T} = -\frac{\partial}{\partial X}\left(\hat{W}_\Lambda(X)G_\Lambda(X,Y;T)\right)} \quad (8.51)$$

The solution is by now standard and generalizes (8.21) and (8.22):

$$G_\Lambda(X,L;T) = \frac{\hat{W}(\bar{X}(T,X))}{\hat{W}(X)} e^{-\bar{X}(T)L}, \quad \frac{d\bar{X}}{dT} = -\hat{W}(\bar{X}), \quad \bar{X}(0) = X. \quad (8.52)$$

One can find $\bar{X}(T;X)$ (and thus $G_\Lambda(X,L;T)$) expressed in terms of elementary functions (see Problem Set 13). Here we will only provide the expression for $X \to \infty$, i.e. when we contract the entrance loop to a point:

$$\bar{X}(T;X=\infty) - \alpha = \frac{\Sigma^2}{\sinh(\Sigma T)(\Sigma \cosh(\Sigma T) + \alpha \sinh(\Sigma T))}, \quad \Sigma = \sqrt{\alpha^2 - \frac{g_s}{2\alpha}}, \quad (8.53)$$

where we, using (8.46), can write

$$\Sigma(\Lambda, g_s) = \sqrt{\Lambda} H\left(\frac{g_s}{\Lambda^{3/2}}\right), \quad H(0) = 1, \quad H\left(\frac{2}{3\sqrt{3}}\right) = 0 \quad (8.54)$$

It can be shown that $\bar{X}(T,X) - \alpha$ falls off exponentially as $e^{-2\Sigma T}$, not only for $X = \infty$ as shown in (8.53), but for all $X > \alpha$ (see Problem Set 13). It follows from (8.52) that the cylinder amplitude falls off as $e^{-2\Sigma T}$, but the coefficient Σ decreases from the CDT value $2\sqrt{\Lambda}$ toward zero when g_s increases to the critical value $g_s^c = 2\alpha_c^3 = 2\Lambda^{3/2}/(3\sqrt{3})$.

We can define the two-point function for GCDT precisely as we did for CDT, eq. (8.34), and we obtain in the same way (only replacing $X^2 - \Lambda$ with $\hat{W}_\Lambda(X)$)

$$\boxed{G_\Lambda(T) = \frac{dW_\Lambda(\bar{X}(T;\infty))}{dT} = \frac{\Sigma^3}{\alpha} \frac{\Sigma \sinh(\Sigma T) + \alpha \cosh(\Sigma T)}{(\Sigma \cosh(\Sigma T) + \alpha \sinh(\Sigma T))^3}} \quad (8.55)$$

It is somewhat tedious, but straight forward, to derive the formula using (8.53) (see Problem Set 13 for some details). The formula itself is remarkable and it looks like a simple generalization of the formula derived for the two-point function in EDT. However, the consequences are very different, and lead to the critical exponents of CDT, so *we have achieved our goal: to find a non-trivial generalization of CDT, which still belongs to the same universality class.* Let us discuss this in the same way as we did for the two-point function of CDT. First, it falls of exponentially as $e^{-2\Sigma T}$ when $T \to \infty$. This indicates critical exponent $\nu_{gcdt} = 1/2$ provided we have a discretized theory where we can write $T = \varepsilon t$ and $\sqrt{\mu - \mu_c} \propto \Sigma \varepsilon$. We will discuss such theories in the next section. Next, we have $G_\Lambda(T \to 0) = 1$, as for CDT, and this indicates an anomalous dimension $\eta_{gcdt} = 1$. Finally

$$\chi(\Lambda) = \int_0^\infty dT\, G_\Lambda(T) = W_\Lambda(\bar{X}(T=\infty;\infty)) - W_\Lambda(\bar{X}(T=0;\infty))$$

$$= W_\Lambda(\alpha) = \frac{\Lambda - \alpha^2}{2g_s} = \frac{1}{2\alpha}, \quad (8.56)$$

and again this indicates a susceptibility exponent $\gamma_{gcdt} = \frac{1}{2}$ provided we can write $\sqrt{\mu - \mu_c} \propto \alpha \varepsilon$ in a discretized theory.

$W_\Lambda(X)$ can be expanded in powers of g_s

$$g_s W_\Lambda(X) = \sum_{n=0}^{\infty} W_\Lambda^{(n)}(X) g_s^{n+1} \qquad (8.57)$$

where $W_\Lambda^{(n)}(X)$ is the disk amplitude with $n+1$ "CDT disk components": $W_\Lambda^{(0)}(X)$ is the CDT amplitude with one component, the CDT universe, $W_\Lambda^{(1)}(X)$ is the universe where at some time T it split in two components which then continue their propagation in time independently without splitting any further, i.e. as two CTD universes, and higher powers of g_s capture the iteration of this splitting process. Let us (for reasons to be clear later) mark one of the components. One can think of the mark as associated with the endpoint in time of that particular CDT component (where the length of the boundary loop is contracted to a point). We obtain the corresponding disk amplitude by differentiating $g_s W_\Lambda(X)$ wrt g_s since the marking of a component in $W_\Lambda^{(n)}(X)$ can be done in $n+1$ ways. A short calculation, using (8.44) and (8.45) leads to the remarkably simple result[4]

$$\tilde{W}_\Lambda(X) := \frac{d(g_s W_\Lambda(X))}{dg_s} = \frac{1}{\sqrt{(X+\alpha)^2 - 2g_s/\alpha}} \qquad (8.58)$$

In some sense $\tilde{W}_\Lambda(X)$ *is* the natural generalization of the CDT disk amplitude $W^{(0)}(X)$. Recall that the CDT disk amplitued was defined by (8.31), and since we now have the cylinder amplitude for GCDT we could use a similar definition of the disk amplitude (which will then differ from $W_\Lambda(X)$). A calculation like (8.31), just using (8.52) instead of (8.21) and (8.22), leads to

$$\int_0^\infty dT\, G_\Lambda(X, L=0; T) = -\int_X^\alpha \frac{d\tilde{X}}{\hat{W}_\Lambda(X)} = \frac{X - \alpha}{\hat{W}_\Lambda(X)} = \tilde{W}_\Lambda(X). \qquad (8.59)$$

If we look at Fig. 8.4, it is seen that contracting L_2 to a point and integrating wrt T can be viewed as labeling one of the CDT components in Fig. 8.5 (the component which contains the contracted loop), and we have indeed agreement between (8.58) and (8.59).

Let us use $\tilde{W}_\Lambda(X)$ as a partition function and calculate the average number of CDT components (baby universes) in this ensemble. Again we obtain this number by differentiating $g_s \tilde{W}_\Lambda(X)$ wrt g_s

$$\langle n \rangle_{g_s} = \frac{1}{\tilde{W}_\Lambda(X)} \frac{d(g_s \tilde{W}_\Lambda(X))}{dg_s} = 1 + \frac{g_s}{3\alpha^2 - \Lambda} \left[\frac{X + 2\alpha}{(X+\alpha)^2 - 2g_s/\alpha} \right] \qquad (8.60)$$

---

[4]This result is the equivalent to the result in EDT that $dW_\Lambda^{edt}(X)/d\Lambda \propto 1/\sqrt{X + \sqrt{\Lambda}}$, which one easily proves by differentiating $W_\Lambda^{edt}(X)$ wrt Λ and which we discuss in more detail in Problem Set 10. The difference is that we in EDT differentiate wrt Λ, not g_s. In the continuum limit of EDT there is no strict equivalent to g_s since an infinite number of baby universes will be created in a continuum time ΔT, even if ΔT is small. Alternatively one can say the baby universes are everywhere and thus marking the "top" of a baby universe will be "proportional" to just marking a point, which is exactly the counting provided in EDT by differentiating wrt Λ.

The Causal Dynamical Triangulation model

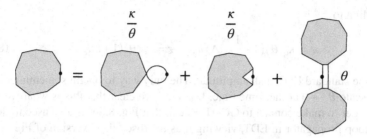

Figure 8.8 The graphic equation for $w(z)$ in the case there $V'(z)$ is given by eq. (8.62). The figure is similar to Fig. 6.6, except that a factor θ is attached to the double link.

It is seen that $\langle n \rangle_{g_s} \to \infty$ for $g_s \to g_s^c$ simply because $d\alpha/dg_s = 1/(3\alpha^2 - \Lambda)$ diverges at that point (see Fig. 8.6, where $dg_s/d\alpha|_{\alpha_c} = 0$). So for $g_s > g_s^c$ the GCDT theory breaks down and since the number of baby universes proliferates, it is natural to conjecture that the EDT picture will take over in any sensible extension of GCDT. This is indeed the case as we will explain in the next section.

8.4 GCDT DEFINED AS A SCALING LIMIT OF GRAPHS

Above we defined GCDT via pictures! It is possible to define GCDT in the same way as we have defined EDT and CDT, via triangulations (and also on much more general classes of graphs), where the links have length ε, and then take the scaling limit $\varepsilon \to 0$. The *continuum* expression for the GCDT disk amplitude was

$$W_\Lambda(X) = \frac{\Lambda - X^2 + (X-C)\sqrt{(X-C_+)(X-C_-)}}{2g_s}, \quad C = \alpha, \ C_\pm = -\alpha \pm \sqrt{\frac{2g_s}{\alpha}}. \tag{8.61}$$

This expression is formally quite similar to the expression for the disk amplitude in EDT *before* one takes the scaling limit:

$$w(z) = \frac{-\kappa + z - \kappa z^2 + \kappa(z-c)\sqrt{(z-c_+)(z-c_-)}}{2\theta}, \quad V'(z) = \frac{1}{\theta}(z - \kappa(1+z^2)) \tag{8.62}$$

A simple shift $z \to z + 1/2\kappa$ will eliminate the linear term, which is present in (8.62) but not present in (8.61). However, the standard scaling limit of EDT is such that c_- does not scale and therefore, in the EDT scaling limit the cut from c_+ to c_- will develop into a cut from a C_+ to $-\infty$, a situation distinctly different from what is seen in (8.61). For this reason we have introduced here a new (allover) coupling constant θ, which was formerly chosen to be 1 in the EDT case. In the defining picture for the EDT disk amplitude this new coupling constant will appear as a coupling contant for the splitting of $w(z)$ in two $w(z)$ as shown in Fig. 8.8, as one can check recalling the arguments leading to the picture in the first place. For a fixed θ the critical EDT point κ_c was the point where $c_+(\kappa_c) = c(\kappa_c)$ and approaching this point, which depends on

θ, according to

$$\kappa = \kappa_c(\theta)\left(1-\frac{1}{2}\varepsilon^2 \Lambda_{edt}\right), \quad z = z_c(\theta)(1+\varepsilon X_{edt}) \qquad (8.63)$$

leads to the standard EDT disk amplitude. The only way to obtain something different is to scale $\theta \to 0$ at the same time. It is intuitive clear that this is what we have to do in order to make contact to GCDT. Recall that Fig. 8.8 was also used to derive the two-loop propagator in EDT, viewing it as the discretized version of Fig. 8.7. It is then clear that if we want to prohibit the creating of baby universes, such that only finite many appear in the continuum limit, we have to scale θ to zero. In fact, if we write $\theta = g_s \varepsilon^3$ then a calculation (the details of which can be found in [32]) leads to

$$w(z) \xrightarrow{\varepsilon \to 0} \frac{1}{\varepsilon} W_\Lambda(X), \quad \kappa = \kappa_c\left(1-\frac{1}{2}\varepsilon^2\Lambda\right), \quad z = z_c(1+\varepsilon X), \quad \kappa_c = \frac{1}{2}, \; z_c = 1. \quad (8.64)$$

κ_c and z_c are just the critical values for CDT. It is now clear from (8.46) and (8.54) that if we have a scaling limit $\theta = g_s \varepsilon^3$ and $\kappa_c - \kappa = \varepsilon^2 \Lambda$ then we indeed can write

$$\sqrt{\kappa_c - \kappa} \propto \alpha\varepsilon \propto \Sigma\varepsilon \qquad \text{for } \varepsilon \to 0 \text{ and fixed } \Lambda \text{ and } g_s. \qquad (8.65)$$

This shows, as already remarked, that the critical exponents in this scaling limit are the ones of CDT and not the EDT exponents.

The graphs generated by $V'(z)$ given in (8.62) will be made of triangles and "one-gons". Let T_1 and T_3 denote the number of one-gons and triangles in a connected planar graph (a "triangulation" T) constructed from these objects. The factor θ associated with T will be $\theta^{-T_1+T_3}$. It follows from writing $z = \theta\tilde{z}$ in (8.62), i.e. we obtain the standard form $V'(\tilde{z}) = \tilde{z} - \kappa(\theta^{-1}+\theta\tilde{z}^2)$ and thus a factor κ/θ and a factor $\kappa\theta$ associated with one-gons and triangles, respectively. Let us consider the graph dual to T, i.e. the graph $\tilde{T}$ where a vertex is put in the center of each triangle and each one-gon, and the vertices in neighboring triangles or one-gons are connected by (dual) links, crossing the links separating the polygons in the original triangulation T. In this way we obtain a planar graph which consists of vertices of order 3 or 1. Such a graph is shown in Fig. 8.9. Let F denote the number of faces in $\tilde{T}$, L the number of links and $V = T_1 + T_3$ the number of vertices in $\tilde{T}$. From Euler's relation we can write

$$F - L + V = 1, \quad \text{where} \quad 2L = T_1 + 3T_3, \quad \text{i.e.} \quad 2F - 1 = -T_1 + T_3. \qquad (8.66)$$

The factor θ associated with T can then be written as the factor

$$\theta^{2F(\tilde{T})-1} \qquad (8.67)$$

associated with the dual graph $\tilde{T}$. Eq. (8.67) shows that the number of faces in the dual graphs will be suppressed when $\theta \to 0$. The following picture then emerges: as long as $g_s < g_s^c$ (for fixed Λ) the criticality of the ensemble of ϕ^3 and ϕ graphs, exemplified in Fig. 8.9, is determined by the criticality of the BPs dressing the ϕ^3 skeleton graphs and the average number of faces, links and vertices in the skeleton graphs will be *finite* in the scaling limit. This is the GCDT limit. However, for g_s

The Causal Dynamical Triangulation model

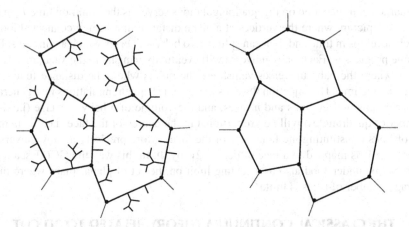

Figure 8.9 A planar graph constructed from vertices of order 1 and 3 (the left graph in the picture) can be viewed as a "skeleton" graph (the right graph) consisting only of vertices of order 3, decorated with tree-graphs (rooted BPs).

larger than g_s^c a different scaling limit will prevail, where the BPs will not be critical (and they are thus not important in the scaling limit), but now the skeleton graphs will define the criticality. This is the EDT limit (see [33] for details).

Seemingly, insisting on only a finite number of baby universes being present in the scaling limit of the triangulated surfaces shows up as a scaling limit where the number of faces is finite on the set of dual graphs. This is not a coincidence. One can formulate a more detailed relationship between such classes of graphs, which also keeps track of graph distances. It is most easily done, not starting with triangulations, but with quadrangulations. We will end this section discussing this, without providing many details, not to mention proofs (they can be found in [34], which also contains a combinatorial definition of GCDT and discusses how to take the scaling limit in detail). Let us define *a planar map* as a connected graph which can be projected on the sphere without any links crossing. It will consist of a number of faces, links and vertices. On such graphs one can mark a vertex and then define the graph distance from the marked vertex to other vertices, i.e. one can define a distance function on the graphs. It can then be shown that there exists a bijection Φ from the planar quadrangulations with a marked vertex[5] to the planar maps with a marked vertex, such that if Q is a planar quadrangulation with N faces and n local maxima of the distance function, then $\Phi(Q)$ is a planar map with N links and n faces, and the distance labelling of Q is mapped to the distance labelling of $\Phi(Q)$. The scaling limit of GCDT is one where we (loosely speaking) keep the number n of baby universes fixed while taking N to infinity. Starting out with the marked vertex, or more generally with a marked entrance loop, the distance function to the marked point or to

---

[5] Strictly speaking one has to make the marking somewhat more precise but we will not go into the technical details of how to do that.

the marked entrance loop on the quadrangulations serves as the common time T, and we have a picture where the vertices at a given distance first form a connected loop which develops in time and then can split in two baby universes which again can split as time progresses. Each baby universe will eventually vanish "in the vacuum". The points where the baby universes vanish are the points where the distance function has local maxima. The bijection Φ sends these GCDT quadrangulations into general planar graphs with N links and n faces, and the volume of a baby universe (i.e. its number of quadrangles) will be proportional to the degree of the face (i.e. the number of links constituting the boundary of the face). More precisely a baby universe of volume V is mapped to a face of degree $2V$ by Φ. In this way the GCDT scaling limit can be understood also as a scaling limit on the set of planar maps where the average number of faces is finite.

8.5 THE CLASSICAL CONTINUUM THEORY RELATED TO 2D CDT

We have now theories, CDT and GCDT, which we have defined as quantum theories. It is natural to ask if there exist *classical* theories, which lead to CDT or GCDT when quantized. For CDT there exists an obvious candidate where the symmetry imposed naively agrees with the symmetry imposed on CDT configurations: the *Hořava-Lifshitz gravity theory* (HLG). It is a modification of General Relativity where time is given a special role. In so-called *projectable* HLG it is assumed that spacetime has a time foliation and that the theory is invariant under spatial diffeomorphisms and time re-definitions, also called foliation preserving diffeomorphisms. This clearly restricts the class of geometries and it agrees with the set of geometries we used to define CDT. In the same way as the Einstein-Hilbert action is characterized as being the unique action invariant under diffeomorphisms and containing at most second derivatives of the metric, one can find the action which contains at most second derivatives and is invariant under foliation preserving diffeomorphisms. In spacetime dimensions larger then 2, what is usually denoted HLG is a theory which actually contains higher spatial derivatives which are added to the theory to make it perturbatively renormalizable. However, in two dimensions renormalizability of gravity is not an issue (we have precisely quantized 2d gravity in these lectures!) and we will not add such terms to two-dimensional HLG. We will not go into any detail, but only mention that the invariance under spatial diffeomorphisms implies that starting out with the metric variables $g_{\mu\nu}(x,t)$, $\nu,\mu = 0,1$, where 0,1 signifies time and space, the only remaining variables will be

$$N(t) = \sqrt{|g_{00}(t)|} \quad \text{and} \quad L(t) = \int dx \sqrt{|g_{11}(x,t)|}. \tag{8.68}$$

$L(t)$ is the length of a spatial universe at time t and it cannot be changed by a spatial diffeomorphism. Further, one assumes in projectable HLG that $g_{00}(x,t)$ is a function only of t. In projectable HLG the so-called proper time:

$$t_p(t) = \int_0^t dt' N(t'), \tag{8.69}$$

The Causal Dynamical Triangulation model

is invariant under time redefinitions, and thus a physical observable and corresponds to $N_p(t_p) = 1$. The classical HLG action rotated to Euclidean signature can now be written (choosing proper time)

$$S_E[L] = \int dt_p \left(\frac{\dot{L}^2(t_p)}{4L(t_p)} + \Lambda L(t_p) \right) \tag{8.70}$$

and the corresponding quantum amplitude will be

$$G_\Lambda(L_1, L_2; T) = \int \mathcal{D}L(t_p)\, e^{-S_E[L]}, \quad L(0) = L_1,\ L(T) = L_2. \tag{8.71}$$

If we write

$$G_\Lambda(L_1, L_2; T) = \langle L_2 | e^{-\hat{H}T} | L_1 \rangle \tag{8.72}$$

$$G_\Lambda(L_2, L_1; \varepsilon) = \langle L_2 | (I - \varepsilon \hat{H} + \mathcal{O}(\varepsilon^{3/2})) | L_1 \rangle, \tag{8.73}$$

we obtain, by discretizing the proper time interval in steps of ε, from (8.70) and (8.71):

$$G_\Lambda(L_2, L_1; \varepsilon) = \frac{L_1}{\sqrt{4\pi\varepsilon L_2}} \exp\left(-\frac{(L_2 - L_1)^2}{4\varepsilon L_2} + \varepsilon \Lambda L_1 \right), \tag{8.74}$$

where the origin of the factor L_1 comes from the marking of the entrance loop of the cylinder amplitude $G_\Lambda(L_1, L_2; T)$. Integrating (8.73) with a wave function $\Psi(L_1)$ we have

$$\int \frac{dL_1}{L_1} G(L_2, L_1; \varepsilon) \Psi(L_1) = \Psi(L_2) - \varepsilon (\hat{H}\Psi)(L_2) + \mathcal{O}(\varepsilon^{3/2}) \tag{8.75}$$

and Taylor expanding $\Psi(L_1)$ appearing in the integral on the lhs of (8.75) around L_2 one finally obtains

$$(\hat{H}\Psi)(L_2) = \left(-L_2 \frac{d^2}{dL_2^2} + \Lambda L_2 \right) \Psi(L_2). \tag{8.76}$$

Thus $\hat{H}$ is precisely the CDT Hamiltonian (8.38). We have thus shown that the classical two-dimensional HLG when quantized leads to CDT.

It is less clear how to associate a classical continuum theory to GCDT. Looking at Fig. 8.4 it seems difficult to associate a classical Hamiltonian to the propagation of space in (proper) time T. The creation of baby universes is not a natural part of a classical theory and the attempts to define a classical theory leading to the GCDT when quantized have so far been forced to put in some kind of "baby universes" by hand in the classical theory, and are at best "unusual" classical theories. Rather, it seems more natural to view GCDT as a quantum generalization of the quantum theory we have denoted CDT, namely a quantum generalization where we allow for the creation of baby universes. In this way one appeals to the idea that in a quantum theory everything, which is not protected by some symmetries (and corresponding conserved charges) should be allowed. That would then lead not to CDT but to GCDT. It is remarkable that we can actually solve this generalized theory explicitly and perform the summation over all possible baby universes. Once we have allowed for the creation of baby universes, it is also natural to allow for the creation of *wormholes*, i.e. a baby universe is created, but rather than vanishing in the vacuum it is allowed to connect back to the "parent universe", in this way changing the spacetime topology (for an illustration see Fig. 5.4 in the case of a propagating string). Even more remarkable than being able to sum over all baby universes is the fact that one can perform this added summation which also includes the summation over all wormholes. We will not go into any detail here, just mention two things. First, the underlying technical reason one is able to perform this summation is the bijection between BPs and CDT, which then can be generalized to a mapping between the generalized surfaces (with baby universes and wormholes) and BPs with loops. As we saw in Problem Set 9, one can sum these BPs with loops. Second, one can write down an "effective Hamiltonian" where the effect of this summation is taken into account and it is a very simple generalization of the CDT Hamiltonian

$$\hat{H}^{\text{eff}} = -L\frac{d^2}{dL^2} + \Lambda L - g_s L^2. \tag{8.77}$$

It is seen that the potential is unbounded from below, a reflection of the fact that the pertubation series in g_s is not even Borel summable because of the numerous wormhole configurations of higher spacetime genus. In this sense the situation is somewhat similar to the situation we described in string theory and also in EDT if one included geometries with arbitrary high genus. Nevertheless, one can make $\hat{H}^{\text{eff}}$ a selfadjoint operator with a discrete energy spectrum where the eigenvalues $E_n \to E_n^{\text{cdt}} = 2n\sqrt{\Lambda}$ for $g_s \to 0$ and $n \geq 1$. In particular, $E_0 = 0$ is still an eigenvalue, and the "disk" function $W_{\Lambda,g_s}(L)$, which now includes all possible wormholes, satisfies

$$\hat{H}^{\text{eff}} W_{\Lambda,g_s}(L) = 0, \quad W_{\Lambda,g_s}(L) = \frac{\text{Bi}\left(\frac{\Lambda - g_s L}{g_s^{2/3}}\right)}{\text{Bi}\left(\frac{\Lambda}{g_s^{2/3}}\right)} + c \cdot \text{Ai}\left(\frac{\Lambda - g_s L}{g_s^{2/3}}\right) \tag{8.78}$$

Bi and Ai are the standard Airy functions (which we also met in Problem Set 9). One can check that for $g_s \to 0$ then $W_{\Lambda,g_s}(L) \to e^{-\sqrt{\Lambda}L}$ for $L \leq \Lambda/g_s$. A couple of solutions is shown in Fig. 8.10. Eq. (8.78) contains the term $\text{Ai}(x/g_s^{2/3})$ which for $x > 0$ falls off like $e^{-2x^{3/2}/3g_s}$. It is thus not part of a perturbative expansion in g_s and undetermined by the requirement that $W_{\Lambda,g_s}(L) \to e^{-\sqrt{\Lambda}L}$ for $g_s \to 0$. $W_{\Lambda,g_s}(L)$ *is the non-perturbative Hartle-Hawking wave function of our quantum GCDT universe.*

8.6 PROBLEM SETS AND FURTHER READING

Problem Set 13 provides the details needed to prove the beautiful formula (8.55). In addition it contains a study of the shape of typical CDT universes. It is pretty boring as long as a boundary cosmological constants are positive and a typical shape is just like shown in Fig. 8.3. However, the situation changes if we allow the cosmological

The Causal Dynamical Triangulation model

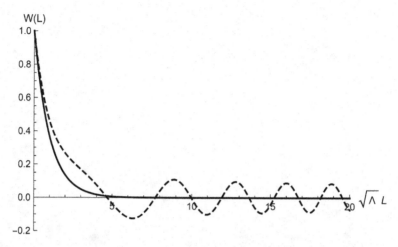

Figure 8.10 The Hartle-Hawking wave function (8.78) which $c=0$ plotted for $g_s = 0$ (full drawn curve and pure CDT, i.e. $W(\Lambda) = e^{-\sqrt{\Lambda}L}$), and for $g_s/\Lambda^{3/2} = 1/3$ (dashed curve). The oscillatory behavior starts only for $\sqrt{\Lambda}L > \Lambda^{3/2}/g_s$ and then the fall off changes from exponential to $1/L^{1/4}$.

constant of one boundary to be negative. This can force the boundary to expand, and if the value is precisely $-\sqrt{\Lambda}$ we obtain a universe expanding in (Euclidean) time precisely such that average universe will have constant negative curvature, i.e. it will be a realization of the so-called pseudosphere. We study some details of this in the last part of Problem Set 13 and further details can be found in the original article [35] on which this part of the Problem Set is based. Similar calculations can also be done in EDT for negative boundary cosmological constants (see [36]).

A very interesting aspect of GCDT is that it can serve as a toy model for *string field theory*. We have described how the bosonic string can split split in two and how two strings can unite into one, and advocated this as a unique feature of string theory compared to an ordinary quantum field theory of particles, where such interaction terms have to be put in "by hand". The multiparticle aspect of quantum field theory is usually described using a so-called Fock space where the basic vectors represent many particles in given quantum states. There have been many attempts to introduce a similar formalism in string theory, but for the closed strings we consider in Chapter 5, it has not been very successful. However, GCDT has a natural formulation as such a string theory. One has free strings propagating. These are just the CDT strings and we know the Hamiltonian for these free strings, it is just (8.38). One can now write down an extended Hamiltonian describing interaction between the strings, i.e. terms allowing the string to split and join and one can construct a Fock space containing the multiple free closed strings of given lengths. Further one can show how the perturbative expansion in the string coupling constant g_s can reproduce a similar expansion for GCDT in g_s. Details of this can be found in [37, 38].

A Preliminary material Part 2: Green Functions

BASICS

The purpose here is to remind the reader about Green functions as they are used in classical physics. No proofs will be given, no mathematical rigor is attempted and the details of calculations are left as exercises. Our starting point will be a simple inhomogeneous second-order differential equation:

$$\left(-\frac{d^2}{dt^2}+V(t)\right)\psi(t) = J(t) \tag{A.1}$$

Assume that we can solve the following equation

$$\left(-\frac{d^2}{dt^2}+V(t)\right)G(t,s) = \delta(t-s) \tag{A.2}$$

This generates a solution to (A.1):

$$\psi(t) = \int ds\, G(t,s) J(s). \tag{A.3}$$

$G(t,s)$ is a Green function for the differential equation (A.1).

Viewed as operators we can write (very formally, no discussion of domains etc.):

$$\hat{D}: \psi \mapsto \hat{D}\psi, \quad (\hat{D}\psi)(t) = -\psi''(t) + V(t)\psi(t)$$

$$\hat{G}: \psi \mapsto \hat{G}\psi, \quad (\hat{G}\psi)(t) = \int ds\, G(t,s)\, \psi(s).$$

Equation (A.2) can now be written as a formal operator identity

$$\hat{D}\hat{G} = \hat{I} \tag{A.4}$$

which suggests that

$$\hat{G} = \hat{D}^{-1}. \tag{A.5}$$

In general $\hat{D}^{-1}$ is not well-defined unless we restrict the function space for $\hat{D}$ since $\hat{D}\psi = 0$ might have many non-trivial solutions, namely the solutions to the homogeneous equation:

$$\left(-\frac{d^2}{dt^2}+V(t)\right)\psi(t) = 0 \tag{A.6}$$

Sometimes we can eliminate these solutions by imposing *boundary conditions*. We usually have to do that anyway if we want $\hat{D}$ to be *Hermitian*. Let us consider $\hat{D}$ defined in the interval $t \in [t_i, t_f]$.

Sturm-Liouville Boundary conditions at t_i, t_f:

$$\alpha \psi(t_i) + \beta \psi'(t_i) = 0 \qquad (A.7)$$
$$\gamma \psi(t_f) + \delta \psi'(t_f) = 0 \qquad (A.8)$$

These boundary conditions ensure that $\hat{D}$ is Hermitian since:

$$\int_{t_i}^{t_f} \phi \frac{d^2}{dt^2} \psi = (\phi \psi' - \phi' \psi)\Big|_{t_i}^{t_f} + \int_{t_i}^{t_f} \left(\frac{d^2}{dt^2} \phi\right) \psi$$

For generic choice of $\alpha, \beta, \gamma, \delta$ (A.6) will *not* have a solution, i.e. $\hat{D}$ will be invertible. Let $\phi_1(t)$ be a solution to (A.6) which satisfies (A.7) (but not (A.8) since we assume that no such solution exists). Let $\phi_2(t)$ be solution to (A.6) which satisfies (A.8) (but not (A.7) ....). Let $\psi_n(t)$ be an *eigenfunction* of the operator $\hat{D}$ (with the given boundary conditions (A.7) and (A.8)):

$$\left[-\frac{d^2}{dt^2} + V(t)\right] \psi_n(t) = \lambda_n \psi_n(t) \qquad (A.9)$$

The functions $\psi_n(t)$ form a complete set since $\hat{D}$ is Hermitian: Any $L^2[t_i, t_f]$ function can be expanded as:

$$f(t) = \sum_n c_n \psi_n(t), \quad c_n = \int_{t_i}^{t_f} \psi_n^*(t) f(t) \qquad (A.10)$$

There are two ways to construct $G(t, s) = \hat{D}^{-1}$:

$$(\text{I}): \quad G(t, s) = \frac{-1}{w} \begin{cases} \phi_1(t)\phi_2(s), & t < s \\ \phi_2(t)\phi_1(s), & t > s \end{cases} \qquad (A.11)$$

$$(\text{II}): \quad G(t, s) = \sum_n \frac{\psi_n^*(s) \psi_n(t)}{\lambda_n} \qquad (A.12)$$

where

$$w(\phi_1, \phi_2) \equiv \phi_1 \phi_2' - \phi_2 \phi_1' \qquad (A.13)$$

is called the *Wronskian* of $\hat{D}$. $w(\phi_1, \phi_2) = const.$ if ϕ_1, ϕ_2 satisfy (A.6).

Exercise 1: Show that (A.11) and (A.12) solves (A.2) and that (A.3) will satisfy the correct boundary conditions.

Let us now consider the special situations where $V(t)$ is *independent* of time and $t_i \to -\infty$, $t_f \to +\infty$, i.e.

$$\text{case (a)}: \quad \left[-\frac{d^2}{dt^2} + \tilde{\omega}^2\right] \psi(t) = J(t) \qquad (A.14)$$

$$\text{case (b)}: \quad \left[-\frac{d^2}{dt^2} - \omega^2\right] \psi(t) = J(t) \qquad (A.15)$$

Case (a)

Impose *the boundary conditions:*

$$\psi(t) \to 0 \text{ for } t \to \pm\infty \tag{A.16}$$

With these boundary condition $\hat{D}$ becomes Hermitean on $L^2(\mathbb{R})$. The complete solution to the homogenous equation corresponding to (A.14) (i.e. $J(t)=0$) is:

$$\psi(t) = a e^{-\tilde{\omega} t} + b e^{\tilde{\omega} t} \tag{A.17}$$

It follows that $\lambda = 0$ is *not* an eigenvalue for (A.16). We see that $\phi_1(t) = e^{\tilde{\omega} t}$ and $\phi_2(t) = e^{-\tilde{\omega} t}$ and this implies by (A.11) that

$$G(t,s) = \frac{1}{2\tilde{\omega}} \left\{ \begin{array}{ll} e^{\tilde{\omega} t} e^{-\tilde{\omega} s}, & t < s \\ e^{-\tilde{\omega} t} e^{\tilde{\omega} s}, & t > s \end{array} \right\} = \frac{e^{-\tilde{\omega}|t-s|}}{2\tilde{\omega}} \tag{A.18}$$

Let us use construction (A.12). The solution to the eigenvalue equation (A.9) is:

$$\left(-\frac{d^2}{dt^2} + \tilde{\omega}^2 \right) \psi_p(t) = \lambda_p \psi_p(t) \quad \Rightarrow \quad \psi_p(t) = e^{ipt}, \quad \lambda_p = p^2 + \tilde{\omega}^2. \tag{A.19}$$

Strictly speaking these are *generalized* eigenfunctions since they do not belong to $L^2(\mathbb{R})$ and do not satisfy the imposed boundary conditions (but they stay bounded at least, contrary to the functions in eq. (A.17)). Obviously, they form a complete set.

$$G(t,s) = \sum_p \frac{\psi_p^*(s) \psi_p(t)}{\lambda_p} \to \int \frac{dp}{2\pi} \frac{e^{ip(t-s)}}{p^2 + \tilde{\omega}^2} = \frac{e^{-\tilde{\omega}|t-s|}}{2\tilde{\omega}} \tag{A.20}$$

Exercise 2: Show this using residue calculus (see Fig. A.1)

Case (b)

Let us impose the boundary condition

$$\psi(t) \sim e^{i\omega t} \text{ for } t \to -\infty, \quad \psi(t) \sim e^{-i\omega t} \text{ for } t \to +\infty \tag{A.21}$$

The functions which enter in these boundary conditions can be viewed as the analytic continuation of solutions $e^{\tilde{\omega} t}$ and $e^{-\tilde{\omega} t}$ to the homogeneous equation of case (a) which satisfy one of the conditions in (A.16) for $t \to -\infty$ and $t \to +\infty$. Rotate the $\tilde{\omega}$ from case (a) as $\tilde{\omega}(\theta) = e^{-i\theta} \tilde{\omega}$ where $\theta \in [0, \pi/2[$. For any such $\tilde{\omega}(\theta)$ we still have that the solutions $\phi_1(t) = e^{\tilde{\omega}(\theta)t}$ and $\phi_2(t) = e^{-\tilde{\omega}(\theta)t}$ to the homogeneous equation of case (a), with $\tilde{\omega}$ replaced by $\tilde{\omega}(\theta)$, go to zero when $t \to \mp\infty$, respectively. We now view the ω of case (b) as the limit of $\tilde{\omega}(\theta)$ for $\theta \to \pi/2$, i.e. we write $\omega \to \omega(1-i\varepsilon)$ and the solutions to the homogeneous equations in case (b) can then be viewed as

$$\phi_1(t) = e^{i(\omega-i\varepsilon)t}, \quad \phi_2(t) = e^{-i(\omega-i\varepsilon)t}$$

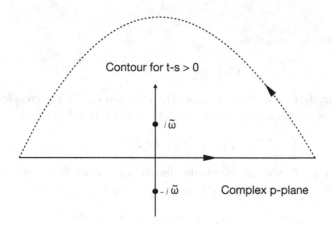

Figure A.1 The integration along the real axis closed by a contour in the upper complex half-plane for $t-s > 0$. For $t-s < 0$ the integration contour should lie in the lower complex half-plane

$$\phi_1(t) \to 0 \quad \text{for} \quad t \to -\infty, \qquad \phi_2(t) \to 0 \quad \text{for} \quad t \to \infty$$

We will use this interpretation of the boundary conditions below, i.e. we will replace ω with $\omega - i\varepsilon$ if ambiguities arise.

The complete solution of the homogeneous equation (A.6) is now in case (b):

$$\psi(t) = ae^{-i\omega t} + be^{i\omega t}, \tag{A.22}$$

i.e. $\lambda = 0$ is *not* eigenvector of (A.9) with boundary condition (A.21). We obtain $\phi_1(t) = e^{i\omega t}$ and $\phi_2(t) = e^{-i\omega t}$, i.e. the Green function constructed according to (A.11) is:

$$G(t,s) = \frac{1}{2\omega i} \left\{ \begin{array}{ll} e^{i\omega t} e^{-i\omega s}, & t<s \\ e^{-i\omega t} e^{i\omega s}, & t>s \end{array} \right\} = \frac{e^{-i\omega|t-s|}}{2i\omega}. \tag{A.23}$$

Can we use (A.12) to construct $G(t,s)$? The answer is yes, with some care! The eigenvalue eq. (A.9) leads to:

$$\left(-\frac{d^2}{dt^2} - \omega^2\right)\psi_p(t) = \lambda_p(t)\psi_p(t) \Rightarrow \psi_p(t) = e^{ipt}, \;\; \lambda_p = p^2 - \omega^2. \tag{A.24}$$

Again these are generalized eigenfunctions in the sense that they do not belong to $L^2(\mathbb{R})$ and do not satisfy the boundary conditions. Compared to (A.20) we have a problem for $p = \pm\omega$ where $\lambda_p = 0$, i.e. for the solutions (A.22) to the homogeneous equation. However, we now resolve this ambiguity by replacing $\omega \to \omega - i\varepsilon$, as mentioned above. With this prescription we have $\lambda_p = p^2 - (\omega - i\varepsilon)^2$ and assuming $\omega > 0$ and ε infinitesimal we can write $\lambda_p = p^2 - \omega^2 + i\varepsilon$ (where we have redefined $2\omega\varepsilon \to \varepsilon$). Finally, we can then write:

$$G_F(t,s) = \sum_p \frac{\psi^*(s)\psi_p(t)}{\lambda_p} \to \int \frac{dp}{2\pi} \frac{e^{ip(t-s)}}{p^2 - \omega^2 + i\varepsilon} = \frac{e^{-i\omega|t-s|}}{2i\omega}. \tag{A.25}$$

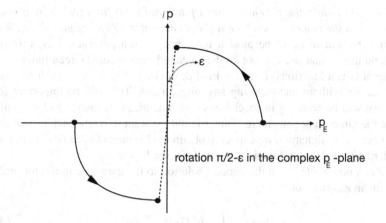

Figure A.2 The rotation $p_E \to e^{i(\pi/2-\varepsilon)} p_E$ in the complex p_E plane. In this way a real value p_E appearing in the propagator $G_E(p_E)$ is transformed into $ip(1-i\varepsilon)$ where p is the real value appearing in the propagator $G_F(p)$.

Exercise 3: Perform the p integration using residue calculus.

By Fourier transformation we get (where we with an abuse of notation will use the same symbol G also for the Fourier transformed Green function):

$$G_F(p) \equiv \int dt\, G_F(t)\, e^{-ipt} = \frac{1}{p^2 - \omega^2 + i\varepsilon} \qquad (A.26)$$

This Green function is called the *Feynman Green function* or the Feynman propagator.

Rather than viewing the $i\varepsilon$ prescription in (A.26) as originating from the analytic continuation in $\tilde{\omega}$ we can view it as an analytic continuation in the complex p-plane. In this way the Green function $G_F(p)$ is seen as (minus[1]) the analytic continuation of the Fourier transformed $G_E(p)$ of (A.20), where

$$G_E(p_E) = \frac{1}{p_E^2 + \omega^2} \qquad (A.27)$$

under a rotation $\pi/2 - \varepsilon$ of in the complex p_E-plane, as show in the Fig. A.2. Explicitly we have, defining $p_E(\theta) = e^{i\theta} p_E$, $\theta \in [0, \pi/2]$ and noting that $p_E(\theta) \to ip(1-i\varepsilon)$ for $\theta \to \pi/2 - \varepsilon$, the follow behavior under the rotation

$$\frac{1}{p_E^2 + \omega^2} \to \frac{1}{(p_E(\theta))^2 + \omega^2} \xrightarrow{\theta = \pi/2-\varepsilon} \frac{1}{-p^2 - i\varepsilon + \omega^2}. \qquad (A.28)$$

---

[1] The minus sign relating the two Fourier transformations is a triviality related to the definitions (A.14) and (A.15). Had we in case (b) used two + signs instead of the two - signs, this minus sign would be absent. And it is not unnatural to use the opposite sign in case (b). It comes if we think of case (b) as arising from case (a) by an analytic continuation $t \to it$ rather than $\tilde{\omega} \to i\omega$.

Thus we can obtain the Feynman Green function (A.26) from (A.27) by a rotation of $\pi/2-\varepsilon$ in the complex p_E-plane without encountering any singularities.. We are spelling this out in detail because it turns out to be a general principle in quantum field theory that one can obtain the so-called time-ordered Green functions (the Feynman Green functions) from the Euclidean Green functions by such an analytic continuation without encountering any singularities. This will be important for us since we will be working in spacetimes with Euclidean signatures and we will calculate the Green functions there. Thus, should we want to rotate back to spacetimes with Lorentzian signature, we expect to obtain the Feynman Green functions, not the standard retarded Green functions we will discuss below.

We can now write down the general solution to the harmonic oscillator problem (b) with an *external* force $J(t)$:

$$\psi(t) = \int_{-\infty}^{\infty} dt'\, G_F(t-t') J(t') \tag{A.29}$$

This is *not* the solution we will usually consider in a classical problem where $\psi(t) = x(t)$ (the coordinate of the particle). In this case we are interested in a *causal* Green function, rather than G_F. The response to $J(t)$ should not influence $x(t)$ at earlier times $t' < t$, i.e.

$$G_R(t,t') = 0 \quad \text{for} \quad t < t' \tag{A.30}$$

We denote this Green function *the retarded Green function*.

$$G_R(p) = \int_{-\infty}^{\infty} dt\, e^{-ipt} G_R(t) = \int_0^{\infty} dt\, e^{-ipt} G_R(t) \tag{A.31}$$

$G_R(p)$ is analytic in the lower complex p-plane if $G_R(t) = 0$ for $t < 0$.

Exercise 4: Show this, assuming $G_R(t)$ is bounded for $t \in \mathbb{R}$.

The retarded Green function is given by

$$G_R(t) = -\theta(t) \frac{\sin \omega t}{\omega}, \quad \theta(t) := \begin{cases} 0 & \text{for } t < 0 \\ 1/2 & \text{for } t = 0 \\ 1 & \text{for } t > 0 \end{cases} \tag{A.32}$$

Exercise 5: Show that $G_R(t)$ satisfies (A.2) with $V(t) = -\omega^2$.

Since $G_R(t)$ and $G_F(t)$ satisfy the same inhomogenous second-order differential equation, the *difference* is a solution of the homogenous equation (A.6):

$$G_F(t) = G_R(t) + \frac{e^{i\omega t}}{2i\omega} \tag{A.33}$$

In the same way we can define *the advanced Green function* $G_A(t)$ such that $G_A(t) = 0$ for $t > 0$

$$G_A(t) = \theta(-t) \frac{\sin \omega t}{\omega} \tag{A.34}$$

If we denote $e^{-i\omega t}$ as *positive frequency oscillations* and $e^{i\omega t}$ as *negative frequency oscillation* we can say that $G_F(t)$ propagate *positive frequencies forward in time* and *negative frequencies backwards in time*. $G_R(t)$ and $G_A(t)$ do not allow a split, but denote by $G_R^{(+)}(t)$ the positive frequency part, $G_A^{(-)}(t)$ the negative frequency part of these functions. then,

$$G_F(t) = G_R^+(t) + G_A^-(t) = \frac{1}{2i\omega}\left[\theta(t)e^{-i\omega t} + \theta(-t)e^{i\omega t}\right]. \qquad (A.35)$$

Is this discussion relevant for higher dimensions? Yes!

Example: Relativistic Free Massive Scalar Particle:

$$\left[-\frac{\partial^2}{\partial t^2} + \frac{\partial^2}{\partial x_i^2} - m^2\right]\phi(x_i,t) = 0 \qquad i = 1,\ldots,n \qquad (A.36)$$

Translational invariance invites to a Fourier transformation:

$$\phi(x_i,t) = \int \frac{d^3k}{(2\pi)^3}\, e^{ik_i x_i}\, \phi(k_i,t) \qquad (A.37)$$

$$\left[-\frac{d^2}{dt^2} - \omega_k^2\right]\phi(k,t) = 0 \qquad \begin{array}{rcl}\omega_k^2 &=& k_i^2 + m^2 \\ \omega_k &=& \sqrt{k_i^2 + m^2}\end{array} \qquad (A.38)$$

Eq. (A.38) is identical to the harmonic oscillator problem already considered, i.e.

$$G_F(t,k_i) = \frac{1}{2i\omega_k}\left[\theta(t)e^{-i\omega_k t} + \theta(-t)e^{i\omega_k t}\right] \qquad (A.39)$$

$$\begin{aligned}G_F(k_0,k_i) &= \int dt\, e^{ik_0 t} G_F(t,k_i) \\ &= \frac{1}{k_0^2 - \omega_k^2 - m^2 + i\varepsilon} = \frac{-1}{k^\mu k_\mu + m^2 - i\varepsilon}.\end{aligned} \qquad (A.40)$$

$$G_R(t,k_i) = -\theta(t)\frac{\sin\omega_k t}{\omega_k}, \quad G_A(t,k_i) = \theta(-t)\frac{\sin\omega_k t}{\omega_k}. \qquad (A.41)$$

SOME OF THE HIGHER DIMENSIONAL GREEN FUNCTIONS:

1. Electrostatics:

$$\nabla \cdot E = \rho, \quad E = -\nabla \phi \quad \Rightarrow \quad -\Delta \phi = \rho \quad \text{(Poisson's eq.)} \tag{A.42}$$

The Green function of this problem satisfies:

$$\left(-\frac{\partial^2}{\partial x_i^2}\right) G(x,y) = \delta^n(x-y), \tag{A.43}$$

where n is the dimension of space and we sum over index $i = 1,\ldots,n$. If $n > 2$ we expect $\phi(x)$ to fall of at infinity for $\rho(x)$ having compact support, i.e. we have the "Euclidean" boundary condition discussed in case (a). In particular, $G(x,y) \to 0$ for $|x-y| \to \infty$, since $G(x,y)$ has the interpretation as $\phi(x)$ for a δ-function source located in y. This is opposite to the expectations for a wave equation. We can solve (A.43) by Fourier transformation:

$$G(x-y) = \int \frac{d^n p}{(2\pi)^n} e^{ip(x-y)} G(p) \qquad \delta^n(x-y) = \int \frac{d^n p}{(2\pi)^n} e^{ip(x-y)}$$

and (A.43) can be written as

$$p^2 G(p) = 1$$

We conclude that the Green function in p space is

$$G(p) = \frac{1}{p^2}, \tag{A.44}$$

and by Fourier transformation we find for $n > 2$

$$G(x) = \int \frac{d^n p}{(2\pi)^n} \frac{e^{ipx}}{p^2} = \frac{\Gamma(\frac{n}{2}-1)}{4\pi^{\frac{n}{2}} |x|^{n-2}}. \tag{A.45}$$

Exercise 6: Show (A.45) by using $\frac{1}{p^2} = \int_0^\infty d\alpha\, e^{-\alpha p^2}$ and interchanging the α and the p integrations (not really allowed, but .....).

From (A.45) we see that $G(x) \to 0$ for $|x| \to \infty$ as desired. For $n = 3$ we get:

$$G(x,y) = \frac{1}{4\pi} \frac{1}{|x-y|} \tag{A.46}$$

$$\phi(x) = \frac{1}{4\pi} \int d^3 y \frac{\rho(y)}{|x-y|} \quad \text{(Coulombs Law!)} \tag{A.47}$$

2. Retarded wave functions in 3d:

$$\left(-\frac{\partial^2}{\partial t^2} + \frac{\partial}{\partial x_i^2}\right)\phi(x_i,t) = -\rho(x_i,t)$$

From (A.47) we can guess the solution since we "know" that disturbances caused by ρ travel with velocity 1 (= c):

$$\phi(x,t) = \frac{1}{4\pi}\int d^3y \frac{\rho(y,t-|x-y|)}{|x-y|} \tag{A.48}$$

Let us derive this from the retarded Green function (A.41):

$$G_R(k,t) = -\theta(t)\frac{\sin|k|t}{|k|}, \quad \omega_k = |k| \text{ for } m = 0 \tag{A.49}$$

$$\begin{aligned}
G_R(x,t) &= -\int \frac{d^3k}{(2\pi)^3} \theta(t)\frac{\sin|k|t}{|k|} e^{ik_i x_i} \\
&= -\theta(t)\frac{1}{4\pi^2}\int_0^\infty d|k||k|^2 \frac{\sin|k|t}{|k|}\int_0^\pi d\theta \sin\theta\, e^{i|k||x|\cos\theta} \\
&= -\frac{\theta(t)}{2\pi^2}\frac{1}{|x|}\int_0^\infty d|k| \sin|k|t \sin|k||x| = -\frac{\theta(t)}{4\pi}\frac{1}{|x|}\delta(t-|x|)
\end{aligned}$$

Conclusion:

$$G_R(x,t) = -\frac{\theta(t)}{4\pi|x|}\delta(t-|x|) = -\frac{\theta(x^0)}{2\pi}\delta(x_\mu x^\mu)$$

and using the retarded Green function we can find any wave propagation caused by the source $\rho(x,t)$':

$$\begin{aligned}
\phi(x,t) &= \int dt' d^3x'\, G_R(x-x',t-t')\,\rho(x',t') \\
&= \frac{1}{4\pi}\int d^3y \frac{\rho(y,t-|x-y|)}{|x-y|}
\end{aligned} \tag{A.50}$$

3. Classical electrodynamics

Maxwell's equations can be written as

$$\partial^\mu F_{\mu\nu} = -j_\nu; \quad F_{\mu\nu} = \partial_\mu A_\nu - \partial_\nu A_\mu, \tag{A.51}$$

or (imposing the Lorentz gauge condition $\partial^\nu A_\nu = 0$ on A_ν):

$$\left(-\frac{\partial^2}{\partial t^2} + \frac{\partial^2}{\partial x_i^2}\right)A_\nu = -j_\nu, \quad \partial^\nu A_\nu = 0 \tag{A.52}$$

The electromagnetic field triggered by a current distribution $j_V(x,t)$ is then given by

$$A_V(x,t) = \int d^4y \, G_R(x^\mu - y^\mu) \, j_V(y^\lambda) \tag{A.53}$$

Exercise 7: Why is $\partial^V A_V = 0$ if A_V is given by (A.53)? (what is the physical requirement for $j^V(x,t)$?)

4. Heat conduction or diffusion

The equation for heat conduction (or diffusion) in a medium with a source $J(x,t)$ is

$$\frac{\partial \varepsilon(x,t)}{\partial t} = b^2 \nabla^2 \varepsilon(x,t) + J(x,t), \quad x_i = 1, \ldots, n \tag{A.54}$$

The Green function satisfies

$$\left(\frac{\partial}{\partial t} - b^2 \nabla^2 \right) G(x,t;y,t') = \delta^n(x-y)\delta(t-t') \tag{A.55}$$

and the boundary conditions are: causal propagation and the requirement that $G(x,t) \to 0$ for $|x| \to \infty$.

Again we find the Green function by Fourier transformation:

$$G_R(k,t) = \int d^n x \, e^{-ik_i x^i} G_R(x,t). \tag{A.56}$$

If we use this in eq. (A.55), we get

$$\left(\frac{\partial}{\partial t} + b^2 k_i^2 \right) G(t,k) = \delta(t) \tag{A.57}$$

which has the solution

$$G(k,t) = \theta(t) \, e^{-b^2 k_i^2 t} \tag{A.58}$$

Exercise 8: Show (A.58).

We can now transform back to x variables

$$G(x,t) = \int \frac{d^n k}{(2\pi)^n} e^{ik_i x^i} G(k,t) = \frac{\theta(t)}{(4\pi b^2 t)^{n/2}} e^{-\frac{|x|^2}{4b^2 t}} \tag{A.59}$$

Exercise 9: Show (A.59).

Note

$$G(x, 0^+) = \delta^n(x). \tag{A.60}$$

We can now solve for the heat conduction from a source $J(x,t)$:

$$\varepsilon(x,t) = \int d^n x' dt' \, G_R(x-x', t-t') J(x',t'). \tag{A.61}$$

5. The Schrodinger equation

Note that $t_{heat} \to it_{schr}$ turns the heat equation into the Schrödinger equation:

$$i\frac{\partial}{\partial t}\psi = -\frac{\hbar}{2m}\nabla^2\psi, \qquad b = \sqrt{\frac{\hbar}{2m}} \qquad (A.62)$$

The Green function and the solutions are obtained by the same analytic continuation of (A.58)–(A.61). In particular, we find

$$\psi(x,t) = \left(\frac{m}{2\pi i\hbar t}\right)^{\frac{n}{2}} \exp\left(\frac{i|x-x_0|^2 m}{2\hbar t}\right). \qquad (A.63)$$

$\psi(x,t)$ is the solution for $t > 0$ to (A.62) such that

$$\psi(x,0^+) = \delta^n(x-x_0)$$

(see (A.60)), i.e. $\psi(x,0^+) = \langle x|x_0\rangle$ and has the quantum mechanical interpretation as the eigenvector of the operator $\hat{x}$ corresponding to the eigenvalue x_0 and $\psi(x,t)$, $t \geq 0$, is the wave function of the free particle located at x_0 at time $t = 0$, i.e.

$$\psi(x,t) \equiv \langle x|e^{-it\hat{H}/\hbar}|x_0\rangle = \left(\frac{m}{2\pi i\hbar t}\right)^{\frac{n}{2}} \exp\left(\frac{i|x-x_0|^2 m}{2\hbar t}\right). \qquad (A.64)$$

SOLUTIONS TO PRELIMINARY MATERIAL 2: GREEN FUNCTIONS

Exercise 1
Recall the following:

$$\frac{d}{dx}|x| = 2\theta(x) - 1, \qquad \frac{d}{dx}\theta(x) = \delta(x). \tag{A.65}$$

where $\theta(x)$ is the function which is 1 for $x > 0$ and 0 for $x < 0$. Thus, differentiating a continuous function $f(x)$ where $f'(x_0^-) = a$ and $f'(x_0^+) = b$ one obtains

$$\frac{d^2}{dx^2}f(x) = (b-a)\delta(x-x_0) + f''(x) \tag{A.66}$$

where $f'(x)$ and $f''(x)$ denote the first and second derivative of f for $x \neq x_0$.

Assume now that ϕ_1 and ϕ_2 both satisfy (A.6). Then we find that the Wronskian $w(\phi_1, \phi_2)$ is constant:

$$\frac{d}{dt}w = \phi_1\phi_2'' - \phi_1''\phi_2 = \phi_1 V(t)\phi_2 - V(t)\phi_1\phi_2 = 0$$

In (A.11) we can then consider w as a constant. $G(t,s)$ is a continuous function of t, but it is not differentiable in $t=s$. For $t \neq s$ it satisfies (A.2) since both $\phi_1(t)$ and $\phi_2(t)$ satisfy (A.6). Thanks to (A.66) the second derivative of $G(t,s)$ wrt t for $t = s$ gives a contribution

$$\frac{-1}{w}\left[\phi_2'(s_+)\phi_1(s) - \phi_1'(s_-)\phi_2(s)\right]\delta(t-s) = -\delta(t-s)$$

Thus $G(t,s)$ satisfies (A.2).

Let us now turn to (A.12). Acting with $D(t)$ on $G(t,s)$ we obtain from (A.9), assuming that we can commute differentiation and summation (we will not deal with such subtleties....) that

$$D(t)G(t,s) = \sum_n \frac{\psi_n^*(s)D(t)\psi_n(t)}{\lambda_n} = \sum_n \psi_n^*(s)\psi_n(t) = \delta(t-s),$$

where we have used that the eigenfunctions $\psi_n(t)$ are assumed to be a complete set of normalized basis vectors in $L^2[t_i, t_f]$. Such a choice can be made since $\hat{D}$ is an Hermitian operator.

Finally, let us look at the solution (A.3) and use the representation (A.11) for $G(t,s)$. We have

$$\psi(t_i) = \phi_1(t_i)\int ds\, \phi_2(s)J(s), \qquad \psi(t_f) = \phi_2(t_f)\int ds\, \phi_1(s)J(s),$$

and it is then clear that $\psi(t)$ satisfies the boundary conditions since $\phi_1(t)$ and $\phi_2(t)$ were chosen to satisfy the boundary conditions at t_i and t_f, respectively.

Exercise 2

In its simplest form the residue theorem states that

$$\oint_C \frac{dz}{2\pi i} \frac{f(z)}{z - z_0} = f(z_0) \tag{A.67}$$

where C is a simple closed curve, oriented anti-clockwise, enclosing z_0, and $f(z)$ is a holomorphic function in an open simple connected region of the complex plane containing C. In the case where $t-s > 0$ we apply the theorem as shown in Fig. A.1, writing

$$\frac{e^{iz(t-s)}}{z^2 + \tilde{\omega}^2} = \frac{f(z)}{z - i\tilde{\omega}}, \qquad f(z) = \frac{e^{iz(t-s)}}{z + i\tilde{\omega}},$$

and using that the part of line-integral in the upper half-plane will vanish when the curve is at infinity and $t-s > 0$ since the function will then vanish exponentially there because $\text{im}\, z > 0$. If $t-s < 0$ we use a contour integral as in Fig. A.1, only with the arc in the lower complex plane, such that $\text{im}\, z < 0$ on the arc.

Exercise 3

We choose the same contours as in Exercise 2, only are the poles now located at $p = \pm(\omega - i\varepsilon)$. For $t-s > 0$ we have to choose the upper half-plane contour and the pole enclosed by the contour is located at $z_0 = -\omega + i\varepsilon$, i.e. we have

$$\frac{e^{iz(t-s)}}{z^2 - (\omega - i\varepsilon)^2} = \frac{f(z)}{z + \omega - i\varepsilon}, \qquad f(z) = \frac{e^{iz(t-s)}}{z - \omega + i\varepsilon}.$$

The contour integral leads to

$$\oint \frac{dz}{2\pi i} \frac{e^{iz(t-s)}}{z^2 - (\omega - i\varepsilon)^2} = \frac{e^{-i\omega|t-s|}}{-2\omega} = \frac{1}{i} G(t,s).$$

The calculation for $t-s < 0$ and the choice of contour in the lower half-plane leads to the same result.

Exercise 4

Let us assume $G_R(t) = 0$ for $t < 0$ and bounded for $t \geq 0$. Eq. (A.41) then shows that $G_R(p_R + ip_I)$ is well-defined and holomorphic for $p_I < 0$. $G_R(p_R + ip_I)$ is holomorphic since

$$i \frac{\partial G_R(p)}{\partial p_R} = \frac{\partial G_R(p)}{\partial p_I} \quad \left(= \int_0^\infty G_R(t)\, t\, e^{-i(p_R + ip_I)t} \right)$$

Exercise 5

Recall (A.65) which tells us that differentiating $\theta(x)$ we obtain $\delta(x)$. We need yet another rule:

$$\delta(x) f(x) = \delta(x) f(0), \quad \text{i.e.} \quad \delta(x) f(x) = 0 \quad \text{if} \quad f(0) = 0. \tag{A.68}$$

We now use this rule differentiating $G_R(t)$:

$$-\frac{dG_R(t)}{dt} = \theta(t)\cos\omega t + \delta(t)\frac{\sin\omega t}{\omega} = \theta(t)\cos\omega t.$$

$$-\frac{d^2 G_R(t)}{dt^2} = -\theta(t)\omega\sin\omega t + \delta(t)\cos\omega t = \omega^2 G_R(t) + \delta(t)$$

$$\left(-\frac{d^2}{dt^2} - \omega^2\right) G_R(t) = \delta(t).$$

Exercise 6

Let us use that we have the one-dimensional integrals

$$\int \frac{dp}{2\pi} e^{-\alpha p^2} = \frac{1}{2\sqrt{\alpha\pi}}, \qquad \int \frac{dp}{2\pi} e^{-\alpha p^2 - ixp} = \frac{e^{-x^2/4\alpha}}{2\sqrt{\alpha\pi}}, \tag{A.69}$$

The first integral we will calculate in a separate exercise dealing with Gaussian integrals. The second integral follows from the first by completing the square: $\alpha(p^2 + ipx) = \alpha(p + ix/2\alpha)^2 + x^2/4\alpha$. The p-integral can now be performed since the integration along the horizontal line with imaginary coordinate $ix/2\alpha$ gives the same result as integrating along the real axis with $x = 0$, again because of the residue theorem (no poles between the two horizontal lines).

From (A.69) we obtain, simply by choosing the p_1-axis parallel to the vector x:

$$\int \frac{d^n p}{(2\pi)^n} e^{-\alpha p^2 - ix\cdot p} = \frac{e^{-x^2/4\alpha}}{2^n (\alpha\pi)^{n/2}}, \tag{A.70}$$

since we have $n-1$ integrals like the left hand integral in (A.69) and one integral (the p_1 integral) like the right hand side of (A.69).

We can now write, assuming that we can change the order of integration:

$$\int \frac{d^n p}{(2\pi)^n} \frac{e^{-ipx}}{p^2} = \int_0^\infty d\alpha \int \frac{d^n p}{(2\pi)^n} e^{-\alpha p^2 - ix\cdot p} = \int_0^\infty d\alpha \frac{e^{-x^2/4\alpha}}{2^n (\pi\alpha)^{n/2}} = \int_0^\infty d\beta \frac{\beta^{n/2-2} e^{-\beta}}{4\pi^{n/2} |x|^{n-2}},$$

where we in the last integral has made the substitution $|x|^2/4\alpha = \beta$. The wanted result now follows from the following formula for the Γ-function:

$$\Gamma(x) = \int_0^\infty d\beta\, \beta^{x-1} e^{-\beta}. \tag{A.71}$$

Exercise 7

A physical current will satisfy $\partial^\mu j_\mu(x) = 0$. This is just the continuity equation. Let now $A_\mu(x)$ be given by (A.53). We then have

$$\frac{\partial}{\partial x_\mu} A_\mu(x) = \int d^4 x' \frac{\partial}{\partial x_\mu} G_R(x-x') j_\mu(x') = \int d^4 x' \left(-\frac{\partial}{\partial x'_\mu}\right) G_R(x-x') j_\mu(x')$$

$$= \int d^4 x'\, G_R(x-x') \frac{\partial}{\partial x'_\mu} j_\mu(x') = 0,$$

Preliminary Material Part 2: Green Functions

where we have assumed that boundary terms vanish at spatial or temporal infinity when we perform the partial integration to go from line one to line two in the equations.

Exercise 8

From the rules (A.65) and (A.68) we obtain

$$\frac{\partial}{\partial t}G(k,t) = -b^2k^2\theta(t)e^{-b^2k^2t} + \delta(t)e^{-b^2k^2t} = -b^2k^2G(k,t) + \delta(t).$$

Exercise 9

The calculation is identical to the one done in eq. (A.70).

B Problem Sets 1–13

B.1 PROBLEM SET 1

In this Problem Set, we treat the free non-relativistic particle in the framework of the path integral.

Gaussian integrals

Prove the following identities (assume the integration range to be $(-\infty, \infty)$ unless specified otherwise):

1. $$\int e^{-(x^2+y^2)} dxdy = \pi.$$

2. $$\int e^{-\frac{1}{2}ax^2} dx = \sqrt{\frac{2\pi}{a}}.$$

3. Let A_{ij} be a real symmetric $n \times n$ matrix with positive eigenvalues.
$$\int d^n x\, e^{-\frac{1}{2}x_i A_{ij} x_j} = \frac{(2\pi)^{\frac{n}{2}}}{\sqrt{\det A}}.$$

   Note that summation over repeated indices is implied all throughout the Problem Set.

4. Let $z_i \in \mathbb{C} = x_i + iy_i$ and let A be a Hermitian matrix with positive eigenvalues.
$$\int \prod_{i,j=1}^n dx_i dy_j\, e^{-\frac{1}{2}z^\dagger A z} = \int \prod_{i,j=1}^n dx_i dy_j\, e^{-\frac{1}{2}z_k^* A_{kl} z_l} = \frac{(2\pi)^n}{\det A}.$$

   Here
$$z = \begin{pmatrix} z_1 \\ \vdots \\ z_n \end{pmatrix}, \quad z^\dagger = (z_1^*, \cdots, z_n^*).$$

5. Let x, b be real n-vectors, and A a real symmetric matrix with positive eigenvalues. Furthermore, let
$$S(x) = \frac{1}{2} x^T A x + b^T x = \frac{1}{2} x_i A_{ij} x_j + b_i x_i.$$

   We now define the so-called "classical solution" x_c, for which
$$\frac{\partial}{\partial x_i} S(x) = 0, \quad i = 1, \cdots, n.$$

Show that
$$S(x_c) = -\frac{1}{2}b^T A^{-1} b, \quad x_c = -A^{-1} b.$$
Then write $x = x_c + \Delta x$, where Δx denotes the "fluctuations" around the classical solution. Then show that
$$S(x) = S(x_c) + S(\Delta x, b = 0).$$
Subsequently, show that
$$\int d^n x \, e^{-S(x)} = e^{-S(x_c)} \int d^n(\Delta x) \, e^{-S(\Delta x, b=0)} = e^{-S(x_c)} \frac{(2\pi)^{\frac{n}{2}}}{\sqrt{\det A}},$$
and thus
$$\int d^n x \, e^{-(\frac{1}{2}x^T A x + b^T x)} = \frac{e^{\frac{1}{2}b^T A^{-1} b} (2\pi)^{\frac{n}{2}}}{\sqrt{\det A}}.$$
This result is still valid if we make the substitutions $A \to iA$, $b \to ib$. We then encounter the Fresnel integrals
$$\int dx \, e^{-\frac{iax^2}{2}} = \sqrt{\frac{2\pi}{ia}}, \quad \int d^n x \, e^{-i(\frac{1}{2}x^T A x + b^T x)} = \frac{e^{\frac{i}{2}b^T A^{-1} b}(2\pi)^{\frac{n}{2}}}{\sqrt{i^n \det A}}.$$

6. The Fourier transformed of a function $f(x)$ is defined by
$$\mathscr{F}(f)(k) := \int d^n x \, f(x) \, e^{-ik^T x} \quad \text{and then} \quad f(x) = \int \frac{d^n k}{(2\pi)^n} e^{ix^T k} \mathscr{F}(f)(k).$$
Show that the Fourier transformation of a Gaussian function is still a Gaussian:
$$\mathscr{F}(e^{-\frac{1}{2}x^T A x}) = \frac{(2\pi)^{\frac{n}{2}}}{\sqrt{\det A}} e^{-\frac{1}{2}k^T A^{-1} k}.$$

7. The convolution $(f*g)(x)$ of two functions $f(x)$ and $g(x)$ is defined as
$$(f*g)(x) = \int dy \, f(x-y) g(y) = \int dy \, g(x-y) f(y).$$
It is a property of convolutions that the Fourier transformed of a convolution is the product of the Fourier transformed of the functions:
$$\mathscr{F}(f*g)(k) = \mathscr{F}(f)(k) \cdot \mathscr{F}(g)(k).$$
Let $f_A(x)$ denote the Gaussian function
$$f_A(x) = \frac{1}{\sqrt{\det(2\pi A)}} e^{-\frac{1}{2}x^T A^{-1} x} \quad \text{i.e.} \quad \mathscr{F}(f_A)(k) = e^{-\frac{1}{2}k^T A k}.$$
Use this to show that
$$(f_A * f_B)(x) = f_{A+B}(x), \quad f_A^{*N}(x) = f_{NA}(x), \quad f^{*N} = f * f * \cdots * f \quad N \text{ times}.$$

The free non-relativistic particle

In the lectures, we saw that

$$\langle x_{n+1}|\hat{O}_\varepsilon^{n+1}|x_0\rangle = \left(\frac{m}{2\pi i\varepsilon\hbar}\right)^{\frac{n+1}{2}} \int \prod_{i=1}^{n} dx_i \, e^{\frac{i}{\hbar}\sum_{i=0}^{n}\varepsilon\left[\frac{m}{2}\left(\frac{x_{i+1}-x_i}{\varepsilon}\right)^2 - V(x_i)\right]}.$$

For the free particle, we have $V(x_i) = 0$. In that case, we can perform the integrals for all the x_i successively, starting with x_1 since they are all Gaussian integrals. In addition they are convolutions. We can thus use what we have just learned about convolutions of Gaussians.

8. Show, using the convolution of Gaussians mentioned above, that

$$\int dx' \frac{1}{\sqrt{2\pi i a}} e^{\frac{i}{2a}(x''-x')^2} \frac{1}{\sqrt{2\pi i b}} e^{\frac{i}{2b}(x'-x)^2} = \frac{1}{\sqrt{2\pi i(a+b)}} e^{\frac{i}{2(a+b)}(x''-x)^2}$$

and use this result to prove, by successive convolutions, that for $V(x) = 0$ we have

$$\langle x_{n+1}|\hat{O}_\varepsilon^{n+1}|x_0\rangle = \sqrt{\frac{m}{2\pi i\hbar t}} \, e^{\frac{i}{\hbar}\frac{m}{2}\frac{(x_b-x_a)^2}{t}} = \Psi(x_b-x_a,t),$$

where

$$t = (n+1)\varepsilon, \qquad x_b = x_{n+1}, \qquad x_a = x_0.$$

Why is this result independent of n?
$\Psi(x,t) \equiv \langle x|\Psi(t)\rangle$ is the wave function of the free particle.

9. Show that the Fourier transformed of $\Psi(x,t)$ wrt x is

$$\mathcal{F}(\Psi)(k) = e^{-i\frac{k^2\hbar}{2m}t} = e^{-\frac{i}{\hbar}\frac{p^2}{2m}t} = \tilde{\Psi}(p,t), \quad p = \hbar k.$$

Here $\tilde{\Psi}(p,t) \propto \langle p|\Psi(t)\rangle$ is of course the solution to the Schrödinger equation for the free particle in momentum basis:

$$i\hbar\frac{\partial}{\partial t}\tilde{\Psi}(p,t) = \frac{p^2}{2m}\tilde{\Psi}(p,t)$$

B.2 PROBLEM SET 2

In this Problem Set, we treat the next-simplest case of the path integral: the harmonic oscillator. The action is written

$$S[x] = \int_{t_a}^{t_b} dt\, L(x(t)), \qquad L(x(t)) = \frac{m}{2}\left(\dot{x}^2(t) - \omega^2 x^2(t)\right).$$

1. Show that the classical solution to the eom with boundary conditions

$$x(t_a) = x_a \quad \text{and} \quad x(t_b) = x_b$$

takes the form

$$\begin{aligned} x_c(t) &= \frac{x_b \sin\omega(t-t_a) + x_a \sin\omega(t_b-t)}{\sin\omega(t_b-t_a)}, \\ S[x_c(t)] &= \frac{m\omega}{2\sin\omega(t_b-t_a)}\left[(x_a^2+x_b^2)\cos\omega(t_b-t_a) - 2x_a x_b\right] \end{aligned} \qquad (\text{B.2.1})$$

2. Make the decomposition

$$x(t) = x_c(t) + \Delta x(t), \qquad \Delta x(t_a) = \Delta x(t_b) = 0.$$

and show that one has

$$S[x(t)] = S[x_c(t)] + S[\Delta x(t)] \qquad (\text{B.2.2})$$

Now recall that the propagation amplitude can be expressed in terms of the path integral as

$$\langle x_b | e^{-i(t_b-t_a)\hat{H}} | x_a \rangle = \int_{\substack{x(t_a)=x_a \\ x(t_b)=x_b}} \mathcal{D}x(t)\, e^{\frac{i}{\hbar}S[x(t)]}.$$

3. Show that

$$\int_{\substack{x(t_a)=x_a \\ x(t_b)=x_b}} \mathcal{D}x(t)\, e^{\frac{i}{\hbar}S[x(t)]} = e^{\frac{i}{\hbar}S[x_c(t)]} \int_{\substack{\Delta x(t_a)=0 \\ \Delta x(t_b)=0}} \mathcal{D}\Delta x(t)\, e^{\frac{i}{\hbar}S[\Delta x(t)]}$$

Set $y(t) = \Delta x(t)$. We now have to compute

$$\int_{\substack{y(t_a)=0 \\ y(t_b)=0}} \mathcal{D}y(t)\, e^{\frac{i}{\hbar}S[y(t)]} = \lim_{\varepsilon \to 0} \langle y(t_b)=0 | \hat{O}^{n+1} | y(t_a)=0 \rangle, \qquad t_b - t_a = (n+1)\varepsilon.$$

As shown in the Chapter 1, this corresponds to the $n \to \infty, \varepsilon \to 0$ limit of the integral

$$\left(\frac{m}{2\pi i \varepsilon \hbar}\right)^{\frac{n+1}{2}} \int \prod_{i=1}^{n} dy_i\, \exp\left[\frac{i}{\hbar}\sum_{i=0}^{n}\varepsilon\left(\left(\frac{y_{i+1}-y_i}{\varepsilon}\right)^2 - \omega^2 y_i^2\right)\right], \qquad y_{n+1} = y_0 = 0.$$

We can write the exponent in terms of a matrix product:

$$\exp\left[\frac{i}{\hbar}\sum_{i=0}^{n}\varepsilon\left(\left(\frac{y_{i+1}-y_i}{\varepsilon}\right)^2 - \omega^2 y_i^2\right)\right] = \exp\left[\frac{i}{\hbar\varepsilon}y^T A_{n\times n} y\right].$$

4. Show that the matrix $A_{n\times n}$ can chosen as the following symmetric matrix

$$A_{n\times n} = \begin{pmatrix} 2-\omega^2\varepsilon^2 & -1 & 0 & \cdots & 0 \\ -1 & 2-\omega^2\varepsilon^2 & -1 & & 0 \\ 0 & -1 & \ddots & \ddots & \vdots \\ \vdots & & \ddots & & -1 \\ 0 & 0 & \cdots & -1 & 2-\omega^2\varepsilon^2 \end{pmatrix}$$

5. Let $D_n \equiv \det A_{n\times n}$. Prove that

$$D_n = (2-\varepsilon^2\omega^2)D_{n-1} - D_{n-2}, \qquad D_0 = 1, \ D_{-1} = 0 \qquad \text{(B.2.3)}$$

Generating functions

We will use the method of generating functions to solve this recursion relation for the determinant. Let a_n be a sequence of numbers. We call

$$f(x) = \sum_n a_n x^n$$

the *generating function* for that sequence. In combinatorics and probability theory, generating functions are very useful since a given problem is often formulated in a less restrictive way using the generating function, and consequently easier solved. Once $f(x)$ is known, we can recover the coefficients a_n straightforwardly as $a_n = \frac{1}{n!}\left(\frac{d}{dx}\right)^n f(x)\Big|_{x=0}$. We will use generating functions all the time!

For D_n we found the recursion relation (B.2.3). Define the generating function

$$D(x) = \sum_{n=0}^{\infty} D_n x^n.$$

6. Show that it satisfies the equation

$$D(x) = 1 + (2-\omega^2\varepsilon^2)xD(x) - x^2 D(x), \quad \text{i.e.} \quad D(x) = \frac{1}{1-(2-\omega^2\varepsilon^2)x+x^2}.$$

Now introduce a new variable $\tilde{\omega}$ such that $\omega\varepsilon/2 = \sin(\tilde{\omega}\varepsilon/2)$.

7. Show that, in terms of this new variable, we have

$$D(x) = \frac{1}{1-2x\cos(\tilde{\omega}\varepsilon)+x^2} = \frac{1}{e^{i\tilde{\omega}\varepsilon}-e^{-i\tilde{\omega}\varepsilon}}\left(\frac{1}{(e^{-i\tilde{\omega}\varepsilon}-x)} - \frac{1}{(e^{i\tilde{\omega}\varepsilon}-x)}\right)$$

8. Next, show that

$$D_n = \frac{\sin((n+1)\varepsilon\tilde{\omega})}{\sin\varepsilon\tilde{\omega}}$$

9. Finally, use this to show that

$$\int_{\substack{y(t_a)=0 \\ y(t_b)=0}} \mathcal{D}y(t)\, e^{\frac{i}{\hbar}S[y(t)]} = \lim_{\varepsilon \to 0} \sqrt{\frac{m}{2\pi i\varepsilon\hbar} \frac{\sin\tilde{\omega}\varepsilon}{\sin((n+1)\varepsilon\tilde{\omega})}}$$

$$= \sqrt{\frac{m\omega}{2\pi i\hbar \sin(\omega(t_b-t_a))}}$$

To summarize: $\boxed{\langle x_b| e^{-\frac{i}{\hbar}(t_b-t_a)\hat{H}} |x_a\rangle = \sqrt{\frac{m\omega}{2\pi i\hbar \sin\omega(t_b-t_a)}}\, e^{\frac{i}{\hbar}S[x_c]}}$

$$e^{\frac{i}{\hbar}S[x_c]} = e^{\frac{i}{\hbar}\frac{m\omega}{\sin\omega(t_b-t_a)}\left((x_b^2+x_a^2)\cos\omega(t_b-t_a)-2x_ax_b\right)}.$$

Now recall that

$$Z = \operatorname{tr} e^{-\beta\hat{H}} = \int dx\, \langle x| e^{-\beta\hat{H}} |x\rangle, \qquad \beta = \frac{1}{k_B T}.$$

Define

$$z(x) := \langle x| e^{-\beta\hat{H}} |x\rangle, \qquad \rho(x) := \frac{z(x)}{Z}$$

10. Show from the previously found result that

$$z(x) = \sqrt{\frac{m\omega}{2\pi\hbar\sin\hbar\beta\omega}}\, e^{-\frac{m\omega}{\hbar}\tanh\frac{\beta\hbar\omega}{2}x^2} \quad \text{and} \quad Z = \int dx\, z(x) = \frac{1}{2\sinh\frac{\beta\hbar\omega}{2}}.$$

11. Let $\psi_0(x)$ denote the ground state wave function of the harmonic oscillator. Show that

$$\rho(x) \to |\psi_0^2(x)|, \quad \text{for } T \to 0 \text{ (i.e. } \beta \to \infty)$$

and explain why we can obtain this result without any detailed calculation.

B.3 PROBLEM SET 3

In this Problem Set we discuss the lattice propagator represented as a random walk (RW) on the lattice.

The lattice propagator

The continuum Laplace operator in $\mathbb{R}^D$ takes the form

$$\Delta f(x) = \sum_{i=1}^{D} \frac{\partial^2}{\partial x_i^2} f(x).$$

Now we replace the continuum by a D-dimensional hypercubic lattice, $(a\mathbb{Z})^D$. The lattice sites are given by

$$x_i(\vec{n}) = a \cdot n_i, \quad \vec{n} = (n_1, \cdots, n_D).$$

Functions on the lattice are functions of the sites, i.e. $f(x_i(n))$. The lattice Laplacian then takes the form

$$(\Delta_L f)(x(n)) = \sum_{j=1}^{D} \frac{f(x(n) + e_j a) + f(x(n) - e_j a) - 2f(x(n))}{a^2}.$$

Here e_j is a unit vector in the $\hat{j}$ direction. We will first derive the lattice propagator in momentum space. The Fourier transform $\hat{F}(p)$ of a function $F(x)$ on the lattice is written

$$\hat{F}(p) = \sum_{x_n} a^D \cdot e^{ip \cdot x_n} F(x_n).$$

1. Prove that

$$\hat{F}(p_i) = \hat{F}\left(p_i + \frac{2\pi}{a}\right).$$

Thus, $\hat{F}$ is periodic with period $\frac{2\pi}{a}$. Therefore, we assume from now on that $p_i \in \left[-\frac{\pi}{a}, \frac{\pi}{a}\right]$. This is called the (first) *Brillouin zone*.

2. Now "prove" the *inversion formula* (use basics from Fourier series):

$$F(x_n) = \int_{-\frac{\pi}{a}}^{\frac{\pi}{a}} \frac{d^D p}{(2\pi)^D} e^{-ip x_n} \hat{F}(p).$$

In order to obtain the lattice propagator, we want to solve

$$(-\Delta_L(x_n) + m^2) G(x_n - x_m) = \delta(x_n - x_m). \tag{B.3.1}$$

The lattice delta function is defined as

$$\delta(x_n - x_m) \equiv \frac{1}{a^D} \delta_{nm} \quad \text{thus}: \quad \sum_{x_n} a^D \delta(x_n - x_m) f(x_n) = f(x_m),$$

where the sum is over all lattice sites. Note that this precisely corresponds to the "shifting" property of the continuum Dirac delta function.

3. Now transform equation (B.3.1) to momentum space, and show that it results in

$$\left[\sum_{i=1}^{D}\frac{2}{a^2}(1-\cos ap_i)+m^2\right]G(p)=1.$$

4. Show that this is equivalent to

$$G(p)=\frac{a^2}{4\sum_{i=1}^{D}\sin^2\frac{ap_i}{2}+m^2a^2}.$$

5. Subsequently show that in the limit $a\to 0$ this reduces to

$$G(p)=\frac{1}{p^2+m^2},$$

consistent with the analogous result in continuum field theory.

Calculation of the lattice propagator in x_n-space

We write Δ_L in matrix form $(\Delta_L)_{nm}$ as follows:

$$(\Delta_L f)(x_n)=\sum_m (\Delta_L)_{nm}f(x_m).$$

6. Show that

$$-(\Delta_L)_{nm}=a^{-2}\left[2D\,\delta_{nm}-Q_{nm}\right],$$

where $Q_{nm}=1$ if n,m are neighboring lattice sites, and $Q_{nm}=0$ otherwise.

7. Now show that $((-\Delta_L)+m^2)$ is invertible for $m^2>0$ (e.g. by using the Fourier transformed operator), and that $\left(-\Delta_L+m^2\right)^{-1}$ allows an expansion in the Neumann series

$$\frac{a^{2-D}}{2D+m^2a^2}\sum_{k=0}^{\infty}\left(\frac{Q}{2D+m^2a^2}\right)^k.$$

Next, show that

$$\left(-\Delta_L+m^2\right)^{-1}_{nm}=\frac{a^{2-D}}{2D+m^2a^2}\sum_{P(x_n,x_m)}(2D+m^2a^2)^{-L(P)/a},$$

where $P(x_n,x_m)$ is a lattice path from x_m to x_n and $L(P)$ is the length of this path. Such a lattice path follows the links of the lattice, and the path length is equal to its number of links times the link length a.

We now consider the path integral for the free relativistic particle and we use the classical action

$$S(P)=m_0 L(P).$$

We want to calculate
$$G(x,y) = \int \mathcal{D}P(x,y)\, e^{-S[P(x,y)]}$$
and we provide a regularization by restricting the paths to a hypercubic lattice:
$$G_a(x_n, x_m) = \sum_{P(x_n, x_m)} e^{-m_0(a) L[P(x_n, x_m)]}.$$

As a side note: by this definition, $G_a(x_n, x_m)$ is dimensionless, contrary to the "real" $G(x,y)$.

8. Show that we can choose $m_0(a)$ as a function of the lattice spacing a, such that
$$\frac{a^{2-D}}{2D + m^2 a^2} G_a(x_n, x_m) = (-\Delta_L + m^2)^{-1}_{nm}.$$
i.e. the (dimensionless) path integral Green function G_a goes to the continuum Green function G_{cont} as follows for $a \to 0$
$$\frac{a^{2-D}}{2D} G_a(x_m, x_n) \to G_{cont}(x_m, x_n)$$

9. Then compare $m_0(a)$ with the form of $m_0(a)$ for the free particle regularized by piecewise linear paths constructed from building blocks of length a as in Chapter 2:
$$m_0(a) = \frac{\log f(0)}{a} + c^2 a m^2 + O(a^3).$$
Here on the lattice we find
$$m_0(a) = \frac{\log 2D}{a} + \frac{1}{2D} a m^2 + O(a^3)$$

10. Give a simple interpretation of $\log 2D$ in terms of the number of paths on the lattice.

B.4 PROBLEM SET 4

This Problem Set aims to calculate critical exponents for the simplest ferromagnetic model of classical spins. The calculation is done in the so-called mean field approximation where we only consider small fluctuations around a dominant spin configuration, and we obtain in this approximate calculation what is called mean field exponents.

The Simplest Lattice Spin Model

Consider a hypercubic lattice in $\mathbb{R}^D$. We set the link length $a = 1$ (it can always straightforwardly be re-introduced later on if necessary). The Hamiltonian is written

$$H = -J \sum_{\langle ij \rangle} S_i S_j - h \sum_i S_i. \tag{B.4.1}$$

The sum over $\langle ij \rangle$ indicates that we sum over all pairs of lattice sites i, j (all links of the lattice). For a hypercubic lattice in D dimensions the number of links is D times the number of vertices. The external magnetic field strength is given by h and $S_i \in \mathbb{R}$ is the spin at lattice site i.

The partition function is given by

$$Z(\beta; h) = \int \prod_i (dS_i \rho(S_i, \beta)) \, e^{-\beta H(S_i)}. \tag{B.4.2}$$

The function $\rho(S, \beta)$ describes the spin properties of the individual lattice sites (or "atoms"). We will assume

$$\rho(S, \beta) = e^{-\kappa(\beta) S^2 - \lambda(\beta) S^4}, \qquad \kappa(\beta) > 0. \tag{B.4.3}$$

Here $\kappa(\beta)$ and $\lambda(\beta)$ are "material" constants that have only a weak β-dependence, i.e. $\kappa(\beta) = \kappa_0$, $\lambda(\beta) = \lambda_0$ as a first approximation (which we will use). We thus have an "effective" Hamiltonian

$$\beta H_{eff} = -\beta J \sum_{\langle ij \rangle} S_i S_j + \kappa_0 \sum_i S_i^2 + \lambda_0 \sum_i S_i^4 - \beta h \sum_i S_i. \tag{B.4.4}$$

In addition we have a "single atom" partition function

$$Z_{s.a.}(\beta, h) = \int dS \, e^{-\kappa_0 S^2 - \lambda_0 S^4 + \beta h S}. \tag{B.4.5}$$

For the expectation value of the spin we have for this "single atom" partition function

$$\langle S(h) \rangle_{s.a.} = \frac{1}{Z_{s.a.}(\beta, h)} \int dS \, S \, e^{-\kappa_0 S^2 - \lambda_0 S^4} e^{\beta h S} \tag{B.4.6}$$

Problem Sets 1–13

$$\langle S(h=0) \rangle_{s.a.} = 0.$$

We will now use $Z(\beta;h)$ in an approximation where we write

$$S_i = \langle S(h) \rangle + \delta S_i, \qquad \langle S_i \rangle = \langle S(h) \rangle \tag{B.4.7}$$

and assume that terms of order $(\delta S_i)^3$ can be ignored.

1. Show that in this approximation we can write

$$Z(\beta,h) = \left(\prod_{i=1}^{V} Z_{\mathrm{mf}}(\beta,h) \right) \times \int \left(\prod_i d(\delta S_i) \right) e^{-\beta H_F(\delta S_i)}. \tag{B.4.8}$$

In (B.4.8) V denotes the "volume" of $\mathbb{R}^D$, i.e. the number of lattice sites. Furthermore, Z_{mf} is the "mean field" partition function per "atom site" i:

$$Z_{\mathrm{mf}}(\beta,h) = e^{-\beta f_{\mathrm{mf}}(\beta,h)} \tag{B.4.9}$$

$$\beta f_{\mathrm{mf}}(\beta,h) = (\kappa_0 - DJ\beta)\langle S(h) \rangle^2 + \lambda_0 \langle S(h) \rangle^4 - \beta h \langle S(h) \rangle \tag{B.4.10}$$

$$\beta H_F(\delta S_i) = \frac{\beta J}{2} \sum_{\langle ij \rangle} (\delta S_i - \delta S_j)^2 + \left(\kappa_0 - DJ\beta + 6\lambda_0 \langle S(h) \rangle^2 \right) \sum_i (\delta S_i)^2 \tag{B.4.11}$$

Finally, $\langle S(h) \rangle$ is determined by the equation

$$2(\kappa_0 - DJ\beta)\langle S(h) \rangle + 4\lambda_0 \langle S(h) \rangle^3 = \beta h. \tag{B.4.12}$$

This is the condition that ensures that $\langle \delta S_i \rangle = 0$ when we only keep terms up to quadratic order in (δS_i). It is obtained by inserting $S_i = \langle S(h) \rangle$ in (B.4.4) and finding the minimum of $H_{eff}(S)$, $S = \langle S(h) \rangle$:

$$\beta H_{eff}(S) = V\beta f_{\mathrm{mf}}(\beta,h;S), \qquad \frac{d\beta f_{\mathrm{mf}}(\beta,h;S)}{dS} = 0, \tag{B.4.13}$$

which leads to (B.4.12).

2. Prove that

$$\frac{df_{\mathrm{mf}}}{dh} = -\langle S(h) \rangle$$

It is the standard result for the free energy density. Why is it called the "mean field approximation"? To see this, consider the partition function for this system with interactions turned off, i.e.

$$Z_{J=0}(\beta,h) = \prod_{i=1}^{V} Z_{s.a.}(\beta,h).$$

where

$$Z_{s.a.}(\beta,h) = e^{-\beta f_{\mathrm{free}}(h)} \tag{B.4.14}$$

$$\beta f_{\mathrm{free}}(h) = \kappa_0 S^2 + \lambda_0 S^4 - \beta h S \tag{B.4.15}$$

If we then ignore the fluctuations, i.e. put $\delta S = 0$, we see that the only difference between $Z(\beta,h,J)$ and $Z_{J=0}(\beta,h)$ is the shift

$$\kappa_0 \to \kappa_0 - D\beta J.$$

This can be understood in the following way. Let us assume $S_i = \langle S(h) \rangle$. Then we can write

$$H = -J\sum_{\langle ij \rangle} S_i S_j - h\sum_i S_i = -(h + DJ\langle S(h)\rangle)\sum_i S_i$$

Thus each spin feels not only the external field h, but also the local field from the neighbors, and we can formally write H as sum of single spins interacting with an effective magnetic field h_{eff}:

$$h \to h_{\text{eff}} = h + DJ\langle S(h)\rangle.$$

3. Check the consistency of this picture by showing that $f_{\text{free}}(h_{\text{eff}}, S)$ in eq. (B.4.15) agrees with $f_{\text{mf}}(h, \langle S(h)\rangle)$ in eq. (B.4.10) when we identify $S = \langle S(h)\rangle$

4. Assume now that $h = 0$ in (B.4.12). For given J draw $\langle S(h=0)\rangle_\beta$ as a function of β.

Spontaneous magnetization starts at

$$\beta_c = \frac{\kappa_0}{DJ}, \qquad T_c = \frac{1}{k_b \beta_c}.$$

Define the critical exponent for magnetization by

$$\langle S(h=0)\rangle_T \sim (T_c - T)^\beta, \qquad T \approx T_c, \quad T < T_c.$$

Note that the exponent β is *not* the inverse of T here.

5. Now convince yourself that $\beta = \frac{1}{2}$.

The susceptibility is defined as

$$\chi(T) = \left.\frac{d\langle S(h)\rangle}{dh}\right|_{h=0}.$$

6. Use (B.4.12) to show that

$$\frac{\partial \langle S(h)\rangle}{\partial h} = \frac{1}{2}\frac{\beta}{(\kappa_0 - DJ\beta) + 6\lambda \langle S(h)\rangle^2},$$

and thus:

$$\chi(T) = \frac{1}{2}\frac{\beta}{\kappa_0 - DJ\beta} \quad T > T_c, \qquad \chi(T) = \frac{1}{4}\frac{\beta}{DJ\beta - \kappa_0} \quad T < T_c.$$

The critical exponent of susceptibility is defined as

$$\chi(T) \to \frac{c}{|T-T_c|^\gamma} \quad \text{for } T \to T_c.$$

7. Convince yourself that $\gamma = 1$ in the mean-field approximation.

Finally, consider the spin-spin correlation function:

$$\langle (S_i - \langle S_i \rangle)(S_j - \langle S_j \rangle) \rangle = \langle \delta S_i \delta S_j \rangle$$

for $h = 0$. The correlation length $\xi(T)$ is defined as the exponential fall-off of $\langle \delta S_i \delta S_j \rangle$, i.e. by

$$\xi(T) = -\frac{\log(\langle \delta S_i \delta S_j \rangle)}{|i-j|} \quad \text{for } |i-j| \to \infty$$

From (B.4.11) we know:

$$\langle (\delta S)_i (\delta S)_j \rangle = \frac{\int \prod_k d(\delta S_k) \, \delta S_i \delta S_j \, e^{-\beta H_F(\delta S_k)}}{\int \prod_k d(\delta S_k) \, e^{-\beta H_F(\delta S_k)}},$$

where $H_F(\delta S)$ is quadratic in δS. Recall from the Gaussian integrations discussed in Problem Set 1:

$$\frac{\int \prod_k dx_k \, e^{-\frac{1}{2} x_i A_{ij} x_j + J_i x_i}}{\int \prod_k dx_k \, e^{-\frac{1}{2} x_i A_{ij} x_j}} = e^{\frac{1}{2} J_i A_{ij}^{-1} J_j}.$$

Furthermore, by definition

$$\langle x_i x_j \rangle = \frac{\int \prod_k dx_k \, (x_i x_j) \, e^{-\frac{1}{2} x_l A_{lm} x_m}}{\int \prod_k dx_k \, e^{-\frac{1}{2} x_l A_{lm} x_m}}.$$

8. Use the last two equations to prove that

$$\langle x_i x_j \rangle = \left(A^{-1}\right)_{ij}.$$

9. Let Δ_L denote the lattice Laplacian. Use (B.4.11) to show that

$$\langle \delta S_i \delta S_j \rangle = \left(A^{-1}\right)_{ij}, \quad A_{ij} = \beta J \left[\Delta_L + m^2\right]_{ij} \quad \text{(B.4.16)}$$

$$m^2 = \frac{1}{\beta J} \left[\kappa_0 - DJ\beta + 6\lambda_0 \langle S(0) \rangle^2\right] \quad \text{(B.4.17)}$$

We know the long distance behavior of $[\Delta_L + m^2]_{ij}^{-1}$. It is

$$-\frac{\log \left[\Delta_L + m^2\right]_{ij}^{-1}}{|i-j|} = m \quad \text{for } |i-j| \to \infty$$

10. Show that
$$m(T) \propto \sqrt{|T - T_c|} \quad \text{for } T \to T_c.$$

For a spin system we have for $|i-j| \to \infty$:

$$\langle \delta S_i \delta S_j \rangle \sim e^{-\frac{|i-j|}{\xi(T)}} \quad \text{i.e.} \quad -\frac{\log \langle \delta S_i \delta S_j \rangle}{|i-j|} \to \frac{1}{\xi(T)} \quad \text{(B.4.18)}$$

which defines the correlation length $\xi(T)$. Close to the phase transition it might have a non-analytic behavior:

$$\xi(T) \sim \frac{1}{|T - T_c|^\nu},$$

signifying long-range correlations.

11. Convince yourself that $\nu = \frac{1}{2}$.

Finally, we know

$$(\Delta_L + m^2)^{-1}_{ij} \sim \frac{1}{|i-j|^{D-2}}. \quad \text{for} \quad 1 \ll |i-j| \ll \xi(T).$$

For our spin system one defines the *anomalous scaling exponent* η:

$$\langle \delta S_i \delta S_j \rangle \sim \frac{1}{|i-j|^{D-2+\eta}}, \quad 1 \ll |i-j| \ll \xi(T).$$

Thus, $\eta = 0$ in mean-field theory.

B.5 PROBLEM SET 5
Rooted planar trees
Recall first:
$$z(\mu) = \sum_{BP} e^{-\mu|BP|} \prod_{v \in BP} w_v \qquad (B.5.1)$$
$$= \sum_L e^{-\mu L} \sum_{\{BP:|BP|=L\}} \prod_{v \in BP} w_v \qquad (B.5.2)$$

Write
$$g = e^{\mu}, \quad f(z) = \sum_{n=2}^{\infty} w_n z^{n-1}, \quad w_1 = 1$$
and $w_v \equiv w_{n(v)}$, where $n(v)$ is the order of the vertex v.

Furthermore, let us write
$$\mathcal{N}(L) := \sum_{\{BP:|BP|=L\}} \prod_{v \in BP} w_v = g_c^L \cdot h(L), \quad \text{i.e.} \quad z(g) = \sum_L h(L) \left(\frac{g_c}{g}\right)^L.$$

where for L large we have $\frac{\log h(L)}{L} \to 0$. We call $\mathcal{N}(L) = g_c^L h(L)$ the number of branched polymers with weights $w_{n(v)}$ and length L. We have
$$g = \frac{1 + f(z)}{z}$$
and $g_c \equiv e^{\mu_c}$ is determined by
$$\left.\frac{dg}{dz}\right|_{z_c} = 0.$$

1. Show that for $0 < \gamma < 1$ we have
$$\boxed{h(L) \propto L^{\gamma-2}(1 + \cdots) \iff z(g) - z(g_c) \propto \left(1 - \frac{g_c}{g}\right)^{1-\gamma} + \cdots}$$

2. Assume all $w_n = 1$ for $n = 2, 3, \cdots$, i.e. all branchings are allowed and have the same weight. Show that
$$g_c = 4 \quad \text{and} \quad \mathcal{N}(L) \to 4^L L^{-\frac{3}{2}} \quad \text{for} \quad L \to \infty$$

3. Show that the explicit expression for $z(g)$ in this case is
$$z(g) = \frac{1 - \sqrt{1 - \frac{4}{g}}}{2}$$

This ensemble of trees is called the ensemble of *uniform random rooted trees*.

4. Assume $w_3 = 1$ and all other $w_n = 0$. Show that
$$g_c = 2 \quad \text{and} \quad \mathcal{N}(L) \to 2^L L^{-\frac{3}{2}} \quad \text{for} \quad L \to \infty.$$
5. Show that for $w_n = 1, w_k = 0, k \neq n, n > 2$ we have
$$g_c(n) = (n-2)^{\frac{1}{n-1}} + (n-2)^{-\frac{n-2}{n-1}}, \qquad \mathcal{N}(L) \to g_c^L(n) L^{-\frac{3}{2}} \quad \text{for} \quad L \to \infty$$
Why does $g_c(n) \to 1$ for $n \to \infty$?
6. Discuss the case $w_2 = 1, w_n = 0, n > 2$. (Solve for z)

Note that in all the cases we have discussed so far we have $\gamma = \frac{1}{2}$. (except for the case discussed in question 6, which was not really a BP). We will now discuss when $\gamma = 1/2$ and how to obtain BPs with $\gamma \neq 1/2$.

Criticality of branched polymers

The basic equation is
$$e^\mu = \frac{1+f(z)}{z}, \quad f(z) = \sum_{m=2}^\infty w_m z^{m-1}$$

Let us assume $w_2 = 0$ and denote $e^\mu = g$. Also, let us assume that $f(z)$ is a polynomial of order n.

7. Show from the very definition of z in terms of μ that for $g \to \infty$ we have $z \to 0$.

For decreasing g, the value of z will increase. Criticality is encountered at the *first* extremum of $g(z)$ for increasing z. Let this point be z_c. By assumption: $w_{n+1} \neq 0$ and $w_{n+2}, w_{n+3}, \cdots = 0$. Assume that
$$g'(z_c) = \cdots = g^{(n-1)}(z_c) = 0, \quad g^{(n)}(z_c) \neq 0$$

8. Show that
$$g(z) - g(z_c) = \frac{\left(1-\frac{z}{z_c}\right)^n}{z}, \quad \text{and that } (w_2=0) \quad g(z_c) = \frac{n}{z_c} \quad (B.5.3)$$

Hint: Taylor expand and use that $zg(z)$ is a polynomial of order n.

9. Find the explicit branching weights w_m corresponding to this function. Note that they alternate in sign starting out with $w_3 > 0$.
10. Define $g_c = g(z_c)$. Show that eq. (B.5.3) leads to
$$z(g) = z_c - z_c^{1+\frac{1}{n}}(g-g_c)^{\frac{1}{n}} + O\left((g-g_c)^{\frac{2}{n}}\right) \quad \text{i.e.} \quad \gamma = 1 - \frac{1}{n}.$$

11. Show that if all weights $w_3, \cdots, w_{n+1}$ are positive then the *only* critical behavior is
$$z(g) = z_c - \kappa(g-g_c)^{1/2} + O(g-g_c) \quad \text{i.e.} \quad \gamma = 1/2$$

Let us now consider the situation where we allow arbitrarily high branching, i.e. w_m can be different from zero for arbitrarily high m.

12. Assume that $f(x)$ is at least two times differentiable, that $f(0), f'(0) \geq 0$, that $f''(x) > 0$ and that $f''(x) > 1/x^2$ for large x. Show that $\gamma = 1/2$ for such an $f(x)$.
13. Some examples of such functions: $f(x) = e^x - 1 - x$, $f(x) = x^2 \tanh x$. Find the corresponding w_m for these functions. Note that in the second example we have an oscillating sign of w_m, but nevertheless $\gamma = 1/2$.

We want to generalize (B.5.3) in a non-trivial way:

$$g(z) - g(z_0) = \frac{\left(1 - \frac{z}{z_0}\right)^s}{z}, \quad z < z_0, \quad n-1 < s < n, \qquad \text{(B.5.4)}$$

where we assume $s > 1$ and the weight $w_2 = 0$, i.e. $g(z_0) = \frac{s}{z_0}$ (like in (B.5.3))

14. Show that weights for $m > 2$ are:

$$w_m = \frac{1}{z_0^{m-1}}(-1)^m \frac{\Gamma(s+1)}{\Gamma(s-m+2)\Gamma(m)} \qquad \text{(B.5.5)}$$

$$= \frac{1}{z_0^{m-1}} \cdot \frac{\Gamma(m-1-s)}{\Gamma(-s)\Gamma(m)} \underset{m \to \infty}{\propto} \frac{(-1)^n}{m^{s+1}} \qquad \text{(B.5.6)}$$

15. Show that the sign of w_m is oscillating for $m < s+2$, like in the situation for integer values of s, but is constant for $m > s+2$.
16. Show that for $1 < s < 2$ all weights are positive.
17. Show that for $s > 1$ we have

$$z - z_c \underset{g \to g_c}{\sim} \kappa (g - g_c)^{\frac{1}{s}} \left(1 + O\left((g - g_c)^{\frac{1}{s}}\right)\right)$$

Thus, $\gamma_s = 1 - \frac{1}{s}$.

For $1 < s < 2$ we have an example of a situation where all weights are positive but $\gamma \neq 1/2$. It requires infinite branching *and* that w_n should not be suppressed too much. For example, a power law $\frac{1}{n^{1+s}}$ rather than $\frac{1}{n!}$, and only $1 < s < 2$.

B.6 PROBLEM SET 6

In this Problem Set we will solve a simple combined matter and BP system. The system is important since it can be generalized to two dimensions where it can also be solved and provides the simplest example of matter coupled to 2d gravity.

Branched polymers with "matter"

Consider a regular lattice in two dimensions. On such a lattice one can put down "dimers" (rods), illustrated by wiggly lines in Fig. B.6.1. One can create a statistical model of these dimers on the lattice by associating with each dimer a fugacity ξ. We will be interested in so-called *hard dimers*, where the dimers are not allowed to touch each other. The partition function for these hard dimers is then

$$z(\xi) = \sum_{\{HD\}} \xi^{|HD|}$$

where the summation is over all possible ways one can put down the hard dimers on the lattice, and $|HD|$ is the number of dimers in the particular dimer configuration.

For two-dimensional lattices these hard dimer models play an important role for exactly solvable lattice spin systems (related to the high-temperature expansion of the spin systems) and the interesting critical behavior of a dimer model is actually obtained for a somewhat "unphysical" *negative* value of the fugacity.

Let G be a connected graph. It is now clear how to define a hard dimer model on G. Let us now consider the ensemble of planar trees or BPs. On each of these trees we can put down dimers and we can consider the partition function of the combined system:

$$z(\mu, \xi) = \sum_{BP} \left(\prod_i w_{v(i)} \right) \left(\prod_\rho e^{-\mu} \right) \sum_{HD(BP)} \xi^{|HD(BP)|}$$

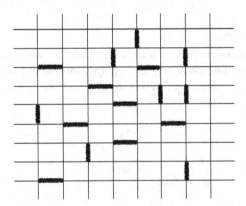

Figure B.6.1 A regular two-dimensional square lattice with (hard) dimers on some of the links (the wiggly lines).

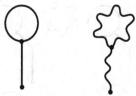

Figure B.6.2 Links without dimers are represented as straight lines and links with dimers as wiggly lines. $z(\mu,\xi)$ where no dimer touches the root is represented as a straight line emergent from the root and connecting to a circular blob (which itself can contain wiggly lines). $\tilde{z}(\mu,\xi)$ where a dimer *does* touch the root is represented as a wiggly line emergent from the root and connecting to a wiggly blob (which itself can contain straight lines).

Here i is a vertex in a BP and $v(i)$ order of the vertex and $w_{v(i)}$ the weight of that vertex. Furthermore, ρ is a link in a BP and $e^{-\mu}$ the usual weight. On each BP we have a statistical system of HDs.

Such an average over both lattices and matter systems on these lattices is called an *annealed average* (contrary to another kind of average: a *quenched average* where we first calculate the free energy of the matter system (i.e. $\log Z$) on a lattice and *then* average over lattices).

Here we consider the simplest BP system:

$$w_v = 1 \quad \text{if } v = 1, 3$$
$$w_v = 0 \quad \text{if } v \neq 1, 3$$

where, as mentioned, $v(i)$ is the order of the vertex i.

Let us consider the equations for *rooted* BPs with dimers. We have two situations: the link touching the root *does not* have a dimer attached and the link touching the root *has* a dimer attached, indicated graphically in Fig. B.6.2.

1. Convince yourself that the graphical representation shown in fig. B.6.3 is correct and leads to the following equations:

$$e^\mu = \frac{1+z^2+2z\tilde{z}}{z}, \quad e^\mu = \xi \frac{1+z^2}{\tilde{z}}$$

Then write $g = e^\mu$ and

2. Show that

$$g = \frac{1+z^2}{z} + \frac{2\xi}{g}(1+z^2). \tag{B.6.1}$$

and that the solution is

$$g(z,\xi) = \frac{1}{2}\left[\frac{1+z^2}{z} + \sqrt{\frac{(1+z^2)^2}{z^2} + 8\xi(1+z^2)}\right] \tag{B.6.2}$$

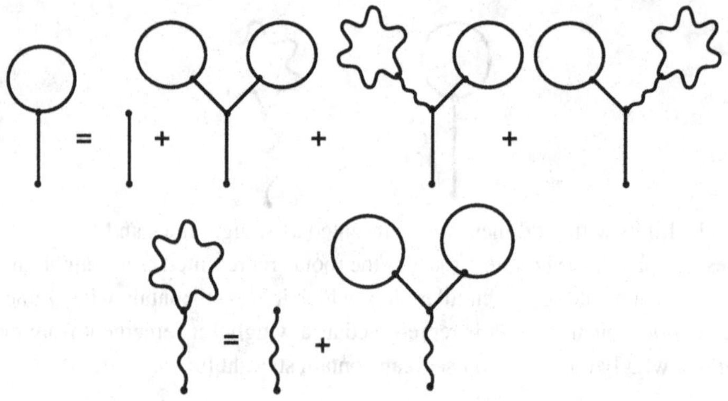

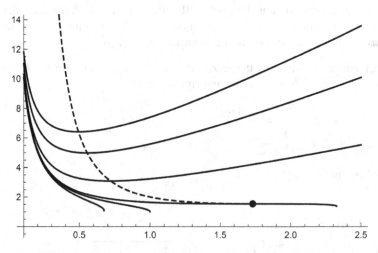

Figure B.6.3 The graphical representation of the equations for rooted hard dimers. The top graphical equation represents $z(\mu)$ where there is no dimer attached to the root link. The first vertex of order 3 met from the root can either split in two $z(\mu)$s or in a $z(\mu)$ and a $\tilde{z}(\mu)$. The bottom graphical equation represents $\tilde{z}(\mu)$ which have a dimer attached to the root link. Here the first vertex of order 3 met from the root can only split in two $z(\mu)$s.

The function is plotted in Fig. B.6.4 for various ξs, and the minima of the curves indicated by the dashed line.

3. Differentiate eq. (B.6.1) with respect to z (while keeping ξ fixed) to find equations for

$$\frac{dg}{dz} = 0 \quad \text{and} \quad \frac{d^2g}{dz^2} = 0$$

Figure B.6.4 The various curves $z \to g(z, \xi)$ for a number of values of ξ. For $\xi > \xi_c = -4/27$ the curves have a minimum, which is also shown as the dashed curve. The last curve having a minimum is the one with $\xi = -4/27$, and the endpoint is indicated by a dot.

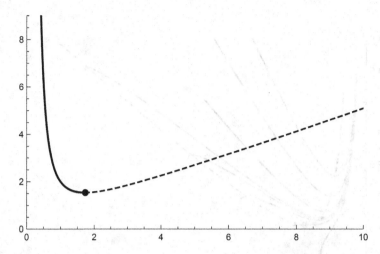

Figure B.6.5 The critical curve is shown in as the thick line in the figure and the endpoint with a dot. The endpoint is (z_c, g_c). This part of the curve is precisely the dashed curve in Fig. B.6.4. The continuation of the curve is the same as in Fig. B.6.6.

There is only one ξ (which we denote ξ_c) where both equations are satisfied.

4. Find ξ_c, g_c, and z_c, and subsequently argue that $\left.\frac{d^3g}{dz^3}\right|_{z_c} \neq 0$.

For a given ξ (larger than $\xi_c = -4/27$) we have a point $(z_k(\xi), g_k(\xi))$ where $dg/dz = 0$. It is the minimum of the curve $z \to g(z, \xi)$ given by eq. (B.6.2). We denote the curve $\xi \to (z_k(\xi), g_k(\xi))$ *the critical curve*. This is where we can take the continuum limit for a given ξ. The critical curve is is shown on Fig. B.6.5. Expanding around the point $z_k(\xi)$ we have ($\xi > \xi_c$ is kept fixed)

$$g(z,\xi) - g(z_k(\xi),\xi) \propto c_2(\xi)(z-z_k(\xi))^2 + O((z-z_k(\xi))^3) \quad \text{i.e.} \quad \gamma = \frac{1}{2}. \quad (B.6.3)$$

However, for $\xi \to \xi_c$ we have $c_2(\xi) \to 0$ since $g''_{zz}(z_c, \xi_c) = 0$ and we obtain for $\xi = \xi_c$

$$g(z,\xi_c) - g(z_c(\xi_c),\xi_c) \propto (z-z_c(\xi_c))^3, \quad \text{i.e.} \quad \gamma = 1 - \frac{1}{3} = \frac{2}{3}. \quad (B.6.4)$$

Thus, the multicritical behavior can actually be reproduced by having a matter system on BPs. The negative weight comes from the matter system.

To summarize: We have a curve of criticality (in this case the curve where $\frac{dg}{dz} = 0$) as the matter coupling constant ξ varies. For all points on the curve we have the same critical behavior, $\gamma = \frac{1}{2}$, except at the endpoint of the critical curve, where $\gamma = \frac{2}{3}$. This is typical in critical phenomena for statistical systems: one has a phase transition line where all points on the line have the same critical behavior except at the endpoint, where the order of the transition, and therefore the corresponding critical exponents, can change.

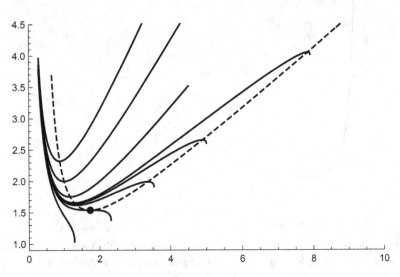

Figure B.6.6 The various curves $z \to g(z,\xi)$ for some values of ξ. For $\xi \in]-4/27, -1/8]$ the curves have a local maximum to the right of the local minimum and the curve $g_k(z_k)$ given by eq. (B.6.5) passes through these local maxima for $z_k > \sqrt{3}$, as seen in the figure.

5. Show that the critical curve is given by

$$g_k(z_k) = \frac{(1+z_k^2)^2}{2z_k^3}. \tag{B.6.5}$$

6. Understand how the critical curve, shown on Fig. B.6.6, is related to the functions $z \to g(z,\xi)$ given by eq. (B.6.2) (also shown on the figure), not only for $z < z_c$ (the *real* critical curve), but also for $z > z_c$.

Let us couple matter to a BP (the dimer model we have just considered is a particular example). We then have a partion function:

$$Z_{\text{matter}}(BP) \equiv \sum e^{-S(\text{matter,BP})}$$

where the sum is over all possible matter configurations on the given BP. We now consider the situation where we sum over all BPs with a given number of vertices V or links L (note that $V = L+1$). We can view L (or V) as the "volume" of the BP, and we have

$$Z_{\text{matter}}(L) \equiv \sum_{BP, |BP|=L} \left(\prod_i w_{v(i)}\right) Z_{\text{matter}}(BP).$$

For large L we expect

$$Z_{\text{matter}}(L) = e^{-L(f(\text{matter})) + o(L)}$$

where f(matter) is the free energy per unit volume (or free energy density). The total (grand canonical) partition function of matter on the ensemble of BPs is then:

$$Z(\mu, \text{matter}) = \sum_{BP} e^{-\mu|BP|} Z_{\text{matter}}(BP)$$
$$= \sum_L e^{-\mu L} Z_{\text{matter}}(L)$$

We thus see that the critical point μ_c is precisely

$$\boxed{\mu_c = -f(\text{matter})}$$

Consider the model $Z(\mu, \text{matter})$ as a BP model where the weight for each BP is

$$\prod_i w_{v(i)} \to \left(\prod_i w_{v(i)}\right) \cdot Z_{\text{matter}}(BP)$$

Matter changes the weight of each BP, but if we can calculate μ_c (which will be a function of the matter couplings) we have automatically calculated the free energy density of the matter on BPs.

Let us now apply this to our dimer model. Our matter coupling is the fugacity. We had $e^\mu = g$, and thus $e^{\mu_c(\xi)} = g_k(\xi)$ where $g_k(\xi)$ was determined by the condition $g'_z(z, \xi) = 0$. Recall from a magnetic system that

$$\frac{df}{dH} = -m(H)$$

It turns out that for the Ising model in two dimensions, at high temperature $f(H)$ has a singularity for an *imaginary magnetic field* $H = i\tilde{H}$ at a certain critical value $\tilde{H}_c(T)$, called the Lee-Yang edge singularity:

$$\left.\frac{\partial f}{\partial \tilde{H}}\right|_{\text{singular}} \propto (\tilde{H} - \tilde{H}_c(T))^\sigma, \quad \left.\frac{\partial^2 f(\tilde{H})}{\partial \tilde{H}^2}\right|_{\text{singular}} \propto (\tilde{H} - \tilde{H}_c(T))^{\sigma-1}, \quad \sigma = -\frac{1}{6}.$$
(B.6.6)

Our fugacity is similar: the coupling of the magnetic field to the spin was

$$e^{HS} \to e^{iHS} \quad \text{when the magnetic field is imaginary}$$

so for aligned spins it is e^{iH} per site. One might even consider diluted spin models where not every lattice site has a spin variable. Similarly, the fugacity is ξ per link where there is a dimer. Let us write $\xi = e^h$. For *negative* fugacity we have

$$h \sim i\pi + \log|\xi|.$$

Let us now ask if we have a singular (i.e. non-analytic) behavior of $f(\xi)$ for $\xi \to \xi_c$. We want to calculate

$$\left.\frac{\partial^2 f(\xi)}{\partial \xi^2}\right|_{\xi \to \xi_c} \sim (\xi - \xi_c)^{\sigma-1}, \quad f(\xi) = -\mu_c(\xi) = -\log g_k(\xi) \quad (B.6.7)$$

First show the following

7. Show that on the critical curve $\xi \to (z_k(\xi), g_k(\xi))$ we have

$$\xi = \frac{(1-z^4)(1+z^2)}{8z^6} \tag{B.6.8}$$

which in principle determines $z(\xi)$.

8. Show that

$$\left.\frac{d\xi}{dz}\right|_{z_c} = 0, \tag{B.6.9}$$

which implies that $\xi(z)$ has a quadratic minimum:

$$\xi - \xi_c = \frac{1}{2}\left.\frac{d^2\xi}{dz^2}\right|_{z_c}(z-z_c)^2 + \cdots, \quad \text{provided} \quad \left.\frac{d^2\xi}{dz^2}\right|_{z_c} \neq 0 \tag{B.6.10}$$

In order to determine the behavior of $g_k(\xi)$ around ξ_c we expand $g(z,\xi)$ around z_c, ξ_c (the reason we do not directly try to expand $g_k(\xi)$ around ξ_c is that if it has a critical behavior, we expect it to be singular around this point. However $g(z,\xi)$ is itself perfectly regular around z_c, g_c)

$$g(z,\xi) = g(z_c,\xi_c) + \left.\frac{\partial g}{\partial z}\right|_{z_c,\xi_c}(z-z_c) + \frac{1}{2}\left.\frac{\partial^2 g}{\partial z^2}\right|_{z_c,\xi_c}(z-z_c)^2 + \left.\frac{\partial g}{\partial \xi}\right|_{z_c,\xi_c}(\xi-\xi_c)$$

$$+ \frac{1}{2}\left.\frac{\partial^2 g}{\partial \xi^2}\right|_{z_c,\xi_c}(\xi-\xi_c)^2 + \left.\frac{\partial^2 g}{\partial \xi \partial z}\right|_{z_c,\xi_c}(z-z_c)(\xi-\xi_c) + \frac{1}{6}\left.\frac{\partial^3 g}{\partial z^3}\right|_{z_c,\xi_c}(z-z_c)^3$$

$$= g(z_c,\xi_c) + \left.\frac{\partial g}{\partial \xi}\right|_{z_c,\xi_c}(\xi-\xi_c) + \frac{1}{2}\left.\frac{\partial^2 g}{\partial \xi^2}\right|_{z_c,\xi_c}(\xi-\xi_c)^2 + \left.\frac{\partial^2 g}{\partial \xi \partial z}\right|_{z_c,\xi_c}(z-z_c)(\xi-\xi_c)$$

$$+ \frac{1}{6}\left.\frac{\partial^3 g}{\partial z^3}\right|_{z_c,\xi_c}(z-z_c)^3 \tag{B.6.11}$$

when we use that

$$\left.\frac{\partial g}{\partial z}\right|_{z_c,\xi_c} = \left.\frac{\partial^2 g}{\partial z^2}\right|_{z_c,\xi_c} = 0$$

9. Use (B.6.10) and (B.6.11) to show that

$$g(z_k(\xi),\xi) = g(z_c,\xi_c) + \left.\frac{\partial g}{\partial \xi}\right|_{z_c,\xi_c}(\xi-\xi_c) + \kappa \cdot (\xi-\xi_c)^{3/2} + O((\xi-\xi_c)^2) \tag{B.6.12}$$

10. show that

$$\frac{d^2 g(z_k(\xi),\xi)}{d\xi^2} \propto \frac{1}{(\xi-\xi_c)^{\frac{1}{2}}} \implies \sigma = \frac{1}{2} \tag{B.6.13}$$

so we indeed have a critical "magnetization" for $\xi \to \xi_c$:

B.7 PROBLEM SET 7

In this Problem Set we consider various aspects of BPs: BPs with infinite Hausdorff dimension, BPs coupled to Ising spins and the relation between BPs with dimers and Ising spins (this last relation is valid for any reasonable lattice system on which one can put Ising spins and dimers).

BPs with infinite Hausdorff dimension

Let us consider BPs with $w_{m+1} = \frac{1}{m^{s+1}}$. Again taking $g = e^{\mu}$, we have:

$$g = \frac{1+f(z)}{z}, \quad f(z) = \sum_{m=1}^{\infty} \frac{z^m}{m^{s+1}} = \text{Li}_{s+1}(z) \tag{B.7.1}$$

Here we encounter the so-called *polylogarithm* function Li. For $s > 0$: $f(1) = \text{Li}_{s+1}(1) = \zeta(s+1)$, which is the Riemann zeta function.

1. Show that $f'(1) = \infty$ for $s \leq 1$, and use this to argue that $z_c < 1$ (the radius of convergence of $f(z)$) and therefore $\gamma = 1/2$ (so we have ordinary BPs). Actually, the argument can be extended and $z_c < 1$ for $s < 1.5915....$ and correspondingly $\gamma = 1/2$

Let $s \in]n, n+1[$, with n an integer ≥ 2. One can show that for $z \to 1$:

$$\text{Li}_s(z) = \text{Li}_s(1) + \text{Li}'_s(1)(1-z) + \cdots + \frac{\text{Li}_s^{(n-1)}(1)}{(n-1)!}(1-z)^{n-1} + c \cdot (1-z)^{s-1} + \cdots$$

Thus we have:

$$g - g_c = c_1(1-z) + \cdots + c_n(1-z)^n + c_s(1-z)^s + \cdots \tag{B.7.2}$$

2. Argue (without giving detailed calculations) that one can add weights

$$\tilde{w}_2, \cdots, \tilde{w}_{n+2} \text{ i.e. } w_{k+1} = \frac{1}{k^{s+1}} + \tilde{w}_{k+1}, \quad k = 1 \cdots, n+1 \tag{B.7.3}$$

such that in this model we have:

$$g - \tilde{g}_c = \tilde{c}_s(1-z)^s + \cdots \tag{B.7.4}$$

This is precisely the same scaling as we encountered before, and an example of universality: in an earlier exercise we found the weights w_{m+1} which produced the relation $g - g_c \propto (z_c - z)^s$, without any corrections and we saw that these coefficients asymptotically behaved like $1/m^{s+1}$. Here we have chosen in (B.7.1) weights w_{m+1} which are exactly $1/m^{s+1}$. When correcting these coefficients in a minimal way (which does not affect the asymptotic behavior), like in (B.7.3) we obtain the critical behavior $g - g_c \propto (z_c - z)^s$, but there are corrections to this expression, as indicated with $+\cdots$ in (B.7.4), but corrections which do not influence the critical behavior. *However:* if we do *not* add $\tilde{w}_2, \cdots, \tilde{w}_{n+2}$ we have (B.7.2) (and note: if we do not allow negative weights (which do not have a straight forward probability interpretation), we cannot get rid of the first n terms in (B.7.2)). Note also that if we do not add the terms we have

$$\frac{dg}{dz} < 0 \quad z \in [0,1], \tag{B.7.5}$$

while if we add the terms $\tilde{w}_m$ in (B.7.3) we obtain $\frac{dg}{dz}\big|_{z=1} = 0$.

3. Show that by inverting (B.7.2) we obtain:

$$z - z_c = z - 1 = \tag{B.7.6}$$
$$d_1(g-g_c) + d_2(g-g_c)^2 + \cdots + d_n(g-g_c)^n + d_s(g-g_c)^s + \cdots$$

The situation is thus very different from

$$g - g_c = (1-z)^s \implies z - 1 = -(g-g_c)^{1/s}$$

How do we define the critical exponent γ for the case (B.7.6)? Recall that γ was defined by

$$\frac{dz}{d\mu} \to \frac{c}{(\mu-\mu_c)^\gamma} \quad \text{for } \mu \to \mu_c.$$

But this definition assumed that $(\mu-\mu_c)^{-\gamma}$ was the dominating term, i.e. $\gamma > 0$. We can write:

$$\frac{d^{n+1}z}{d\mu^{n+1}} \to \frac{1}{(\mu-\mu_c)^{\gamma+n}}.$$

Let us apply this to (B.7.6):

$$\frac{d^{n+1}z}{d\mu^{n+1}} \to \frac{1}{(\mu-\mu_c)^{n+1-s}}, \quad \text{and thus} \quad \boxed{\gamma = 1-s < 0 \text{ for } s > 2}.$$

These BPs are very different from the ones where $\gamma > 0$. They are dominated by configurations where a few vertices have very high order and the rest have order 1. We will not prove that here, but there are simple arguments pointing in that direction.

Recall that if $0 < \gamma < 1$: $dz/d\mu \propto (\mu-\mu_c)^{-\gamma} \to \infty$ and thus $d\mu/dz \to 0$. One important consequence of (B.7.6) is that $d\mu/dz \neq 0$ for $z \to z_c$.

4. Use this in the expression for $G^{(l)}(\mu)$, the intrinsic two-point function, to show that the mass $m_I(\mu) \to c > 0$ as $\mu \to \mu_c$.

Thus the mass does not scale to zero. Recall that we have $m(\mu) \equiv |\mu-\mu_c|^\nu$. If $m(\mu)$ does not scale to zero we formally have $\nu = 0$. Furthermore, for the Hausdorff dimension d_H we had $d_H = \frac{1}{\nu}$, so formally $\nu = 0$ implies that $d_H = \infty$. Effectively, one can reach all vertices in just a few steps! Intuitively this is possible if we have vertices of very high order, such that many vertices can be connected via these high order vertices.

However, there is more to be said, since differentiating the partition function $z(\mu)$ n times, it becomes divergent for $\mu \to \mu_c$. Thus the m-point functions, $m \geq n$ are critical, but we will not discuss the interpretation of this any further here.

The Ising model coupled to BPs

Let BP be a branched polymer:

$$Z_{BP}(\beta,h) = \sum_{\{\sigma_i\}} e^{\beta \sum_{\langle ij \rangle} \sigma_i \sigma_j + h \sum_i \sigma_i}$$

where $\langle ij \rangle$ is the link between neighboring vertices i and j in the BP. We take $\sigma_i = \pm 1$ and h is an external magnetic field[1]. This spin model is the so-called Ising model, and one can put the model on any graph consisting of vertices and links. In particular one can put the Ising model on regular lattices. More physics related to the Ising model is discussed in Problem Set 11. Here we consider the Ising model on BPs. The total partition function is then

$$Z(\mu,\beta,h) = \sum_{BP} e^{-\mu|BP|} \rho(BP) Z_{BP}(\beta,h), \qquad \rho(BP) = \prod_i w_{v(i)}$$

We consider rooted BPs and use the convention that the root vertex has no magnetic field attached. Denote Z_+ the partition function where $\sigma_{root} = 1$ and Z_- the partition function where $\sigma_{root} = -1$. As usual we define $f(Z) = \sum_{n=2}^{\infty} w_n Z^{n-1}$ and assume $w_1 = 1$. We assume until stated differently that $f(Z)$ is such that $\gamma = 1/2$.

5. Show:

$$Z_+ = e^{-\mu}\left[e^{\beta+h} + e^{-\beta-h} + e^{\beta+h}f(Z_+) + e^{-\beta-h}f(Z_-)\right] \qquad (B.7.7)$$

$$Z_- = e^{-\mu}\left[e^{\beta-h} + e^{-(\beta-h)} + e^{-(\beta-h)}f(Z_+) + e^{\beta-h}f(Z_-)\right] \qquad (B.7.8)$$

6. Show that when $h = 0$: $Z_+ = Z_-$ and

$$e^{\mu} = 2\cosh\beta \; \frac{1+f(Z)}{Z} \qquad (B.7.9)$$

and thus that

$$\mu_c(\beta) = \tilde{\mu}_c + \ln(2\cosh\beta), \qquad (B.7.10)$$

where

$$e^{\tilde{\mu}} = \frac{1+f(Z)}{Z}, \quad (Z = Z_+ = Z_-). \qquad (B.7.11)$$

---

[1] Compared to the notation used in Problem Set 4 we have chosen here to put the coupling constant $J = 0$ and we have left out the factor β usually multiplying $h\sum_i \sigma_i$ in $Z_{BP}(\beta,h)$. This is just to simplify some of the formulas below.

So for a fixed β we see that the critical Z_c is determined by the same equation as the BPs without Ising spins, namely

$$\frac{d}{dZ}\left(\frac{1+f(Z)}{Z}\right)\bigg|_{Z=Z_c} = 0. \tag{B.7.12}$$

7. Show that

$$\frac{1+f(Z_c)}{Z_c} = f'(Z_c) \quad \text{and thus} \quad e^{\mu_c(\beta)} = 2\cosh\beta \, f'(Z_c). \tag{B.7.13}$$

For $\beta \to 0$ (or $T \to \infty$): $\mu_c(\beta) = \hat{\mu}_c + \ln 2$.

8. Explain the $\ln 2$ as coming from the entropy of Ising spins.

Recall from the discussion of the dimer model that the partition function for a fixed "volume" L (number of links)

$$Z_L(\beta) = e^{-f_L(\beta)\cdot L} = e^{-F_L(\beta)}$$

and for $L \to \infty$ we have that $f_L(\beta) \to f(\beta) + O(1/L)$, where the free energy per volume, $f(\beta)$ is related to the critical point by

$$f(\beta) = -\mu_c(\beta)$$

A critical temperature in the spin model is a β_c where $f(\beta_c)$ is non-analytic.

*Thus there is **no** critical temperature β_c for Ising models on BPs, since $\mu_c(\beta)$ is analytic for all β*

For the Ising model the situation on a regular lattice is the following:

$d = 1$: no phase transition.
$d = 2$: the Onsager phase transition, the most famous phase transition in physics!
$d \geq 2$: a phase transition

For BPs, we have $d_H = 2$ for $\gamma = \frac{1}{2}$, which was what we assumed above. But contrary to the situation for a regular lattice we have no magnetic phase transition. It is possible to check that also in the cases discussed above, where $\gamma < 0$ and $d_H = \infty$ we have no magnetic phase transition. *Therefore d_H is not a good indicator of dimension in all situations* (the linear structure of the trees seems more important in this case).

Let us finally ask whether we have spontaneous magnetization. Recall for regular lattices:

$d = 1$: no spontaneous magnetization and $d \geq 2$: spontaneous magnetization.

We define spontaneous magnetization as

$$\langle m \rangle = -\lim_{h \to 0^+} \frac{\partial f(\beta, h)}{\partial h} = \lim_{h \to 0^+} \frac{\partial \mu_c(\beta, h)}{\partial h}$$

Problem Sets 1–13

where $\mu_c(\beta, h)$ is the critical value of μ obtained by solving eqs. (B.7.7) and (B.7.8) for μ and then finding the smallest value of μ for given β and h.

First, we consider BPs with $\gamma = \frac{1}{2}$. After that we will analyze $\gamma < 0$ separately.

Case 1: $\gamma = \frac{1}{2}$

$$Z_\pm(\beta, h) = Z_c(\beta) + \Delta Z_\pm \qquad \text{for } h \to 0: \Delta Z_\pm = c_\pm h + O(h^2)$$

9. Show from (B.7.7) and (B.7.8) by expanding to linear order in h that:

$$(\Delta Z_+ + \Delta Z_-)\left(e^{\mu_c(\beta)} - 2\cosh\beta f'(Z_c)\right) + 2Z_c e^{\mu_c(\beta)} \Delta\mu = 0 \qquad (B.7.14)$$

$$(\Delta Z_+ - \Delta Z_-)\left(e^{\mu_c(\beta)} - 2\sinh\beta f'(Z_c)\right) = 4h(1+f(Z_c))\sinh\beta \qquad (B.7.15)$$

10. Show, using (B.7.13) that $\Delta\mu = O(h^2)$ and thus that $\langle m \rangle = 0$.

For $\gamma = \frac{1}{2}$ we have no spontaneous magnetization and the Hausdorff dimension $d_H = 2$ is not a good guidance.

Case 2: $\gamma < 0$

To be specific, let us consider the case where $w_{n+1} = 1/n^{s+1}$, $s > 2$ which we analyzed above and which has $\gamma = 1 - s$. Here we have $Z_c(\beta) = 1$ (the radius of convergence of $\sum_{n=1}^{\infty} w_{n+1} Z^n = f(Z)$). If $h > 0$ one expects that $\Delta Z_+ \geq 0$ since in average each vertex will have more + spins than − spins. Since the spin at the root is fixed to be +, the spin interaction between the root and its neighbor vertex will thus in average contribute positively to Z_+ (and similarly negatively to Z_-) compared to the situation where $h = 0$. However, since $Z_+(\beta, h=0) = 1$, the maximum value, $Z_+(\beta, h)$ cannot increase further, i.e. $\Delta Z_+ = 0$. Also, note that we no longer have $e^{\mu_c(\beta)} = 2\cosh\beta f'(Z_c)$, since $\frac{d}{dZ}\frac{1+f(Z)}{Z} \neq 0$ in (B.7.6) for $s > 2$.

11. Use this to show that based on (B.7.14) we get $\Delta\mu \propto \Delta Z_-$ and from (B.7.14) we get $\Delta\mu \propto h$.

12. Show that we have spontaneous magnetization and find $\langle m \rangle$ as a function of β.

Now for $\gamma < 0$ we have $d_H = \infty$, leading to spontaneous magnetization. Therefore in this case the Hausdorff dimension is a good guidance.

The relation between the Ising model and hard dimers

Let G be a connected graph. It can be a regular lattice, a BP or another kind of random graph (we are later going to consider so-called two-dimensional random graphs). We can place an Ising spin model on this graph by assigning the spins to the vertices, and the interaction between spins will be between neighboring vertices connected by a link in the graph. We have as before:

$$Z_G(\beta,h) = \sum_{\{\sigma_i\}} e^{\beta \sum_{\langle ij \rangle} \sigma_i \sigma_j + h \sum_i \sigma_i}$$

13. Let V be the number of vertices in G and L the number of links. Use the identity
$$e^{\sigma X} = \cosh X + \sigma \sinh X, \qquad \sigma = \pm 1$$
to show that
$$Z_G(\beta,h) = \cosh^V h \cosh^L \beta \cdot \sum_{\{\sigma_i\}} \prod_j (1+\sigma_j \tanh h) \prod_{\langle kl \rangle} (1+\sigma_k \sigma_l \tanh \beta)$$

Expanding the products and summing over σ_i, it is clear that terms with an odd number of $\sigma_{i_1} \sigma_{i_2} \cdots \sigma_{i_{2n+1}}$ will average to zero. We want to use this and let $\beta \to 0$ (the high temperature expansion)

14. Let $\theta(n)$ denote the number of ways one can put down n hard dimers on G. Show that
$$Z(\beta,h) = (2\cosh h)^V \cosh^L \beta \times$$
$$\left(1 + \tanh^2 h \left[\theta(1)\beta + O(\beta^2)\right] + \tanh^4 h \left[\theta(2)\beta^2 + O(\beta^4)\right] + \cdots \right)$$

Define $\xi = \beta \tanh^2 h$. We now take the limit $\beta \to 0$, $h \to i\pi/2$ while ξ is fixed.

15. Show that the partition function for the hard dimer on G with *negative* ξ is
$$\tilde{Z}(\xi) = \lim_{h \to i\frac{\pi}{2},\, \beta \to 0,\, \xi \text{ fixed}} \frac{1}{(2\cosh h)^V} Z(\beta,h)$$

This relates the dimer model and the Ising model with an imaginary magnetic field.

B.8 PROBLEM SET 8

Asymptotic expansions

Most perturbation expansions are only so-called asymptotic expansions. This is true even for the perturbative expansion of the ground state energy E_0 of the quantum mechanical anharmonic oscillator:

$$\hat{H} = \frac{1}{2m}\hat{p}^2 + \frac{1}{2}m\omega^2\hat{x}^2 + g\hat{x}^4, \qquad E_0 = \sum_{n=0}^{\infty} c_n g^n.$$

The coefficients c_n in the expansion can be calculated to any order using textbook perturbation theory. However, the coefficients c_n grow so fast that the radius of convergence in the power series is zero. Note that this is not surprising: if there was a radius of convergence, the theory for g and $-g$ (for small g) would essentially be the same since everything would be analytic in g for small g, but that is clearly not the case. The dynamical system above is well-defined for positive small g, and is just a small deformation of the harmonic oscillator. However, for small negative g it is a very unhealthy system. For sufficient large energy we have classical run-away solutions accelerating to infinity and quantum mechanically there will for every energy always be a finite probability for tunnelling to such a situation. This implies that it is even non-trivial to define $\hat{H}$ as an Hermitian operator for negative g (and the possible definitions are non-unique).

The non-convergence of the perturbative series of E_0 for any g leads to the question: assume that we have calculated all the c_n. Do we have a way to calculate E_0? One method is *Borel summation*. Let $f(x)$ be "defined" by its formal power series. The Borel transform of f, $B(f)$, is then also defined as a formal power series

$$f(x) = \sum_{n=0}^{\infty} a_n x^n, \qquad B(f)(x) := \sum_{n=0}^{\infty} \frac{a_n}{n!} x^n \qquad (B.8.1)$$

Assume now that the power series for $B(f)$ has radius of convergence $r > 0$ and that the corresponding function $B(f)(x)$ can be analytically continued into a wedge region $|\arg z| < \varepsilon$ of the complex plane, and that it grows slower than exponential in this region. Then one can write formally write, interchanging summation and integration, which might or might not be allowed from a mathematical point of view,

$$f(x) = \sum_{n=0}^{\infty} a_n x^n = \sum_{n=0}^{\infty} n! \frac{a_n}{n!} x^n, \qquad n! = \int_0^{\infty} dt\, t^n\, e^{-t}$$

$$f(x) = \int_0^{\infty} e^{-t} \left(\sum_{n=0}^{\infty} \frac{a_n}{n!} (xt)^n \right) = \int_0^{\infty} e^{-t} B(f)(xt).$$

This integral now exists and it is called *the Borel sum of the formal power series* $f(x)$.

1. Assume $x > 0$ and apply this procedure to

$$f(x) = \sum_{n=0}^{\infty} n!(-1)^n x^{n+1} = x \sum_{n=0}^{\infty} n!(-1)^n x^n, \qquad (B.8.2)$$

to obtain

$$f(x) = x \int_0^{\infty} dt \, \frac{e^{-t}}{1+xt} \qquad (B.8.3)$$

The perturbation series of the anharmonic oscillator is divergent like (B.8.2), so it has zero radius of convergence. Two questions arise. (a) Assume the perturbation series can be Borel summed, like the series (B.8.2). Of course we know that $E_0(g)$ exists in quantum mechanics. How can we be sure that the Borel sum actually gives the correct value of $E_0(g)$. To be sure of that one has to appeal to properties of $E_0(g)$, which have to be proven outside perturbation theory, e.g. using general theorems for unbounded Hermitian operators like $\hat{H}$, known from functional analysis. and combine these with other general conditions which a function $f(x)$ has to satisfy in order that the Borel sum of its asymptotic series actually is equal to $f(x)$. We will not discuss these mathematical issues. (b) At a much more mundane level one can ask the following: even if we know that one, by some fancy method, is able to sum a series like (B.8.2) to the correct answer, to what extent does a perturbation expansion, which is only an asymptotic expansion help us at all? Clearly, given a value of x, e.g. the coupling constant g of the anharmonic oscillator, it makes no sense to continue calculating to very high order since $n! x^n \to \infty$ for $n \to \infty$. In fact, in general this observation also contains the practical answer we know from perturbation theory, here formulated for the asymptotic series: for a given x, higher order terms will only improve the approximation to the function $f(x)$ we are looking for if $n < |x|$ (such that $|a_n x^n| < 1$, $|a_n| \sim n!$). Of course there exist many methods by which one can do much better than just naively summing the series up to a given n, but as with the Borel summation, to be sure that they work, one has to know something more about the function $f(x)$.

There are other (less general) methods to sum divergent series:

2. Show that the formal power series $f(x)$ defined in (B.8.2) satisfies the following differential equation:

$$\frac{df}{dx} + \frac{1}{x^2} f(x) = \frac{1}{x}$$

3. Solve this to find $f(x)$ explicitly. (Hint: change variables to $u = \frac{1}{x}$ if you do not remember the general solution to a linear differential equation.)

The exponential-integral function $\mathrm{Ei}(u)$ is defined by

$$\mathrm{Ei}(u) = -\int_{-u}^{\infty} dt \, \frac{e^{-t}}{t} \qquad \text{for} \quad u < 0 \qquad (B.8.4)$$

It is convenient to define the following function

$$\mathrm{Ei}_c(u) = -\int_{-u}^{-c} dt \, \frac{e^{-t}}{t} \qquad \text{for} \quad u,c > 0 \qquad (B.8.5)$$

4. Show that
$$f(x) = -e^{\frac{1}{x}} \operatorname{Ei}\left(-\frac{1}{x}\right), \quad x > 0$$
and that $f(x)$ can be written in the form (B.8.3).

Thus we have seen that starting from the formal power series (B.8.2) we obtain the Borel sum (B.8.3) and solving the differential equation which the formal power series obeys, we obtain the same function. (and it is easy to show that the function $f(x)$ we have found precisely has the asymptotic expansion (B.8.3), e.g. by partial integration in representation (B.8.4) of $\operatorname{Ei}(-1/x)$).

5. Perform the partial integrations and convince yourself that it is correct....

6. Now repeat the same steps for the function
$$g(x) = \sum_{n=0}^{\infty} n! x^{n+1} \tag{B.8.6}$$
and Borel sum to obtain (formally)
$$g(x) = x \int_0^\infty dt \, \frac{e^{-t}}{1-xt} \tag{B.8.7}$$
and show that $g(x)$ satisfies the differential equation:
$$\frac{dg}{dx} - \frac{1}{x^2} g(x) = -\frac{1}{x}, \tag{B.8.8}$$
and finally that for all positive c we have solutions
$$g_c(x) = -e^{-\frac{1}{x}} \operatorname{Ei}_c\left(\frac{1}{x}\right), \tag{B.8.9}$$
to eq. (B.8.8), which *all* have the asymptotic expansion (B.8.6). (Hint for the asymptotic expansion: partial integrate $\operatorname{Ei}_c(1/x)$ and use that the asymptotic expansion of $e^{-1/x}$ is zero!)

The series (B.8.6) is *not* Borel summable ($B(g/x)(x) = 1/(1-x)$ has a singularity on the positive real axis and the integral (B.8.7) does not exist). However there is no problem solving the differential equation for positive x and the corresponding solutions $g_c(x)$ all have the correct asymptotic expansion. Notice also that $e^{-1/x}$ is a solution to the homogeneous differential equation. This is why we found a whole family of solutions:
$$g_{c_1}(x) - g_{c_2}(x) = \text{const.} \, e^{-\frac{1}{x}},$$
and since the Taylor expansion around 0^+ of $e^{-1/x}$ is zero, they all have the same asymptotic expansion. So we here see a simple example where the asymptotic series *does not fix the function uniquely. For that we need more information.* Of course we also have homogeneous solutions we could add to our function $f(x)$ which was

Borel summable. However, in this case the solution to the homogeneous differential equation would be $e^{1/x}$, which blows up at $x \to 0^+$, and there might be good physical arguments to discard this contribution.

We say that the $e^{-1/x}$ is a "non-perturbative" contribution: if $x = g$, the coupling constant, the contribution $e^{-1/g}$ will never be seen in a simple perturbation expansion. Nevertheless, there are many examples in physics where such contributions are important. Maybe the simplest one is the energy shift between the two lowest energy levels for the anharmonic oscillator with double well:

$$V(x) = -\frac{\omega^2 x^2}{4} + \frac{g}{4}x^4$$

where the energy shift is

$$\Delta E \propto \hbar\omega\, e^{-\left(\frac{\omega^4}{3\hbar\omega}\frac{1}{g}\right)}$$

In the Problem Set 9 we will count BPs with loops and see that the partition function is only defined as an asymptotic series, which is *not* Borel summable. Nevertheless, we can find a differential equation for it, which we can subsequently solve.

B.9 PROBLEM SET 9

Branched polymers with loops

We start with the simplest branching, where

$$w_3 = 1, \quad w_1 = 1+j, \quad e^\mu = g.$$

Usually we have always chosen $w_1 = 1$, but for later use it is convenient to choose $w_1 = 1+j$ where we will set $j=0$ in the end. Now the rooted BP partition function satisfies the graphical equation shown in Fig. B.9.1, which when formulated in terms of g and z reads (recall that there is a factor $1/g$ associated to each link):

$$z = \frac{1}{g}(w_1 + w_3 z^2), \quad \text{i.e.} \quad z = \frac{1}{g}(1+j+z^2) \tag{B.9.1}$$

1. Solve this for $z(g,j)$ and determine whether a plus or minus sign should be used in the process.

2. Show that $\gamma = \frac{1}{2}$, $z_c = \sqrt{1+j}$, and $g_c = 2\sqrt{1+j}$.

Define the susceptibility (almost) like in Chapter 4:

$$\chi(g,j) = \frac{dz(g,j)}{dj}. \tag{B.9.2}$$

The difference between this susceptibility and the one we used in Chapter 4 is that there we differentiated wrt g rather than j. The power with which j appears in a graph is $V_1 - 1$, were V_1 is the number of vertices of order 1 (the -1 is the root). The power with which g appears in a graph is $L = V - 1$ in a BP. However, for the graphs we consider we have $V_1 = L/2 - 3/2$. Thus there will not be any difference in the critical behavior of this χ and the one in the Chapter 4. Graphically the difference is that the two marked vertices in the graph for the present χ will be vertices of order 1 (of which one is the root vertex), while for the χ in Chapter 4, it can be any two

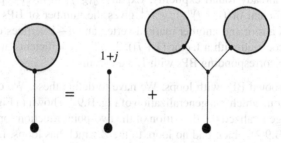

Figure B.9.1 The graphical equation for the rooted branched polymers, represented in the by now standard way. The bottom big dot is the root and the free little dot of order 1 has weight $w_1 = 1+j$. The vertex of order 3 has weight 1.

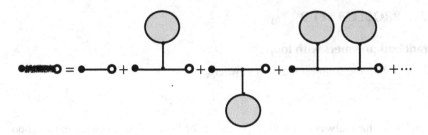

Figure B.9.2 The susceptibility $\chi(g,j)$ defined by eq. (B.9.2). The graph on the lhs of the equality sign, the susceptibility, is defined as the sum of the graphs on the rhs of the equality sign. The large black dot denotes the root (coming from $z(g,j)$ in eq. (B.9.2)) and the large dot to the right is the vertex on which d/dj has acted. The weight $1+j$ of such a vertex is then replaced by weight 1.

vertices in the graph. The graphical representation of the present $\chi(g,j)$ is shown in fig. B.9.2. Note the following: in order to obtain the correct number of graphs it is important to be aware that χ *has two marked vertices, and these marks can be distinguished:* one mark is on the root, while the other marked vertex came from removing a j from one of the vertices of order 1 when differentiating wrt j. The present definition will be more convenient when we consider graphs which are not tree graphs.

Now let us define

$$\Delta \equiv \frac{g^2}{4} - (1+j), \qquad z = \frac{g}{2} - \sqrt{\Delta} \qquad (B.9.3)$$

At the critical point we have $\Delta_c = \Delta(g_c, j) = 0$.

3. Show from the graphical representation of χ in fig. B.9.2 that

$$\chi(g,j) = \frac{1}{2\sqrt{\Delta}} = \frac{1}{g-2z}, \qquad (B.9.4)$$

which is of course what one obtains just by differentiating the $z(g,j)$ which we have already found explicitly. Expand $\chi(g,j)$ in inverse powers of g. The coefficient of $(1+j)^{n-1}/g^{2n-1}$ gives the number of BPs with $2n-1$ links and a root and another marked vertex and $n-1$ vertices of order one (each associated with a factor $(1+j)$). The first coefficients are 1,2 and 6. Draw the corresponding BPs with 1, 3 and 5 lines.

Let us turn to rooted BPs with loops. We have to define these. We choose to use a graphic definition, which is a generalization of Fig. B.9.1, shown in Fig. B.9.3. In Fig. B.9.3 we have generalized the definition of the two-point function (or susceptibility) from z in eq. (B.9.2) which had no loop, to the Z which has loops, i.e. the equation corresponding to Fig. B.9.3 reads:

$$Z = \frac{1}{g}\left((1+j) + Z^2 + \frac{1}{\Lambda}\chi^{(2)}\right), \qquad \chi^{(2)}(g,j,\Lambda) = \frac{dZ(g,j,\Lambda)}{dj}. \qquad (B.9.5)$$

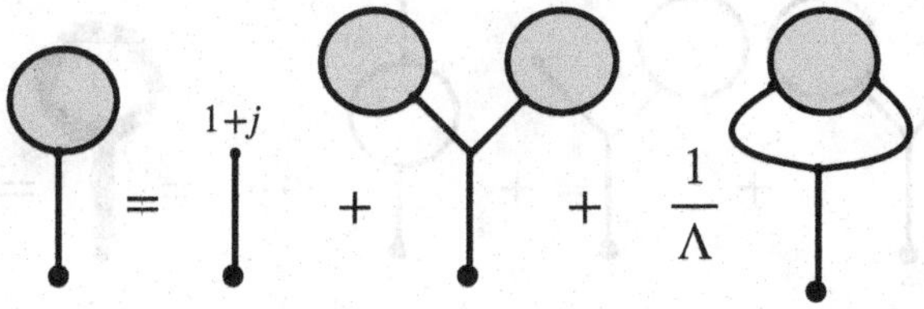

Figure B.9.3 The graphical equation for BPs with loops. It is no longer an algebraically closed equation since it involves the two-point function appearing in the last graph in the figure. This term comes from the vertex of order 3, connected to the root, where the links not connected to the root will be part of a loop. A new coupling constant Λ is introduced, such that Λ^{-n} will be multiplying graphs with n loops when we solve the graphical equation iteratively in the number of loop.

Thus we now have a differential equation in Z, rather than an algebraic equation as for the BPs without loops. However, we can turn eq. (B.9.5) into an infinite set of algebraic equations by first introducing the k-point functions $\chi^{(k)}$:

$$\chi^{(k+1)}(g,j,\Lambda) = \frac{d\chi^{(k)}(g,j,\Lambda)}{dj} = \frac{d^k Z(g,j,\Lambda)}{dj^k}, \quad (B.9.6)$$

and then differentiation eq. (B.9.5) wrt j:

$$\chi^{(2)} = \frac{1}{g}\left(1 + 2Z\chi^{(2)} + \frac{1}{\Lambda}\chi^{(3)}\right), \quad (B.9.7)$$

$$\chi^{(3)} = \frac{1}{g}\left(2Z\chi^{(3)} + 2\chi^{(2)}\chi^{(2)} + \frac{1}{\Lambda}\chi^{(4)}\right), \quad (B.9.8)$$

$$\chi^{(4)} = \cdots\cdots \quad (B.9.9)$$

These equations allow us to make a systematic double expansion in powers of $1/g$ and $1/\Lambda$, such that for $j=0$ the number of graphs of the k-point function with n lines and ℓ loops is the coefficient to the power $g^{-n}\Lambda^{-\ell}$ when we make the expansion

$$Z(g,j,\Lambda) = \sum_{\ell,n} \frac{Z_{n,\ell}(j)}{\Lambda^\ell g^n} = \sum_{\ell=0}^{\infty} \frac{Z_\ell(g,j)}{\Lambda^\ell}, \quad Z_\ell(g,j) = \sum_{n=|3\ell-1|}^{\infty} \frac{Z_{n,\ell}(j)}{g^n}, \quad (B.9.10)$$

$$\chi^{(k)}(g,j,\Lambda) = \sum_{\ell,n} \frac{\chi^{(k)}_{n,\ell}(j)}{\Lambda^\ell g^n} = \sum_{\ell=0}^{\infty} \frac{\chi^{(k)}_\ell(g,j)}{\Lambda^\ell}, \quad \chi^{(k)}_\ell(g,j) = \sum_{n=|3\ell-1|}^{\infty} \frac{\chi^{(k)}_{n,\ell}(j)}{g^n} \quad (B.9.11)$$

Thus $Z_{n,\ell}(j=0)$ is the number of graphs with ℓ loops and n lines. A natural starting point of the iteration of these equations is the BPs, i.e. Z_0 and $\chi^{(k)}_0$ since we know

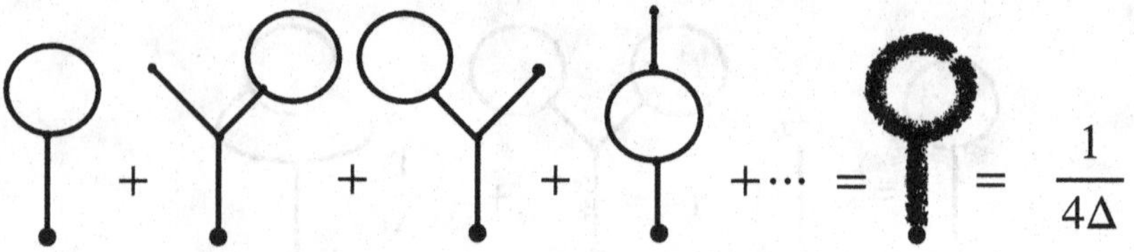

Figure B.9.4 The figure illustrates how the summation of one-loop rooted BPs can be represented as a BP two-point function connected to the root and a another BP two-point function forming a loop meeting at the endpoint of the BP two-point function connected to the root.

these functions explicitly:

$$Z_0 = \frac{g}{2} - \sqrt{\Delta} = \frac{1}{2}\left(g - \sqrt{g^2 - 4(1+j)}\right) \quad (B.9.12)$$

$$\chi_0^{(2)} = \frac{1}{g - 2Z_0} = \frac{1}{2\sqrt{\Delta}} \quad (B.9.13)$$

$$\chi_0^{(k+2)} = 2^k (2k-1)!! \left(\chi_0^{(2)}\right)^{2k+1}. \quad (B.9.14)$$

4. Show that the first iterations are

$$Z_1 = \frac{1}{g}\left(2Z_0 Z_1 + \chi_0^{(2)}\right), \quad (B.9.15)$$

$$(B.9.16)$$

$$Z_2 = \frac{1}{g}\left(2Z_0 Z_2 + Z_1^2 + \chi_1^{(2)}\right), \quad (B.9.17)$$

and show that it can be written

$$Z_1 = \left(\chi_0^{(2)}\right)^2, \quad \chi_1^{(2)} = 4\left(\chi_0^{(2)}\right)^3, \quad Z_2 = 5\left(\chi_0^{(2)}\right)^5. \quad (B.9.18)$$

5. Argue that the one-loop diagrams corresponding to Z_1 can be presented as in Fig. B.9.4. Find the coefficients to $1/g^2$, $1/g^4$ (and, if you are energetic, $1/g^6$ in the expansion of $1/(4\Delta)$. These are the number of one-loop diagrams with 2, 4 and 6 lines. Draw them.

6. Argue (no proof...) that the graphs representing Z_2 are of the form shown in Fig. B.9.5, i.e. all two-loop ϕ^3 graphs with one "external" line and one "external" vertex. Extend the arguments to Z_ℓ with ℓ loops, which can then be represented as dressed ℓ-loop ϕ^3 "tadpole" graphs, and argue that generating function of these behave like

$$Z_\ell(g, j) = \frac{C_\ell}{\Delta^{3\ell/2 - 1/2}} \quad (B.9.19)$$

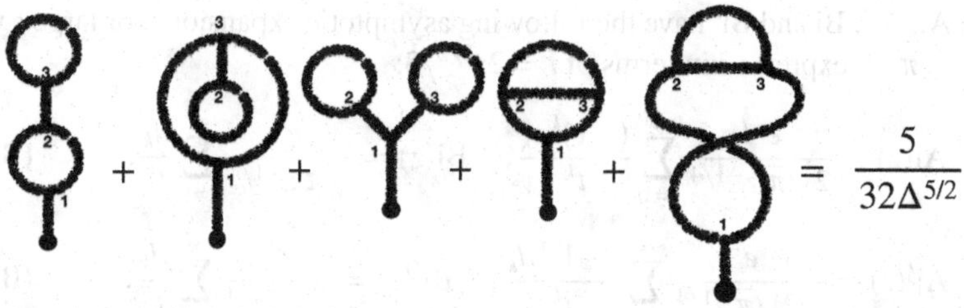

Figure B.9.5 The two-loop rooted BPs decomposed into tadpole diagrams where the vertices (except the root) are of order 3 and the lines represent BP two-point functions. Note that the coefficient here, 5/32, and the coefficient 1/4 in the former figure, are the coefficients reproduced in the asymptotic expansion given by eq. (B.9.31).

Returning to the full generating function for rooted BPs, also including loops, we write

$$Z(\Lambda, g, j) = \sum_{\ell=0}^{\infty} \frac{Z_\ell(g,j)}{\Lambda^\ell} = \frac{g}{2} + \sqrt{\Lambda} \sum_{\ell=0}^{\infty} \frac{C_\ell}{\left(\Lambda\Delta^{\frac{3}{2}}\right)^\ell}, \quad C_0 = -1. \tag{B.9.20}$$

Introduce the notation

$$\Lambda\Delta^{\frac{3}{2}} = \frac{3}{2}t \tag{B.9.21}$$

$$Z(g,j,\Lambda) = \frac{g}{2} - \Delta^{\frac{1}{2}} F(t) \tag{B.9.22}$$

We then insert this Z in the defining equation (B.9.5) for Z. Keeping g fixed we have

$$\Delta = \frac{g^2}{4} - (1+j) \implies \frac{d}{dj} = -\frac{d}{d\Delta}$$

and thus, using F from (B.9.22) instead of Z

$$F^2 + \frac{F}{3t} + \frac{dF}{dt} = 1. \tag{B.9.23}$$

In this equation we finally put $j = 0$.

7. Show that eq. (B.9.5) implies eq. (B.9.23).

Eq. (B.9.23) a so-called *Riccati equation* and it can be solved. The solution can be expressed in terms of Airy functions Ai, Bi and their derivatives. The Airy functions $w(z)$ are solutions to following differential equation

$$\frac{d^2w}{dz^2} = zw(z), \quad w(z) = a\operatorname{Ai}(z) + b\operatorname{Bi}(z), \tag{B.9.24}$$

and the solution to eq. (B.9.23) in terms of Airy functions is then

$$F(t) = \frac{1}{\sqrt{z}} \frac{w'(z)}{w(z)}, \quad t = \frac{2}{3} z^{3/2}. \tag{B.9.25}$$

where Ai, Ai', Bi and Bi' have the following asymptotic expansions for large z where $|\arg z| \leq \pi/3$, expressed in terms of $t = 2z^{3/2}/3$:

$$\text{Ai}(z) = \frac{e^{-t}}{2\sqrt{\pi} z^{-1/4}} \sum_{k=0}^{\infty} \frac{(-1)^k a_k}{t^k}, \quad \text{Bi}(z) = \frac{e^t}{\sqrt{\pi} z^{-1/4}} \sum_{k=0}^{\infty} \frac{a_k}{t^k}. \tag{B.9.26}$$

$$\text{Ai}'(z) = -\frac{e^{-t}}{2\sqrt{\pi} z^{1/4}} \sum_{k=0}^{\infty} \frac{(-1)^k d_k}{t^k}, \quad \text{Bi}'(z) = \frac{e^t}{\sqrt{\pi} z^{1/4}} \sum_{k=0}^{\infty} \frac{d_k}{t^k}. \tag{B.9.27}$$

One has

$$a_k = \frac{\Gamma(3k+\frac{1}{2})}{54^k k! \Gamma(k+\frac{1}{2})}, \quad d_k = -\frac{6k+1}{6k-1} a_k. \tag{B.9.28}$$

These expansions are only asymptotic since the coefficients $a_k, |d_k|$ grow like $k!$.

From the asymptotic expansions (B.9.26) we see that the requirement that the asymptotic expansion of $F(t)$ starts out as $1 + O(1/t)$ only fixes $F(t)$ up to exponential corrections of order e^{-2t}. We have

$$F(t) = \frac{\text{Bi}'(z)}{\text{Bi}(z)} \left(\frac{1 + c\,\text{Ai}'(z)/\text{Bi}'(z)}{1 + c\,\text{Ai}(z)/\text{Bi}(z)} \right) \tag{B.9.29}$$

$$= 1 - \frac{1}{6t} - \frac{5}{72t^2} + \cdots + e^{-2t}\left(k_0 + \frac{k_1}{t} + \cdots\right) + e^{-4t}\left(h_0 + \frac{h_1}{t} + \cdots\right) + \cdots \tag{B.9.30}$$

So there is a one-parameter class of solutions, depending on the constant c, which have the same leading asymptotic expansion

$$F(t) = 1 - \frac{1}{6t} - \frac{5}{72t^2} - \cdots \tag{B.9.31}$$

8. Write

$$F(t) = \sum_{n=0}^{\infty} \frac{c_n}{t^n}$$

and use (B.9.23) to find a recursive relation for the c_n and find the first two coefficients shown in (B.9.31).

9. Show that

$$c_n \propto -\frac{\Gamma(n)}{2^n} \quad \text{for } n \to \infty,$$

up to factors $n^{-\alpha}(1 + O(1/n))$ where α is not determined.

Finally, we have achieved our goal:

$$Z(g, \Lambda) = \frac{g}{2} - \sqrt{\Delta} + \sqrt{\Delta} \sum_{\ell \geq 1} \left(\frac{3}{2}\right)^{\ell} \frac{c_{\ell}}{\left(\Delta^{\frac{3}{2}} \Lambda\right)^{\ell}} \tag{B.9.32}$$

where

$$\sqrt{\Delta} = \sqrt{\frac{g^2}{4} - 1} = \frac{1}{2}\sqrt{g - g_c}, \quad g_c = 2 \tag{B.9.33}$$

Let us discuss if we can associate any critical behavior to this partition function when $g \to g_c = 2$. Our starting point was that $Z_0(g) = g/2 - \sqrt{\Delta}$ was the partition function for BPs and that it has $\gamma = 1/2$, i.e. $Z_0(g) = c_0 + c_1(g-g_c)^{1-\gamma} + \cdots$, where $\gamma = 1/2$. Similarly we have for $Z_\ell(g)$, the partition function for PBs with ℓ loops, that $Z_\ell(g) \propto \Delta^{(1-3\ell)/2}$.

10. Show that $Z_\ell(g) \sim (g-g_c)^{1-\gamma_\ell}$ where $\gamma_\ell = \frac{3}{2}\ell + \frac{1}{2}$.

It is now clear that we cannot associate an ordinary critical behavior to the function $Z(g, \Lambda)$ given by eq. (B.9.32) since for fixed Λ it becomes more and more singular for increasing number of loops when $g \to g_c$. However, one can try to take a so-called *double scaling limit*, where we together with a scaling $g \to g_c$ also scale the "coupling constant for loops", $1/\Lambda$ to zero such that

$$\Lambda \Delta^{3/2} = \frac{3}{2} t \quad \text{is fixed.} \tag{B.9.34}$$

The "physics" of this double scaling limit is the following: all partition functions $Z_\ell(g)$ have the same critical point $g_c = 2$. This is actually quite remarkable. The number $Z_\ell(k)$ of BPs with k links and ℓ loops has a leading asymptotic behavior

$$Z_\ell(g) = \sum_k \frac{Z_\ell(k)}{g^k}, \quad Z_\ell(k) \sim k^{\gamma_\ell - 2} 2^k (1 + O(1/k)) \quad \text{for} \quad k \gg 1. \tag{B.9.35}$$

Thus there is an exponential growth 2^k of the ℓ-loop BPs with the number of links. However, the sub-leading, but universal, factor $k^{\gamma_\ell - 2}$ grows with ℓ, and since ℓ in principle can be of an order proportional to k this factor can actually end up being more important than the exponential growth. This has two consequences: (1) eventually, for large k the number of BPs with a very large number of loops (proportional to k) will completely dominate in numbers those of small ℓ and (2) this rapid growth (factorial, not exponential) is the reason that the partition function $Z(g, \Lambda)$ given by eq. (B.9.32) is only given by an asymptotic expansion which is not convergent, and (as we argued) thus does not uniquely define $Z(g, \Lambda)$. The double scaling limit is an attempt to take a limit which tries to make a compromise between allowing the number of links to go to infinity (which is needed if we want to associate any continuum physics to BPs), i.e. to let $g \to g_c$ and thus $\Delta \to 0$, and at the same time allow BPs of arbitrary high loop number ℓ to play a role. Clearly taking $\Lambda \to \infty$ suppress graphs with a large number of loops, but the double scaling limit is the only one where we in principle can have graphs with an infinite number of links co-existing together with graphs having an infinite number of loops.

11. Finally: show that starting out with any BP (weights $w_3, w_4, \ldots$) where $\gamma = \frac{1}{2}$, the leading higher loop diagrams reduce precisely to the ϕ^3 diagrams we have already considered (just with changed $w_3 \neq 1$).

Lesson: We have given a perturbative definition of $Z(g, \Lambda)$ and we have found the expansion, and even explicit functions (the Airy functions) which reproduce this expansion. However, the $Z(g,\Lambda)$ is not uniquely fixed by its asymptotic expansion. In order to completely fix it we need a *non-perturbative* definition (which we do not have).

This example illustrates in a quite precise way the problem encountered in string theory, where one has a well-defined expansion in genus of the worldsheet (the equivalent to our expansion in loops), but is lacking a non-perturbative definition of string theory itself. We can even go one step further and study the BP equivalence to the attempts in string theory to find a non-perturbative definition of the theory.

BPs with Loops formulated as a ϕ^3 Field theory

The graphs we have studied from a combinatorial point of view are basically ϕ^3-graphs. It should thus not come as a surprise that the defining equation can be derived from a ϕ^3 "field theory". We write "field theory" because we will only keep the zero-dimensional real number ϕ in the path integral, not the real field $\phi(x)$. In this way the path integral will just generate the graphs, but propagators will be trivial equal to a number (which we choose to match the $1/g$ we assigned to each link in the combinatorial approach.

We define the partition function of the ϕ^3-graphs to be

$$\Omega(g,j,\Lambda) = \int d\phi \, e^{-S(\phi)}, \qquad S(\phi) = \Lambda\left(\frac{1}{2}g\phi^2 - \frac{1}{3}\phi^3 - (1+j)\phi\right) \qquad (B.9.36)$$

We are already here facing the problem that the integral is ill-defined if we simply integrate ϕ along the real axis. However, for the moment we will ignore this. We obtain the Feynman graphs corresponding to Ω by expanding $e^{\Lambda(\phi^3/3+(1+j)\phi)}$ in powers of ϕ and performing the remaining Gaussian integral using Wick's theorem. This procedure *is* well-defined to any finite order. If we consider a connected *tadpole* Feynman graph (i.e. a graph coming from the interaction term $\Lambda(\phi^3/3 + (1+j)\phi)$ with one "external" vertex of order 1) where the total number of vertices is V and the number of links is L, it will have ℓ loops, where

$$\ell = L - V + 1 \qquad (B.9.37)$$

Since a factor Λ is associated to each vertex, except the external vertex, and a factor $1/\Lambda$ to each link, we see that the total Λ-factor associated to a connected tadpole Feynman graph with ℓ loops will be $1/\Lambda^\ell$. The other coupling constants are chosen such that the connected *inequivalent* tree-graphs are assigned a weight 1 for each vertex of order 3 and weight $(1+j)$ for each vertex of order 1, while each link is assigned a weight $1/g$, such that we reproduce the standard BPs.

In this field theoretical language we have:

$$\langle \phi \rangle := \frac{1}{\Omega} \int d\phi \, \phi \, e^{-S(\phi)} = \frac{1}{\Lambda} \frac{1}{\Omega} \frac{d\Omega}{dj} := Z \qquad \left(\text{i.e. } \frac{1}{\Lambda} \frac{d\Omega}{dj} = Z\Omega\right) \qquad (B.9.38)$$

Our partition function Z for BPs with one external vertex is precisely $\langle\phi\rangle$. Had we been in higher dimensions the vertex would have a coordinate x and we would have $\langle\phi(x)\rangle$. Similarly we can write

$$\langle\phi^2\rangle := \frac{1}{\Omega}\int d\phi\, \phi^2\, e^{-S(\phi)} = \frac{1}{\Lambda^2}\frac{1}{\Omega}\frac{d^2\Omega}{dj^2} = Z^2 + \frac{1}{\Lambda}\frac{dZ}{dj}. \tag{B.9.39}$$

The so-called Dyson-Schwinger equation states that the expectation value of the classical eom is zero. The classical eom is

$$0 = \frac{dS}{d\phi} = \Lambda\left(g\phi - \phi^2 - (1+j)\right). \tag{B.9.40}$$

Using eqs. (B.9.38) and (B.9.39) we obtain

$$gZ - \left(Z^2 + \frac{1}{\Lambda}\frac{dZ}{dj}\right) - (1+j) = 0, \tag{B.9.41}$$

which is precisely our fundamental graphical equation (B.9.5) for BPs. Let us for completeness derive the DS equation:

$$0 = \int d\phi\, \frac{d}{d\phi}e^{-S(\phi)} = -\int d\phi\, \left(\frac{dS}{d\phi}\right) e^{-S(\phi)}, \quad \text{thus} \quad \left\langle\frac{dS}{d\phi}\right\rangle = 0. \tag{B.9.42}$$

We now have the following situation: we have a perturbative expansion of graphs defined by the integral (B.9.36). The integral itself is ill defined when the integration contour is along the real axis, but the perturbative expansion makes sense to any (finite) order, by expanding the interaction $e^{\Lambda(\phi^3/3 + (1+j)\phi)}$ in powers of ϕ and performing the remaining Gaussian integration (as already remarked above). Is it possible to make the complete integral well-defined and in this way arrive at a non-perturbative definition of the theory? Yes, in fact it is easy: rotate the integration contour by $\pi/6$ in the complex ϕ plane. We thus make the substitution $\phi \to e^{i\pi/6}\phi$. In this way we still integrate over real ϕ, but the action is changed to

$$S_{\text{mod}}(\phi) = \Lambda\left(\frac{g}{4}(1+i\sqrt{3})\phi^2 - \frac{1+j}{2}(\sqrt{3}+i)\phi - \frac{i}{3}\phi^3\right) \tag{B.9.43}$$

and

$$\Omega_{\text{mod}} = e^{i\pi/6}\int_{-\infty}^{\infty} d\phi\, e^{-S_{\text{mod}}(\phi)} \tag{B.9.44}$$

is well-defined. Further it is easy to show that the expectation value of ϕ^n calculated *perturbatively* to a finite order is unchanged, since in such a calculation we have just performed a well-defined rotation of the contour of integration. One can also directly check that the factors of $e^{i\pi/6}$ cancel between vertices and links.

Thus it seems as if we have managed to define the summation over BPs non-perturbatively. However, it turns out that Ω_{mod} defined in this way is complex, and the non-perturbative contributions to $\langle\phi^n\rangle$ will typically be complex. Clearly we do not really want complex contributions and it reflects that the non-perturbative definition mentioned is not really based on any physical principle. Such a principle is presently missing, both for our BPs and for string theory.

B.10 PROBLEM SET 10

The purpose of this Problem Set is to show that the characterization of criticality for ensembles of polygon graphs can be done in a way very similar to what we did for BPs and also to generalize the generic behavior studied in Chapter 6 to so-called multicritical behavior, again as for BPs.

A general even potential V(x)

We use the notation

$$V(x) = \frac{1}{g}\sum_n t_n x^{2n} = \frac{1}{g}\tilde{V}(x) \tag{B.10.1}$$

and we will, for a start, consider the t_n as fixed such that we can only vary g. Note that the notation of this g is somewhat different from the g in Chapter 6. Recall that the generic behavior was obtained if $t_1 > 0$ and $t_n \leq 0$ for $n > 1$ (and at least one of these $t_n < 0$).

Consider

$$\oint_C \frac{d\omega}{2\pi i} \frac{f(\omega)}{\sqrt{\omega^2 - a^2}} \tag{B.10.2}$$

where $f(\omega)$ is analytic in a region Ω including the cut $[-a, a]$ and the contour C encircles the cut and is located in Ω.

1. Show that

$$\oint_C \frac{d\omega}{2\pi i} \frac{f(\omega)}{\sqrt{\omega^2 - a^2}} = \int_{-a}^{a} \frac{dx}{\pi} \frac{f(x)}{\sqrt{a^2 - x^2}} = \int_{-1}^{1} \frac{dy}{\pi} \frac{f(ay)}{\sqrt{1 - y^2}} \tag{B.10.3}$$

Hint: contract C to be just above and below the cut and use that

$$\omega = x \pm i\varepsilon \implies \frac{1}{\sqrt{\omega - a}} = \mp i \frac{1}{\sqrt{a - x}} \quad \text{for} \quad x \in]-a, a[$$

For integer $k > 0$ in

$$\oint_C \frac{d\omega}{2\pi i} \frac{f(\omega)}{(\omega^2 - a^2)^{k+\frac{1}{2}}}$$

one cannot simply contract the contour to the cut because the integral

$$\int_{-a}^{a} \frac{dx}{\pi} \frac{f(x)}{(a^2 - x^2)^{k+\frac{1}{2}}}$$

is singular. However, we have

$$\oint_C \frac{d\omega}{2\pi i} \frac{f(\omega)}{(\omega^2 - a^2)^{k+\frac{1}{2}}} = \frac{1}{k - \frac{1}{2}} \cdots \frac{1}{\frac{1}{2}} \left(\frac{d}{da^2}\right)^k \oint_C \frac{d\omega}{2\pi i} \frac{f(\omega)}{(\omega^2 - a^2)^{\frac{1}{2}}}$$

$$= \frac{1}{k - \frac{1}{2}} \cdots \frac{1}{\frac{1}{2}} \int_{-1}^{1} \frac{dy}{\pi} \frac{\left(\frac{d}{da^2}\right)^k f(ay)}{\sqrt{1 - y^2}}$$

2. Show from the formulas in Chapter 6 for the disk amplitude $W(z)$, that for an even potential we have

$$W(z) = \left(\int_0^a \frac{dx}{\pi} \frac{xV'(x)}{(z^2-x^2)\sqrt{a^2-x^2}}\right)\sqrt{z^2-a^2} \qquad \text{(B.10.4)}$$

3. Show that for $V(x)$ given by (B.10.1) the condition $W(z) \to \frac{1}{z}$ for $|z| \to \infty$ leads to

$$g(a^2) = \int_0^a \frac{dx}{\pi} \frac{x\tilde{V}'(x)}{\sqrt{a^2-x^2}} \qquad \text{(B.10.5)}$$

This determines g as a function of the position of the cut a.

4. Show that

$$g(a^2) = \int_0^a \frac{dx}{\pi} \frac{x\tilde{V}'(x)}{\sqrt{a^2-x^2}} = \sum_n \frac{t_n a^{2n}}{B(n,\frac{1}{2})} \qquad \text{(B.10.6)}$$

Hint: use the integral below and the definition of the beta-function:

$$\int_0^{\frac{\pi}{2}} d\theta \sin^{2n}\theta = \frac{\pi}{2} \cdot \frac{(2n-1)!!}{(2n)!!}, \qquad B(x,y) := \frac{\Gamma(x)\Gamma(y)}{\Gamma(x+y)}$$

This fixes $g(a^2)$ as a polynomial, knowing the poynomial $\tilde{V}(x)$.

Consider the simplest situation where

$$t_1 = \frac{1}{2}, \quad t_2 = -\frac{1}{4} \implies \tilde{V}'(x) = x - x^3$$

5. Show that

$$g(a^2) = \frac{1}{4}a^2 - \frac{3}{16}a^4$$

$$g(a^2) = g(a_c^2) - \frac{3}{16}(a_c^2 - a^2)^2 = \frac{1}{12} - \frac{3}{16}\left(\frac{2}{3} - a^2\right)^2$$

Thus we start at $g = 0$ for $a^2 = 0$ and reach g_c where $\frac{dg}{da^2} = 0$ for $a_c^2 = \frac{2}{3}$. For larger g we have no solution $g(a^2)$ (see Fig. B.10.1). The situation is thus somewhat similar to the BP case (except that the g used here is more like the $e^{-\mu} = 1/g_{BP}$: $g \to 0$ corresponds to the partition function going to zero, as does $\mu \to \infty$ for BPs. This is why the curve turns downwards on the figure rather than upwards as in the BP case): the critical behavior is obtained when g is approaching it maximal value g_c and the continuum limit is obtained by expanding around that maximum.

6. Show that we have the same qualitative behavior for $t_1 > 0, t_n \leq 0$ for $n > 1$ and at least one $t_n < 0$.

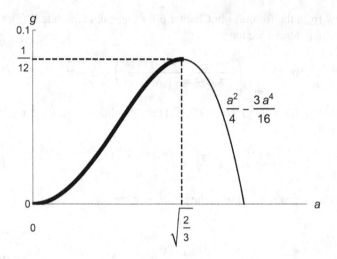

Figure B.10.1 The graph shown is $g(a^2) = \frac{1}{4}a^2 - \frac{3}{16}a^4$. The physical region of the curve is from $(a,g) = (0,0)$ to $(\sqrt{2/3}, 1/12)$, the top of the curve, and is shown as the thick black part of the graph.

Thus: universality! Close to g_c we have $g_c - g \approx (a_c^2 - a^2)^2$, or equivalently $a^2 \approx a_c^2 - c \cdot \sqrt{g_c - g}$.

We see that the situation is very similar to the BP case, and inspired by BPs we can now define multicriticality by dropping the requirement that $t_n \leq 0$ for $n > 1$. We thus lose a strict probabilistic interpretation of the random triangulations. However, like for BPs it is often possible to view the negative weights as coming from some matter interacting with the random geometry.

Let us in the same way as for BPs define a multicritical point by

$$\frac{dg}{da^2}\bigg|_{a^2=a_c^2} = 0, \cdots, \left(\frac{d}{da^2}\right)^{m-1} g\bigg|_{a^2=a_c^2} = 0, \quad \left(\frac{d}{da^2}\right)^m g\bigg|_{a^2=a_c^2} \neq 0 \quad \text{(B.10.7)}$$

To satisfy this we require a polynomial of order at least $2m$ (we consider only even potentials $V(x)$).

7. Show that *if* the polynomial is of order $2m$ and we assume $t_1 = 1/2$, then we have:

$$g(a^2) = g(a_c^2) - c(a_c^2 - a^2)^m, \quad g(a_c^2) = c \cdot a_c^{2m}, \quad c = \frac{1}{4m a_c^{2m-2}}$$
(B.10.8)

The value $a_c^2 > 0$ can be chosen arbitrarily, but after that the coefficients $t_n, n > 1$ are completely fixed (we already assumed $t_1 = 1/2$). We choose $a_c^2 = 1$ from now on.

8. Show that with the choice $a_c = 1$ we have (in the case of the $2m$'th order polynomial)

$$t_n = \frac{(-1)^{n-1}}{4m} \binom{m}{n} B\left(n, \frac{1}{2}\right) \quad \text{for } n \leq m \qquad t_n = 0 \quad \text{for } n > m \quad \text{(B.10.9)}$$

We thus have the multicritical behavior ($a_c^2 = 1$)

$$a^2 = 1 - \left(1 - \frac{g}{g_c}\right)^{\frac{1}{m}}, \qquad g_c = \frac{1}{4m} \qquad \text{(B.10.10)}$$

Define

$$\tilde{M}_k(a^2) = \oint_C \frac{d\omega}{2\pi i} \frac{\omega \tilde{V}'(\omega)}{(\omega^2 - a^2)^{k+\frac{1}{2}}} \qquad \text{(B.10.11)}$$

where $M_k = \frac{1}{g}\tilde{M}_k$, with M_k as in Chapter 6.

9. Show that

$$\tilde{M}_0(a^2) = 2g(a^2) \qquad \text{(B.10.12)}$$

Define $\varepsilon = a_c^2 - a^2 \sim \left(1 - \frac{g}{g_c}\right)^{\frac{1}{m}}$.

10. Show that $\varepsilon \to 0$

$$M_k(a^2) = \mu_k \varepsilon^{m-k} + O\left(\varepsilon^{m-k+1}\right), 0 < k < m, \qquad M_m(a^2) \neq 0,$$
$$\text{(B.10.13)}$$

The potential corresponding to the choice (B.10.9) of t_n (where only g is allowed to vary) is called the *Kazakov potential*, and varying g we have a behavior like (B.10.13). Taking $\varepsilon \to 0$ (or $g \to g_c$) we approach the *mth-multicritical point* in a specific way.

A more general approach to the mth-multicritical point is obtained by also allowing $t_n \to t_n + \delta t_n$ but in such a way that (B.10.13) is satisfied (with μ_k depending on δt_n). We say that the choice $\mu_1, \cdots, \mu_{m-1}$ defines the approach to the mth-multicritical point. One can show that the μ_k's are related to so-called intersection indices on Riemann surfaces.

We have now defined the so-called multicritical behavior, if we have a situation like (B.10.10), with $m = 2, 3, \ldots$:

$$a^2 = 1 - \left(1 - \frac{g}{g_c}\right)^{1/m}, \quad g_c = \frac{1}{4m}, \quad \text{or} \quad g(a^2) = \frac{1}{4m} - \frac{1}{4m}(1-a^2)^m. \quad \text{(B.10.14)}$$

Let us now generalize the critical behavior, for $m < s < m+1$ to

$$a^2 = 1 - \left(1 - \frac{g}{g_c}\right)^{1/s}, \quad g_c = \frac{1}{4s}, \quad \text{or} \quad g(a^2) = \frac{1}{4s} - \frac{1}{4s}(1-a^2)^s. \quad (B.10.15)$$

From eq. (B.10.6) we can now find potential $\tilde{V}(x)$ by an expansion of $(1-a^2)^s$ in powers of a^2. We have already made this expansion in problem 5, dealing with multicritical BPs.

11. Show that we have

$$t_n \sim \frac{1}{n^{s+3/2}} \quad \text{for} \quad n \to \infty. \quad (B.10.16)$$

Thus the potential $\tilde{V}(x)$ is given by an infinite power series in x^2. The corresponding function is a hypergeometric function. So whenever s is non-integer we need triangulations which have vertices of arbitrarily high order, if the model shall reproduce a critical behavior like (B.10.15)

We end this exercise by proving an amazing universal result:

$$\frac{dgW(z)}{dg} = \frac{1}{\sqrt{z^2 - a^2}} \quad (B.10.17)$$

This result is true for *any* even potential $V(x)$ of the form (B.10.1). We call it an amazing universal result, but it should maybe not come as big surprise, considering that we have already in Chapter 6 proven that the two-loop function is universal in the sense that it only depends on a, and eq. (B.10.17) is essential the disk amplitude, differentiated after g. This differentiation corresponds to putting a mark everywhere on the disk, and this is combinatorially the same as contracting one of the loops to a point. Thus one should be able to obtain eq. (B.10.17) from the expression for the two-loop function and one can indeed do that (after some work...). In Chapter 6 we mainly dealt with positive probabilities, i.e. $t_n \leq 0$ for $n \geq 2$, but the combinatorial arguments are still valid if we drop that restriction on the t_n's. Eq. (B.10.17) has a direct translation to conformal field theories coupled to 2d quantum gravity in the scaling limit $g \to g_c$, $a \to a_c$ as we will discuss later.

12. Show that

$$\tilde{M}_1(a^2) = 4\frac{dg}{da^2}, \quad \tilde{M}_{k+1}(a^2) = \frac{1}{k+\frac{1}{2}} \frac{d}{da^2} \tilde{M}_k(a^2) \quad (B.10.18)$$

13. Show that we can write

$$gW(z) = \frac{1}{2}\left[\tilde{V}(z) - \tilde{M}(z)\sqrt{z^2 - a^2}\right], \quad \tilde{M}(z) = \sum_{k=1}^{m} \tilde{M}_k(z^2 - a^2)^{k-1}. \quad (B.10.19)$$

14. Use (B.10.18) to show (B.10.17). Hint:

$$\frac{dgW}{dg} = \frac{da^2}{dg}\frac{dgW}{da^2} = -\frac{1}{2}\frac{da^2}{dg}\left(\frac{d\tilde{M}(z)}{da^2}\sqrt{z^2-a^2} - \frac{\tilde{M}(z)}{2\sqrt{z^2-a^2}}\right)$$
$$= \frac{1}{\tilde{M}_1}\frac{1}{\sqrt{z^2-a^2}}\left(\tilde{M}(z) - 2(z^2-a^2)\frac{d\tilde{M}(z)}{da^2}\right) \qquad (B.10.20)$$

The continuum limit of (B.10.17) reads

$$\frac{dW(Z)}{d\Lambda} = \frac{1}{\sqrt{Z+\sqrt{\Lambda}}} \qquad (B.10.21)$$

where

$$z^2 = a_c^2 + \varepsilon Z, \quad a^2 = a_c^2 - \varepsilon\sqrt{\Lambda}, \quad g - g_c \sim (a_c^2 - a^2)^m \sim \varepsilon^m \Lambda^{m/2} \qquad (B.10.22)$$

$W(z)$ is the disk amplitude with one marked point at the boundary and Z the boundary cosmological constant. Differentiating with respect to the cosmological constant Λ corresponds to an insertion anywhere. It turns out that one obtains the same result in so-called quantum Liouville theory coupled to a (p,q) conformal field theory where the insertion is a specific so-called conformal operator, namely the so-called primary operator with largest negative dimension in the (p,q) conformal theory.

B.11 PROBLEM SET 11

MULTI-ISING SPINS COUPLED TO 2D GRAVITY

Physics of the Ising model on a regular lattice

The partition function of the Ising model is[2]

$$Z(\beta) = \sum_{\{\sigma_i\}} e^{\frac{\beta}{2}\sum_{\langle ij \rangle}(\sigma_i\sigma_j - 1)}, \quad \sigma = \pm 1. \tag{B.11.1}$$

where the Ising spins are placed at the vertices i of the lattice, and neighboring spins interact. $\langle ij \rangle$ denotes the link between site i and j if they are neighbors. $\{\sigma_i\}$ denotes the set of all spin configurations and the summation in the action is over all links.

For an infinite lattice there exists a so-called critical β_0, such that for $\beta > \beta_0$ we have magnetization while for $\beta < \beta_0$ we have no magnetization. The phase transition at β_0 is a second-order phase transition.

For very large β (small temperatures, $T = 1/\beta$) almost all $\sigma_i = 1$ (or almost all $\sigma_i = -1$). The excitations around the configuration where all $\sigma_i = 1$ are *small* spin clusters with $\sigma_i = -1$. The reason that the spin clusters are small is that the energy $E = \beta/2\sum_{\langle ij \rangle}(\sigma_i\sigma_j - 1)$ is only different from zero when two neighboring spins are different, i.e. along the boundaries between regions of $\sigma_i = 1$ and $\sigma_i = -1$. For a regular lattice, starting with all spins $\sigma_i = 1$, a large region with $\sigma_i = -1$ will also have a long boundary (at least like $\sqrt{A}$, A being the area of a region where $\sigma_i = -1$). Thus large regions of $\sigma_i = -1$ will be suppressed for large β if we start out in a state (the ground state) where all $\sigma_i = 1$.

As β decreases toward β_0 the size of $\sigma_i = -1$ spin clusters as well as the number of them will grow and at β_0 there is an equal number of $\sigma_i = 1$ and $\sigma_i = -1$ spins, and the distributions of $\pm$ spin clusters will be the same and the cluster sizes can be large.

The Ising model on dynamical triangulations

Let us now consider Ising spins coupled to dynamical triangulations (DT). We put the spins at the center of the triangles (this is not essential, but convenient here) as illustrated in fig. B.11.1. The partition function is defined as

$$Z(\mu, \beta) = \sum_T e^{-\mu N_T} Z_T(\beta). \tag{B.11.2}$$

---

[2] We have already defined the Ising model several times, and each time the definition has been slightly different because we have chosen the version most suited for our purpose. The physics associated with these slightly different models is of course the same.

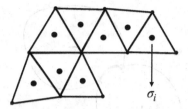

Figure B.11.1 The figure shows part of a triangulation. The Ising spins will be located at the centers of the triangles, marked as dots on the figure, and the index i in σ_i in eq. (B.11.1) now refers to these centers. The interaction is between neighboring triangles, i.e. each spin interacts with three neighboring spins and the links $\langle ij \rangle$ in eq. (B.11.1) now refer to the so-called dual links connecting the centers of neighboring triangles.

where the summation is over a suitable class of triangulations. $Z_T(\beta)$ refers to the partition function (B.11.1), defined as mentioned on the graph corresponding to the triangulation T.

A few facts about this model (which we are not going to prove). For each β there is a critical $\mu_0(\beta)$. In addition there is a β_0 such that for $\beta > \beta_0$ there is magnetiztion and for $\beta < \beta_0$ there is no magnetiztion. The phase transition at β_0 is *third order* on the ensemble of DT. Also the critical exponents $\alpha, \beta_m, \gamma_m$ for the magnetic system at β_0 are different from the famous Onsager exponents on a regular lattice. *So the DT ensemble of geometries influences the critical properties of the spin system.* One can express the size of the spin clusters at the critical β_0 as a function of the critical exponents, i.e. the fractal properties of spin clusters of the Ising model change on dynamical triangulations. *In addition the spin system influences the critical properties of geometry, but only for $\beta = \beta_0$, i.e. when the spin system itself is critical.* We know that for pure gravity (DT without matter), the susceptibility exponent is $\gamma = -1/2$. For $\beta \neq \beta_0$ this is still true for the combined system. However, for $\beta = \beta_0$ one finds $\gamma(\beta_0) = -1/3$. Thus the long range interactions of the large spin clusters also change the fractal structure of the dynamical triangulations.

The mean field model

The purpose of this Problem Set is to understand this interplay between geometry and matter in a simple "mean-field" model. The starting point is that the minimal boundaries separating $\pm$ spin clusters can be very different from those on a regular lattice as illustrated in fig. B.11.2. So for *some* geometries it is possible to have huge spin clusters separated by small boundaries, i.e. small energy. Conversely, the relative weight of these geometries will be enhanced in the combined matter-geometry ensemble relative to more regular triangulations which do not have such a "pinching", simply because the energy of the matter part will be small.

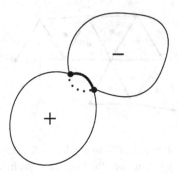

Figure B.11.2 A triangulation with the topology of the sphere belonging to the class $\mathcal{T}^{(2)}$, which can be cut in two almost equal parts along a minimal loop consisting of two links. On such a triangulation one can have macroscopic spin clusters separated by a minimal boundary consisting of the two links.

Let us now consider the following toy model designed to capture this: we allow two links to be connected to the same two vertices, but only if cutting the triangulation along the two links separates the triangulation in two disconnected parts (as illustrated in Fig. B.11.2). We now consider triangulations which have one boundary, and this boundary consists only of two links. We can now make a decomposition of the triangulation by peeling away "baby universes" connected to the rest of the surface by only two links and then closing the links. The fact that we have a boundary makes this a systematic procedure, the ultimate "parent" universe being connected to this boundary (see Fig. B.11.3). This class of triangulations is denoted $\mathcal{T}^{(2)}$ in Chapter 5. We now only sum over spin configurations which are such that the spin of a baby universe component is either $+$ or $-$. This approximation is inspired by fig. B.11.2 and is expected to be a good approximation for large β. It is also expected to be good for somewhat smaller β if we have many "independent" Ising models coupled to the ensemble $\mathcal{T}^{(2)}$. We write "independent" because different copies of the Ising spins do not interact directly, but they interact indirectly via the common geometry which they influence. The model is expected to be correct all the way down to a critical β_0 if n, the number of independent Ising spins, is sufficiently large. For $n=1$ it is only an approximation, which does not give the correct critical exponents (like ordinary mean-field theory in 2 and 3 dimensions for spin systems).
We can now write down the one-loop function

$$G(\mu,\beta) = \sum_{T \in \mathcal{T}^{(2)}(2)} e^{-\mu N_T} \sum_{\{\sigma_i\}}{}' e^{\frac{\beta}{2} \sum_{\langle ij \rangle} (\sigma_i \sigma_j - 1)} \qquad (B.11.3)$$

where $\mathcal{T}^{(2)}$ refers to the class of triangulations discussed and $\mathcal{T}^{(2)}(2)$ refers to this class with a boundary consisting of 2 links. The $\sum'$ refers to the summation over the restricted class of spin configurations we mentioned above. Eq. (B.11.3) is illustrated in Fig. B.11.4.

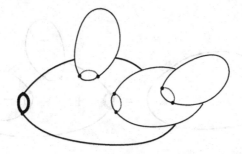

Figure B.11.3 A triangulation belonging to the class $\mathscr{T}_1^{(2)}$, where the boundary consists of two links. Relative to this boundary one can now define a hierarchical structure of baby universes sharing with their parent universe only two links as illustrated in the figure.

For a given β the model (B.11.3) has a critical $\mu_0(\beta)$, such that the sum is convergent for $\mu > \mu_0(\beta)$ and divergent for $\mu < \mu_0(\beta)$. We now define:

$$\chi(\mu,\beta) = -\frac{\partial G(\mu,\beta)}{\partial \mu} \qquad (\text{B.11.4})$$

If there is no Ising spin, i.e. we have our original pure gravity model from Chapter 6, we define

$$G_0(\mu) = \sum_{T \in \mathscr{T}^{(2)}(2)} e^{-\mu N_T}, \qquad \chi_0(\mu) = -\frac{\partial G_0(\mu)}{\partial \mu} \qquad (\text{B.11.5})$$

and we denote the corresponding critical μ by μ_0. We know that since $\gamma_0 = -1/2$ for the model without Ising spin that

$$\chi_0(\mu) = a_1 + a_2(\mu - \mu_0)^{1/2} + \cdots \qquad (\text{B.11.6})$$
$$G_0(\mu) = b_1 + b_2(\mu - \mu_0) + b_3(\mu - \mu_0)^{3/2} + \cdots \qquad (\text{B.11.7})$$

and we write

$$\chi(\mu,\beta) = c_1 + c_2(\mu - \mu_0(\beta)) + \cdots + c(\mu - \mu_0(\beta))^{-\gamma(\beta)} + \cdots \qquad (\text{B.11.8})$$

We want to determine $\gamma(\beta)$

1. Show, by first summing over the spin of baby universes, that

$$G(\mu,\beta) = \sum_{T \in \mathscr{T}^{(2)}(2)} e^{-\mu N_T} \left(1 + e^{-2\beta} G(\mu,\beta)\right)^{N_L(T)-2}$$

$$= \sum_{T \in \mathscr{T}^{(2)}(2)} e^{-\mu N_T} \left(1 + e^{-2\beta} G(\mu,\beta)\right)^{3N_T/2-1} \qquad (\text{B.11.9})$$

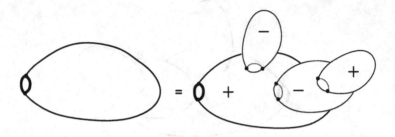

Figure B.11.4 Illustration of the spin configurations which respect the baby universe structure shown in Fig. B.11.3 and therefore can be included in the summation of the spin configurations in eq. (B.11.3).

We will ignore the -2 and -1 in the powers, since we will solve the model close to criticality where N_T and N_L are large.

2. show using (B.11.9) that we have

$$G(\mu,\beta) = \sum_{T\in\mathcal{T}^{(2)}(2)} e^{-\bar{\mu}N_T} = G_0(\bar{\mu}), \qquad (B.11.10)$$

$$\bar{\mu} = \mu - \frac{3}{2}\log\left(1+e^{-2\beta}G(\mu,\beta)\right), \qquad (B.11.11)$$

3. Show that

$$\mu = \bar{\mu} + \frac{3}{2}\log\left(1+e^{-2\beta}G_0(\bar{\mu})\right). \qquad (B.11.12)$$

4. Show that

$$\frac{\partial \mu}{\partial \bar{\mu}} = \frac{1-\frac{3}{2}e^{-2\beta}\chi_0(\bar{\mu})+e^{-2\beta}G_0(\bar{\mu})}{1+e^{-2\beta}G_0(\bar{\mu})} = \frac{e^{2\beta}-\left(\frac{3}{2}\chi_0(\bar{\mu})-G_0(\bar{\mu})\right)}{e^{2\beta}+G_0(\bar{\mu})} \qquad (B.11.13)$$

and thus that

$$\frac{\partial \bar{\mu}}{\partial \mu} = \frac{e^{2\beta}+G_0(\bar{\mu})}{e^{2\beta}-\left(\frac{3}{2}\chi_0(\bar{\mu})-G_0(\bar{\mu})\right)} \qquad (B.11.14)$$

5. Show, using $G(\mu,\beta)=G_0(\bar{\mu})$ (see (B.11.10)), that

$$\chi(\mu,\beta) = \chi_0(\bar{\mu})\frac{\partial \bar{\mu}}{\partial \mu} \qquad (B.11.15)$$

6. Show from the definition of $G_0(\mu)$ and $\chi_0(\mu)$ that $\frac{3}{2}\chi_0(\bar{\mu})-G_0(\bar{\mu})$ is a decreasing function of $\bar{\mu}$, with its maximum for the smallest possible value one can use in G_0 and χ_0, namely the critical value μ_0, i.e. $\bar{\mu}=\mu_0$. Note that both $\chi_0(\mu_0)$ and $G_0(\mu_0)$ are finite at μ_0, according to (B.11.6) and (B.11.7).

7. Show that this implies that there exists a β_0 such that the denominator in (B.11.14) is > 0 for all $\mu > \mu_0(\beta)$ provided $\beta > \beta_0$.

8. Let now $\beta > \beta_0$ be fixed. We have a critical point $\mu_0(\beta)$. Let us now decrease μ toward this critical point. The susceptibility satisfies (B.11.15), which we write in detail as

$$\chi(\mu,\beta) = \chi_0(\bar\mu(\mu,\beta)) \frac{\partial \bar\mu(\mu,\beta)}{\partial \mu}. \qquad (B.11.16)$$

Use this formula and the arguments above to argue that $\chi(\mu,\beta)$ can only be critical if

$$\bar\mu(\mu_0(\beta),\beta) = \mu_0, \qquad (B.11.17)$$

and that this implies that

$$\boxed{\gamma(\beta) = \gamma_0 = -1/2 \text{ for } \beta > \beta_0.} \qquad (B.11.18)$$

β_0 is the largest β for which

$$e^{2\beta} = \frac{3}{2}\chi_0(\bar\mu) - G_0(\bar\mu) \qquad (B.11.19)$$

has a solution. For $\beta \leq \beta_0$ we now define $\bar\mu_0(\beta)$ by

$$\left.\frac{\partial \mu}{\partial \bar\mu}\right|_{\bar\mu_0(\beta)} = 0. \qquad (B.11.20)$$

Thus, according to (B.11.13), $\bar\mu_0(\beta)$ satisfies eq. (B.11.19).

9. Show that we have

$$\bar\mu_0(\beta_0) = \mu_0, \qquad \bar\mu_0(\beta) > \mu_0 \text{ for } \beta < \beta_0. \qquad (B.11.21)$$

10. Show that (B.11.20) implies that

$$\mu - \mu_0(\beta) = c\left(\bar\mu - \bar\mu_0(\beta)\right)^2 + O((\bar\mu - \bar\mu_0(\beta))^3). \qquad (B.11.22)$$

11. *Assume now that $\beta < \beta_0$*. Use (B.11.15), (B.11.21) and (B.11.22) to show, by Taylor expanding around $\bar\mu_0(\beta)$ ($> \mu_0$, so one *can* Taylor expand), that for $\mu \to \mu_0(\beta)$ we have

$$\chi(\mu,\beta) \sim \frac{1}{\bar\mu - \bar\mu_0(\beta)} \sim \frac{1}{\sqrt{\mu - \mu_0(\beta)}} \qquad \beta < \beta_0. \qquad (B.11.23)$$

Thus

$$\boxed{\gamma(\beta) = 1/2 \text{ for } \beta < \beta_0} \text{ like BPs!} \qquad (B.11.24)$$

We have now seen that there is a phase transition at β_0, where the critical exponent $\gamma(\beta)$ jumps from $-1/2$, the value for pure gravity without Ising spins, for $\beta > \beta_0$, to $\gamma(\beta) = 1/2$, the value for BPs, for $\beta < \beta_0$.

We will finally determine $\gamma(\beta_0)$. We can no longer Taylor expand for $\mu \to \mu_0(\beta)$ because $\bar{\mu}_0(\beta_0) = \mu_0$ (see (B.11.21)), and $\chi_0(\mu_0)$ and $G_0(\mu_0)$ are not analytic in that point. However, we know their behavior there, see (B.11.6) and (B.11.7).

12. Show that
$$\frac{\partial \mu}{\partial \bar{\mu}} \sim (\bar{\mu} - \bar{\mu}_0(\beta_0))^{-\gamma_0} + O(\bar{\mu} - \mu_0) \qquad \text{(B.11.25)}$$

13) Show
$$\mu - \mu_0(\beta_0) = c(\bar{\mu} - \mu_0)^{1-\gamma_0} + O((\bar{\mu} - \mu_0)^2) \qquad \text{(B.11.26)}$$

14) Show, using (B.11.15), that this implies
$$\chi(\mu, \beta_0) \sim \frac{1}{(\bar{\mu} - \mu_0(\beta_0))^{-\gamma_0}} \sim \frac{1}{(\mu - \mu_0(\beta_0))^{-\gamma_0/(1-\gamma_0)}} \qquad \text{(B.11.27)}$$

Thus
$$\boxed{\gamma(\beta_0) = \frac{\gamma_0}{\gamma_0 - 1} = \frac{1}{3}} \qquad \text{(B.11.28)}$$

Summary: For large β ($\beta > \beta_0$), i.e. for low temperature, we have $\gamma(\beta) = \gamma_0$ ($= -1/2$). This is the magnetized phase where spin fluctuations are small, and the geometry is not affected by the spin. At $\beta = \beta_0$ there is a phase transition. At the transition $\gamma(\beta)$ jumps to $1/3$. For $\beta < \beta_0$ (high temperature) there are many baby universes and $\gamma(\beta) = 1/2$, like for BPs. In this phase there is no spontaneous magnetization in accordance with the fact that BPs have no spontaneous magnetization.

General remarks: the high temperature phase of our model, where $\beta < \beta_0$, does not represent well a single Ising spin coupled to DT (as already mentioned). In the real, full model one has $\gamma(\beta) = -1/2$ for $\beta < \beta_0$. Also, in the full model $\gamma(\beta_0) = -1/3$ (and not $1/3$). However, again as already mentioned, the model represents very well many Ising spins coupled to DT. The models with many Ising spins coupled to DT cannot be solved analytically, but have been studied by computer simulations.

Finally: Note that the physics of the magnetized baby universes seems amazingly similar to the physics of real magnets, the baby universes playing the role of magnetized domains.

B.12 PROBLEM SET 12

Deriving the Multiloop Formulas

The purpose of this problem set is to derive the the multiloop formulas (6.73), (6.77) and (6.78) using (6.72). We will simply use the representation (6.61) for the loop insertion operator and act on the disk function $w(\vec{g}, z)$ written in the form (6.52), using the results (6.57)–(6.60). Let us write (6.73) in the following way

$$w(\vec{g}, \omega, z) = \frac{1}{(z^2 - \omega^2)^2} \left(-2z\omega + \frac{2z^2\omega^2 - c^2(z^2 + \omega^2)}{(z^2 - c^2)^{1/2}(\omega^2 - c^2)^{1/2}} \right) \quad \text{(B.12.1)}$$

1. Show that

$$\frac{2}{\tilde{M}_1} \frac{d}{dc^2} \sum_{k=1}^{\infty} \tilde{M}_k (\omega^2 - c^2)^{k-1/2} = -\frac{1}{(\omega^2 - c^2)^{1/2}}. \quad \text{(B.12.2)}$$

2. Show that

$$\frac{\partial}{\partial V(z)} \sum_{k=1}^{\infty} \tilde{M}_k (\omega^2 - c^2)^{k-1/2} = \frac{d}{dz} \left[\left(\frac{\omega^2 - c^2}{z^2 - c^2} \right)^{1/2} \frac{z}{z^2 - \omega^2} \right]. \quad \text{(B.12.3)}$$

3. Use now (6.57) to write

$$\frac{dw(\vec{g}, \omega)}{dV(z)} = \frac{-2\omega z}{(z^2 - \omega^2)^2} + \frac{c^2}{(z^2 - c^2)^{\frac{3}{2}}(\omega^2 - c^2)^{\frac{1}{2}}} - \frac{d}{dz} \left[\left(\frac{\omega^2 - c^2}{z^2 - c^2} \right)^{\frac{1}{2}} \frac{z}{z^2 - \omega^2} \right]$$
(B.12.4)

and show that the last two terms, after differentiation, can be reorganized in the following way:

$$\frac{1}{(z^2 - \omega^2)^2} \left(\frac{(z^2 - \omega^2)(z^2 - c^2)\omega^2}{(z^2 - c^2)^{3/2}(\omega^2 - c^2)^{1/2}} + \frac{(\omega^2 - c^2)(z^2 + \omega^2)}{(z^2 - c^2)^{1/2}(\omega^2 - c^2)^{1/2}} \right) \quad \text{(B.12.5)}$$

4. Use the above to prove formula (B.12.1).

We now turn to the proof of the three-loop formula (6.77). Since the two-loop function only depends on the coupling constants $\vec{g}$ via the position of the cut, $c(\vec{g})$, the loop insertion operator becomes very simple in the form (6.52) when acting on the two-loop function

5. Prove that

$$\frac{d}{dc^2} \left(\frac{2z^2\omega^2 - c^2(z^2 + \omega^2)}{(z^2 - c^2)^{1/2}(\omega^2 - c^2)^{1/2}} \right) = \frac{1}{2} \frac{c^2(z^2 - \omega^2)^2}{(z^2 - c^2)^{3/2}(\omega^2 - c^2)^{3/2}} \quad \text{(B.12.6)}$$

6. Use this to prove formula (6.77) for the three-loop function

Let us next prove the 4-loop formula. What we have to show is that

$$\frac{d}{dV(z)} \frac{f(c)}{\tilde{M}_1} = \frac{2}{\tilde{M}_1} \frac{d}{dc^2} \frac{f(c)}{\tilde{M}_1(z^2-c^2)^{3/2}}. \tag{B.12.7}$$

7. Show that

$$\frac{\partial \tilde{M}_1}{\partial V(z)} = -\frac{d}{dc^2} \frac{2c^2}{(z^2-c^2)^{3/2}} \tag{B.12.8}$$

and use this to show (B.12.7).

Finally, let us turn to the n-loop formula, which we have just proven for $n = 3, 4$. Assume it is correct up to $n-1 \geq 4$.

8. Use (B.12.8) to prove the following

$$\left[\frac{d}{dV(z)}, \frac{2}{\tilde{M}_1} \frac{d}{dc^2}\right] = 0 \tag{B.12.9}$$

9. Use this and (B.12.7) to prove the multiloop formula (6.78)

B.13 PROBLEM SET 13

In this Problem Set we will solve the characteristic equation (8.52), use the solution to find the two-point function as well as to calculate the "average shape" of the quantum universe, also in a situation where the universe is "expanding" to infinity for $T \to \infty$.

The Characteristic Function and the Two Point Function

1. Show that the solution to

$$\frac{d\bar{X}(T)}{dT} = -\hat{W}(\bar{X}), \qquad \bar{X}(T=0;X) = X, \qquad (B.13.1)$$

can be written as

$$T = \int_{\bar{X}(T;X)}^{X} \frac{dx}{\hat{W}(x)} = \int_{\bar{X}(T;X)}^{X} \frac{dx}{(x-\alpha)\sqrt{(x+\alpha)^2 - \frac{2g_s}{\alpha}}} \qquad (B.13.2)$$

Here $\bar{X}(T)$ and X are both larger than α and it is seen that $T \to \infty$ implies that $\bar{X}(T) \to \alpha$. Let us now introduce the notation

$$\Sigma = \sqrt{\alpha^2 - \frac{g_s}{2\alpha}}, \qquad \cosh\beta = \frac{\alpha}{\sqrt{g_s/2\alpha}}, \qquad \sinh\beta = \frac{\Sigma}{\sqrt{g_s/2\alpha}}, \qquad (B.13.3)$$

The integral in (B.13.2) can be written as

$$T = -\frac{1}{\Sigma} \sinh^{-1}\sqrt{F(x)}\Big|_{\bar{X}(T)}^{X}, \qquad F(x) = \sinh^2(\beta/2) + \sinh^2\beta \frac{\sqrt{g_s/2\alpha}}{x-\alpha} \qquad (B.13.4)$$

2. Shown that (B.13.4) is the integral appearing in (B.13.2) by differentiation.

3. Show that (B.13.4) leads to

$$\bar{X}(T) - \alpha = \frac{\Sigma^2}{\sqrt{\frac{g_s}{2\alpha}}} \frac{1}{\sinh^2\left(\Sigma T + \sinh^{-1}\sqrt{F(X)}\right) - \sinh^2(\beta/2)} \qquad (B.13.5)$$

4. Show that the large T behavior of $\bar{X}(T)$ is

$$\bar{X}(T) - \alpha \to \frac{4\Sigma^2}{\sqrt{\frac{g_s}{2\alpha}}} \left(\sqrt{1+F(X)} - \sqrt{F(X)}\right)^2 e^{-2\Sigma T} \quad \text{for } T \to \infty \qquad (B.13.6)$$

5. Show that for $X \to \infty$ we have

$$\bar{X}(T) - \alpha = \frac{\Sigma^2}{\sqrt{\frac{g_s}{2\alpha}}} \frac{1}{\sinh(\Sigma T) \sinh(\Sigma T + \beta)}$$

$$= \frac{\Sigma^2}{\sinh(\Sigma T)\Big(\Sigma \cosh(\Sigma T) + \alpha \sinh(\Sigma T)\Big)}. \qquad (B.13.7)$$

6. Show that for $g_s \to 0$ (B.13.7) agrees with the corresponding CDT expression (8.23) in Chapter 8 (for $X \to \infty$).

We now turn to the two-point function $G_\Lambda(T)$. We have seen that it can be expressed as

$$G_\Lambda(T) = \frac{dW_\Lambda(\bar{X}(T, X=\infty))}{dT}, \qquad 2g_s W_\Lambda(x) = \Lambda - x^2 + \hat{W}_\Lambda(x). \qquad (B.13.8)$$

7. Show, using the notation $\bar{X}(T) \equiv \bar{X}(T, X=\infty)$, that

$$2g_s G_\Lambda(T) = -2\bar{X}(T)\frac{d\bar{X}(T)}{dT} - \frac{d^2\bar{X}(T)}{dT^2} \qquad (B.13.9)$$

$$= \frac{4\Sigma^3(\alpha^2 - \Sigma^2)\Big(\alpha \cosh(\Sigma T) + \Sigma \sinh(\Sigma T)\Big)}{\Big(\Sigma \cosh(\Sigma T) + \alpha \sinh(\Sigma T)\Big)^3} \qquad (B.13.10)$$

You should not try to perform the (trivial) differentiations leading from (B.13.9) to (B.13.10) by hand unless you really love calculations. Rather, convince yourself that the end result is precisely the formula for $G_\Lambda(T)$ given in Chapter 8.

The Average Shape of CDT and GCDT Universes

Until now we have mainly considered the two-loop function in the form $G_\Lambda(X, L; T)$, given by (8.52). However, here it will convenient to consider the situation where we have a boundary cosmological constant Y associated with the exit loop at T. The corresponding two-loop function $G_\Lambda(X, Y; T)$ is obtained by a Laplace transformation:

$$G_\Lambda(X, Y; T) = \int_0^\infty dL\, e^{-LY} G_\Lambda(X, L; T) = \frac{\hat{W}(\bar{X}(T; X))}{\hat{W}(X)} \frac{1}{\bar{X}(T; X) + Y} \qquad (B.13.11)$$

We thus have an ensemble of universes which start start out at $T=0$ with boundaries with a length distribution determined by the boundary cosmological constant X and which at time T have boundaries with a lengths distribution monitored by the boundary cosmological constant Y. It is natural to ask about the average length of a spatial universe at time t between 0 and T. We view $G_\Lambda(X,Y;T)$ as the partition function for the ensemble of universes and then the average length at t is defined as

$$\langle L(t)\rangle_{X,Y} = \frac{1}{G_\Lambda(X,Y;T)} \int_0^\infty dL\, G_\Lambda(X,L;t)\, L\, G_\Lambda(L,Y;T-t). \quad (\text{B.13.12})$$

We will show that

$$\langle L(t)\rangle_{X,Y} = \frac{\hat{W}'(\bar{X}(t;X)) - \hat{W}'(\bar{X}(T;X))}{\hat{W}(\bar{X}(t;X))} + \frac{\hat{W}(\bar{X}(T;X))}{\hat{W}(\bar{X}(t;X))} \frac{1}{\bar{X}(T;X)+Y} \quad (\text{B.13.13})$$

where $\hat{W}'(X)$ denotes the derivative of $\hat{W}(X)$ wrt X.

Before deriving (B.13.13) let us discuss some implication of the formula. First note that we have

$$\hat{W}(X) = \frac{(X-\alpha)}{\tilde{W}(X)}, \quad (\text{B.13.14})$$

where $\tilde{W}(\alpha) \neq 0$. This is true both in EDT, CDT and GCDT, just with slightly different[3] α and $\tilde{W}(X)$.

The smaller Y, the larger we expect length of the exit loop to be, and correspondingly also $\langle L(t)\rangle_{X,Y}$. In particular a negative Y will try expand the exit loop, to the extent it is possible (such an expansion will also typically result in an enlarged area, which is suppressed by the action). Let us assume $Y > -\alpha$. Now take $T \to \infty$. Recall that $\bar{X}(T;X) \to \alpha$ for $T \to \infty$

8. Show that

$$\langle L(t)\rangle_{X,Y} = \frac{\alpha}{\Sigma^2} + O(e^{-2\Sigma t}) \quad \text{for large } t. \quad (\text{B.13.18})$$

In the case of CDT (i.e. $g_s = 0$) this is just $1/\sqrt{\Lambda}$.

We are here considering a situation where T is infinity or very large, and when we are far away from from the entrance loop (and by construction very far away from the exit loop) and the average length of the boundary loop is then constant (and simply $1/\sqrt{\Lambda}$ in the case of CDT). Of course there are fluctuations and the situation is basically the one shown in Fig. 8.3 and captured in eq. (8.42). If we choose $Y < -\alpha$

---

[3] In all case $\tilde{W}(X)$ can be viewed as related to the disk amplitude, as discussed in connection with formula (8.58). We make here a list:

$$\tilde{W}(X) = \frac{1}{\sqrt{X+\sqrt{\Lambda}}}, \qquad \alpha = \sqrt{\Lambda}/2 \quad \text{EDT} \quad (\text{B.13.15})$$

$$\tilde{W}(X) = \frac{1}{X+\sqrt{\Lambda}}, \qquad \alpha = \sqrt{\Lambda} \quad \text{CDT} \quad (\text{B.13.16})$$

$$\tilde{W}(X) = \frac{1}{\sqrt{(X+\alpha)^2 - 2g_s/\alpha}}, \qquad \alpha = \alpha \quad \text{GCDT} \quad (\text{B.13.17})$$

the system becomes unstable and the exit boundary will expand to infinity in a finite time. However, exactly when $Y = -\alpha$ we have a situation where the length of the exit boundary expands to infinity when $T \to \infty$.

9. Show that for $Y = -\alpha$ and $T \to \infty$ eq. (B.13.13) becomes

$$\langle L(t) \rangle_{X,Y=-\alpha} = \frac{1}{\bar{X}(t;X) - \alpha} - \frac{\tilde{W}'(\bar{X}(t;X))}{\tilde{W}(\bar{X}(t;X))}, \qquad (B.13.19)$$

and show that $\langle L(t) \rangle_{X,Y=-\alpha}$ grows exponentially like $e^{2\Sigma t}$ for large t. Finally, show that the corrections to (B.13.19) if we keep T finite but much larger that t is of order $e^{-2\Sigma(T-t)}$.

Let us now for simplicity study eq. (B.13.19) in the CDT case, where $\alpha = \sqrt{\Lambda}$ and $\tilde{W}_\Lambda(X) = 1/(X + \sqrt{\Lambda})$. $\bar{X}(t;X)$ is given by (8.23).

10. Show that

$$\bar{X}(t;X)) = \sqrt{\Lambda} \coth \sqrt{\Lambda}(t + t_0(X)), \qquad X = \sqrt{\Lambda} \coth t_0 \qquad (B.13.20)$$

and

$$\langle L(t) \rangle_{X,Y=-\sqrt{\Lambda}} = \frac{1}{\sqrt{\Lambda}} \sinh 2\sqrt{\Lambda}(t + t_0(X)). \qquad (B.13.21)$$

This shows that if we view t as the geodesic distance from the entrance boundary with boundary cosmological constant X and $\langle L(t) \rangle_{X,Y=-\sqrt{\Lambda}}$ as the length of the curve a geodesic distance t from the boundary, this average geometry can be viewed as belonging to the *hyperbolic plane*, also called the *pseudosphere*. Recall that for a sphere of radius R the infinitesimal geodesic distance between points with (spherical) coordinates (θ, ϕ) and $(\theta + d\theta, \phi + d\phi)$ is given by

$$ds^2 = R^2(d\theta^2 + \sin^2\theta\, d\phi^2) = dt^2 + R^2 \sin^2(t/R)\, d\phi^2, \qquad t = R\theta. \qquad (B.13.22)$$

Here t is the geodesic distance on the sphere from the north pole where $\theta = 0$ to a point with coordinates (θ, ϕ). Also the curve at geodesic distance t from the north pole (curve of fixed latitude θ) has the length $\ell(t) = 2\pi R \sin(t/R)$. The intrinsic curvature (the Gaussian curvature) is constant on the sphere, and equal $R_g = 1/R^2$. We obtain the geodesic distance on the pseudo-sphere by formally rotating $R \to iR$ in the above line element,

$$ds^2 = dt^2 + R^2 \sinh^2(t/R)\, d\phi^2, \qquad R_g = -\frac{1}{R^2}, \qquad L(t) = 2\pi R \sinh(t/R). \qquad (B.13.23)$$

In the (t, ϕ) coordinate system, t is also the geodesic distance of point (t, ϕ) to the point with coordinate $t = 0$ and $L(t)$ the length of the curve of points with geodesic distance t to the point with $t = 0$. All points on the pseudo-sphere has intrinsic curvature $R_g = -1/R^2$. It is now seen that we can view (B.13.21) as corresponding to the part of the pseudo-sphere where $t \in [t_0(X), \infty[$ if we identify $R = 1/2\sqrt{\Lambda}$ and $\ell(t) = \pi \langle L(t) \rangle_{X,Y=-\sqrt{\Lambda}}$. It is remarkable that different choices of X lead to the "same" pseudo-sphere, with $t + t_0(X)$ being the geodesic distance to the "origin" (which of course is arbitrary as a point on the pseudo-sphere, like the north-pole being an "arbitrary" point on the sphere). For $X \to \infty$ the entrance boundary loop will in average contract to a point, and the whole pseudo-sphere is covered. The fact that we have to change the length assignment of $L(t)$ relative to t should not be a course of worry. From the beginning in the CDT model there was an arbitrariness in the relative length assignment of spatial links and temporal links.

Let us now return to (B.13.13) and prove the formula.

1. Use (B.13.11) to show that eq. (B.13.12) can be written as

$$\langle L(t) \rangle_{X,Y} = \frac{(Y + \bar{X}(T;X))\hat{W}(\bar{X}(t;X))}{\hat{W}(\bar{X}(T;X))} \times \qquad \text{(B.13.24)}$$
$$\left(-\frac{d}{d\bar{X}(t;X)}\right)\left[\frac{\hat{W}(\bar{X}(T-t;\bar{X}(t;X)))}{\hat{W}(\bar{X}(t;X))} \frac{1}{\bar{X}(T-t;\bar{X}(t;X)) + Y}\right]$$

2. Show that (B.13.2) implies that

$$\frac{\partial}{\partial X}\bar{X}(t;X) = \frac{\hat{W}(t;X)}{\hat{W}(X)} \qquad \text{(B.13.25)}$$

3. Show (if you do not feel it is trivial) that

$$\bar{X}(T-t; \bar{X}(t;X)) = \bar{X}(T;X) \qquad \text{(B.13.26)}$$

4. Show (B.13.13) using (B.13.25) in (B.13.24). Note that you are only allowed to use (B.13.26) *after* having performed the differentiation in (B.13.24).

C Solutions to Problem Sets 1–13

C.1 SOLUTIONS TO PROBLEM SET 1

Gaussian integrals

1. Switch to polar coordinates. We have

$$\int e^{-(x^2+y^2)}\,dxdy = \int_0^{2\pi} d\theta \int_0^\infty r dr e^{-r^2} = 2\pi \times \frac{1}{2}\int_0^\infty dr^2 e^{-r^2} = \pi$$

2. From question 1. it is clear that

$$\int e^{-x^2} dx = \sqrt{\pi},$$

and thus (by changing integration variable to $y = \sqrt{\frac{a}{2}}x$) that

$$\int e^{-\frac{ax^2}{2}} dx = \sqrt{\frac{2}{a}} \int e^{-y^2} dy = \sqrt{\frac{2\pi}{a}}$$

3. We write the exponent in vector notation as

$$x_i A_{ij} x_j = x^T A x, \quad x \in \mathbb{R}, \quad A \in \mathrm{Sym}_{n \times n}(\mathbb{R}).$$

Now any real, symmetric matrix can be decomposed as

$$A = Q^T \Lambda Q,$$

such that Λ is diagonal and Q is an orthonormal matrix, meaning that

$$Q^{-1} = Q^T, \quad \det Q = 1.$$

Therefore we have

$$x^T A x = x^T Q^T \Lambda Q x,$$

so it is convenient to make the change of variables

$$y = Qx, \quad d^n y = \det(Q) d^n x = d^n x.$$

The exponent then simplifies considerably since Λ is diagonal. We have

$$x^T A x = y^T \Lambda y = \lambda_i y_i^2,$$

DOI: 10.1201/9781003320562-C

where the λ_i are the elements on the diagonal of Λ (and therefore, the eigenvalues of A). Our integral can now be performed for all the y_i separately. We see that

$$\int \prod_{i=1}^{n} dx_i e^{-\frac{1}{2} x_i A_{ij} x_j} = \int \prod_{i=1}^{n} dy_i e^{-\frac{\lambda_i}{2} y_i^2} = \prod_{i=1}^{n} \sqrt{\frac{2\pi}{\lambda_i}}$$

by our result from question 2. Now note that

$$\prod_{i=1}^{n} \lambda_i = \det \Lambda = \det Q^T \det \Lambda \det Q = \det \left(Q^T \Lambda Q \right) = \det A,$$

which completes the proof.

4. We know that we can decompose a Hermitian matrix A as the product

$$A = Q^{\dagger} \Lambda Q,$$

again with Λ diagonal with *real* matrix elements, but now Q is a unitary matrix, meaning that

$$Q^{-1} = Q^{\dagger}, \quad |\det Q| = 1.$$

Question 4. is then reduced to question 3. by realizing that a unitary transformation $w = Qz$ in the n-dimensional complex vector space becomes an orthogonal transformation in the $2n$-dimensional real vector space obtained by writing $z_k = x_k + i y_k$, $w_k = u_k + i v_k$ and $w = Qz$ as

$$\begin{pmatrix} u \\ v \end{pmatrix} = \begin{pmatrix} \text{Re } Q & -\text{Im } Q \\ \text{Im } Q & \text{Re } Q \end{pmatrix} \begin{pmatrix} x \\ y \end{pmatrix}$$

where

$$Q^{\dagger} Q = I_{n \times n} \implies \begin{pmatrix} \text{Re } Q & -\text{Im } Q \\ \text{Im } Q & \text{Re } Q \end{pmatrix}^T \begin{pmatrix} \text{Re } Q & -\text{Im } Q \\ \text{Im } Q & \text{Re } Q \end{pmatrix} = \begin{pmatrix} I_{n \times n} & 0 \\ 0 & I_{n \times n} \end{pmatrix}$$

By the orthogonal change of variables $(x, y) \to (u, v)$ we obtain

$$z^{\dagger} A z \to \sum_{k=1}^{n} \lambda_k (u_k^2 + v_k^2)$$

and as in question 3. we obtain

$$\int \prod_{i=1}^{n} (dx_i dy_i) e^{-\frac{1}{2} z_i^* A_{ij} z_j} = \int \prod_{i=1}^{n} (du_i dv_i) e^{-\frac{\lambda_i}{2} (u_i^2 + v_i^2)}$$

$$= \prod_{i=1}^{n} \left(\sqrt{\frac{2\pi}{\lambda_i}} \sqrt{\frac{2\pi}{\lambda_i}} \right) = \frac{(2\pi)^n}{\det A}.$$

Solutions to Problem Sets 1–13

5. Use the rules for differentiation:

$$\frac{\partial S(x)}{\partial x_i} = \frac{1}{2}(\delta_{ij}A_{jk}x_k + x_j A_{jk}\delta_{ik}) + b_j\delta_{ij} = A_{ij}x_j + b_i$$

$$\frac{\partial^2 S(x)}{\partial x_i \partial x_j} = A_{ij}.$$

We then find

$$\left.\frac{\partial S(x)}{\partial x_i}\right|_{x_c} = 0 \implies x_c = -A^{-1}b \text{ and thus } S(x_c) = -\frac{1}{2}b^T A^{-1}b,$$

Making the expansion $x = x_c + \Delta x$, we now compute

$$S(x_c + \Delta x) = S(x_c) + \left.\frac{\partial S(x)}{\partial x_i}\right|_{x_c} \Delta x_i + \frac{1}{2}\left.\frac{\partial^2 S(x)}{\partial x_i \partial x_j}\right|_{x_c} \Delta x_i \Delta x_j$$

$$= S(x_c) + S(\Delta x, b=0).$$

Since x_c is a constant vector we have:

$$d^n \Delta x = d^n x.$$

As a result, we can pull $e^{-S(x_c)}$ out of the integral and obtain

$$\int d^n x \, e^{-S(x)} = e^{-S(x_c)} \int d^n \Delta x \, e^{-S(\Delta x, b=0)} = e^{-S(x_c)} \int d^n \Delta x \, e^{-\frac{1}{2}\Delta x^T A \Delta x}$$

Thus we have obtained the desired result

$$\int d^n x \, e^{-S(x)} = e^{-S(x_c)} \frac{(2\pi)^{\frac{n}{2}}}{\sqrt{\det A}} = e^{\frac{1}{2}b^T A^{-1}b} \frac{(2\pi)^{\frac{n}{2}}}{\sqrt{\det A}}.$$

6. We just use the above derived formula with $b = -ik$, then we obtain

$$\mathscr{F}(e^{-\frac{1}{2}x^T A x}) = \int d^n x \, e^{-\frac{1}{2}x^T A x} e^{-ik^T x} = \frac{(2\pi)^{\frac{n}{2}}}{\sqrt{\det A}} e^{-\frac{1}{2}k^T A^{-1}k}$$

7. It is clear from the explicit formula for the Fourier transformed $\mathscr{F}(f_A)$ that

$$\mathscr{F}(f_A) \cdot \mathscr{F}(f_B) = \mathscr{F}(f_{A+B})$$

Thus we obtain

$$f_A * f_B = \mathscr{F}^{-1}(\mathscr{F}(f_A * f_B)) = \mathscr{F}^{-1}(\mathscr{F}(f_A) \cdot \mathscr{F}(f_B))$$
$$= \mathscr{F}^{-1}(\mathscr{F}(f_{A+B})) = f_{A+B}$$

The free non-relativistic particle

8. The integral given depends (by translational invariance) only on $x'' - x$ since we can shift the integration variable $x' \to x' + x$, and it then becomes a standard convolution with "external variable" $x'' - x$. Thus the formula listed is just the one-dimensional version of our general formula for convolution of Gaussians with $A = ia$ and $B = ib$.

   By translational invariance the matrix element will depend only on $x_b - x_a = x_{n+1} - x_0$. We can thus choose $x_0 = 0$ and $x_{n+1} = x_b - x_a$. By choosing $a = \varepsilon\hbar/m$ the expression for $\langle x_{n+1}|\hat{\mathcal{O}}_\varepsilon^{n+1}|0\rangle$ becomes successive convolutions of the gaussian function f_{ia}

$$\langle x_{n+1}|\hat{\mathcal{O}}_\varepsilon^{n+1}|0\rangle = (f_{ia}*f_{ia}*\cdots*f_{ia})(x_{n+1}) = f_{i(n+1)a}(x_{n+1}) = \frac{e^{-\frac{x_{n+1}^2}{2i(n+1)a}}}{\sqrt{2\pi i(n+1)a}}.$$

   Now, using $x_{n+1} = x_b - x_a$ and $t = (n+1)\varepsilon$ we obtain the desired formula for $\langle x_s|\hat{\mathcal{O}}_\varepsilon^{n+1}|x_0\rangle$.

   The result is independent of n. The reason is that the Hamiltonian in the case of a free particle only depends on the momentum operator. Recall that the n-dependence entered because the concept of a path integral was introduced via the Trotter-Kato theorem, where we actually changed the operator $e^{-i\varepsilon\hat{H}/\hbar}$ into an exponential depending only on the operator $\hat{p}$ and another exponential depending only on $\hat{x}$. In this procedure one only got back $e^{-it\hat{H}/\hbar}$ in the limit where $n \to \infty$. However, if $\hat{H}$ only depends on $\hat{p}$ the subdivision in n is exact and one is always calculating the matrix element of $\langle x|e^{-it\hat{H}/\hbar}|y\rangle$ independent of how many subdivisions one makes.

9. Again this is a simple application of our Fourier formula for Gaussians $f_{ia}(x)$, with $a = \hbar t/m$.

C.2 SOLUTIONS TO PROBLEM SET 2

1. The Euler-Lagrange equation is

$$\frac{\delta L}{\delta x(t)} = \frac{d}{dt}\frac{\delta L}{\delta \dot{x}(t)}, \quad \text{i.e.} \quad -\omega^2 x(t) = \ddot{x}(t).$$

Clearly, the given solution $x_c(t)$ satisfies this differential equation as well as the boundary conditions. To calculate the action one can just insert the solution, or (slightly easier) perform a partial integration

$$\frac{m}{2}\int_{t_a}^{t_b} dt(\dot{x}^2 - \omega^2 x^2) = -\frac{m}{2}\int_{t_a}^{t_b} dt(\ddot{x} + \omega^2 x)x + \frac{m}{2}x\dot{x}\Big|_{t_a}^{t_b}.$$

The integrand will be zero if $x(t)$ satisfies the classical equation, and inserting the classical solution in the boundary term produces the wanted expression for $S[x_{cl}]$.

2. We simply insert the decomposition $x_c(t) + \Delta x(t)$ in the action:

$$S[x(t)] = S[x_c(t) + \Delta x(t)]$$
$$= \frac{m}{2}\int dt\left((\dot{x}_c + \dot{\Delta x})^2 - \omega^2(x_c + \Delta x)^2\right)$$
$$= \frac{m}{2}\int dt\left(\dot{x}_c^2 + 2\dot{x}_c\dot{\Delta x} + \dot{\Delta x}^2 - \omega^2 x_c^2 - 2\omega^2 x_c\Delta x - \omega^2 \Delta x^2\right)$$
$$= S[x_c] + S[\Delta x] + \frac{m}{2}\int dt(2\dot{x}_c\dot{\Delta x} - 2\omega^2 x_c\Delta x)$$
$$= S[x_c] + S[\Delta x] + m\dot{x}_c\Delta x\Big|_{t_a}^{t_b} + \frac{m}{2}\int dt\left[-2(\ddot{x}_c + \omega^2 x_c)\Delta x\right]$$
$$= S[x_c] + S[\Delta x].$$

Here we performed a partial integration in the fifth line, and used the fact that the variation vanishes at the endpoints. Further, the integral also vanishes since $x_c(t)$ satisfies the eom.

3. The "measure" $\mathscr{D}x(t)$ is invariant under the decomposition $x(t) = x_c(t) + \Delta x(t)$. $x_c(t)$ acts like a translation of the "vector" $x(t)$, and if we represent $\mathscr{D}x(t)$ as a kind of limit of $\prod_{i=1}^{n} dx(t_i)$ we have:

$$\mathscr{D}x(t) = \mathscr{D}\Delta x(t),$$

meaning that we can move the factor $e^{\frac{i}{\hbar}S[x_c(t)]}$ outside the functional integral.

4. It is easily checked by direct calculation. Since $y_0 = y_{n+1} = 0$, we can relabel the sum over y_{i+1}^2 to y_i^2, so that we pick up this term twice. The off-diagonal elements -1 provide the cross terms $y_i y_{i+1}$.

Note that this matrix A is not a unique solution - however, we are looking for a symmetric matrix in order to be able to use our results for Gaussian integrals, and the symmetric matrix is unique.

5. We compute the determinant by expanding D_n in the top row:

$$D_n = (2-\varepsilon^2\omega^2)D_{n-1} - (-1) \begin{vmatrix} -1 & 2-\omega^2\varepsilon^2-1 & -1 & & 0 \\ 0 & -1 & \ddots & & \vdots \\ \vdots & & \ddots & & -1 \\ 0 & 0 & \cdots & -1 & 2-\omega^2\varepsilon^2 \end{vmatrix}$$

$$= (2-\varepsilon^2\omega^2)D_{n-1} - D_{n-2}.$$

The second term in the first line was expanded in the first column. The values of D_0 and D_{-1} follow by consistency for the recursion relations for D_2 and D_1.

6. We now find the generating function using the recursion relation:

$$D(x) - D_0 = \sum_{n=1}^{\infty} D_n x^n = 1 + \sum_{n=1}^{\infty}((2-\varepsilon^2\omega^2)D_{n-1} - D_{n-2})x^n$$
$$= (2-\varepsilon^2\omega^2)xD(x) - (x^2 D(x) - xD_{-1})$$

Thus

$$D(x) - 1 = (2-\varepsilon^2\omega^2)xD(x) - x^2 D(x) \quad \text{or} \quad D(x) = \frac{1}{1-(2-\varepsilon^2\omega^2)x + x^2}$$

7. In terms of the new variable $\tilde{\omega}$, $1 - \frac{\omega^2\varepsilon^2}{2} = 1 - 2\sin^2\frac{\tilde{\omega}\varepsilon}{2} = \cos\tilde{\omega}\varepsilon$, and therefore

$$D(x) = \frac{1}{1-2x\cos(\tilde{\omega}\varepsilon) + x^2} = \frac{1}{(e^{i\tilde{\omega}\varepsilon}-x)(e^{-i\tilde{\omega}\varepsilon}-x)}$$
$$= \frac{1}{e^{i\tilde{\omega}\varepsilon}-e^{-i\tilde{\omega}\varepsilon}}\left(\frac{1}{(e^{-i\tilde{\omega}\varepsilon}-x)} - \frac{1}{(e^{i\tilde{\omega}\varepsilon}-x)}\right)$$

8. Using $(a-x)^{-1} = \sum_{n=0}^{\infty} x^n/a^{n+1}$ we obtain

$$D_n = \frac{1}{e^{i\tilde{\omega}\varepsilon}-e^{-i\tilde{\omega}\varepsilon}}\left(e^{i\tilde{\omega}\varepsilon(n+1)} - e^{-i\tilde{\omega}\varepsilon(n+1)}\right) = \frac{\sin\tilde{\omega}\varepsilon(n+1)}{\sin\tilde{\omega}\varepsilon}$$

9. Using that $(n+1)\varepsilon = t_b - t_a$, $\tilde{\omega} = \omega + O(\varepsilon)$ and $\sin\tilde{\omega}\varepsilon = \omega\varepsilon + O(\varepsilon^2)$ we obtain the required formula for the amplitude.

10. It is readily seen that if we make the replacement $i(t_b - t_a) = \hbar\beta$ and $x_a = x_b = x$ in the formula for the amplitude, and use that

$$\frac{\cosh\omega\beta\hbar - 1}{\sinh\omega\beta\hbar} = \tanh(\omega\beta\hbar/2),$$

we obtain the wanted formula

$$z(x) = \langle x|e^{-\beta\hat{H}}|x\rangle = \sqrt{\frac{m\omega}{2\pi\hbar\sin\omega\beta\hbar}}\, e^{-\frac{m\omega}{\hbar}\tanh\frac{\omega\beta\hbar}{2}x^2}.$$

The integral $\int dx\, z(x)$ is a Gaussian integral, which we know how to calculate and we obtain

$$Z = \int_{-\infty}^{\infty} z(x) = \sqrt{\frac{m\omega}{2\pi\hbar\sin\omega\beta\hbar}}\sqrt{\frac{\pi\hbar}{m\omega\tanh\frac{\omega\beta\hbar}{2}}} = \frac{1}{2\sinh\frac{\omega\beta\hbar}{2}}.$$

11. From the expressions for $z(x)$ and Z we obtain, taking the limit $\beta \to \infty$,

$$\frac{z(x)}{Z} \to \frac{\sqrt{\frac{m\omega}{2\pi\hbar}}\, e^{-\frac{\hbar\beta\omega}{2}} e^{-\frac{m\omega}{\hbar}x^2}}{e^{-\frac{\hbar\beta\omega}{2}}} = \sqrt{\frac{m\omega}{2\pi\hbar}}\, e^{-\frac{m\omega}{\hbar}x^2} = |\psi_0(x)|^2.$$

That the $\beta \to \infty$ limit leads to the square of the ground state wave function should be clear without this detailed calculation and is valid for an arbitrary system with a discrete spectrum bounded from below since we have

$$Z = \sum_n \langle E_n|e^{-\beta\hat{H}}|E_n\rangle \to e^{-\beta E_0} + O(e^{-\beta E_1})$$

and similarly

$$\begin{aligned} z(x) &= \sum_{n,m} \langle x|E_n\rangle\langle E_n|e^{-\beta\hat{H}}|E_m\rangle\langle E_m|x\rangle = \sum_n \langle x|E_n\rangle e^{-\beta E_n}\langle E_n|x\rangle \\ &\to e^{-\beta E_0}|\langle x|E_0\rangle|^2 + O(e^{-\beta E_1}) \end{aligned}$$

C.3 SOLUTIONS TO PROBLEM SET 3

1.
$$\hat{F}(p) = \sum_{x_n} a^D e^{ip \cdot x_n} F(x_n).$$

Prove that

$$\hat{F}(p_i) = \hat{F}\left(p_i + \frac{2\pi}{a}\right).$$

Follows trivially from $\frac{2\pi}{a} x_n^i = 2\pi n$. Thus $e^{ip_i \cdot x_n^i} = e^{i(p_i + \frac{2\pi}{a})x_n^i}$.

2. The standard formulas for Fourier series of functions $\hat{F}$ periodic with period 2π are

$$\hat{F}(q) = \sum_n e^{iq \cdot n} F(n), \qquad F(n) = \int_{-\pi}^{\pi} \frac{d^D q}{(2\pi)^D} e^{-iq \cdot n} \hat{F}(q).$$

The formulas in problem 3 are the same, just introducing the dimensionful parameter a (the length of a lattice link).

3. One has, from the definition of Δ_L,

$$-a^2 \Delta_L e^{-ipx_n} = \sum_{j=1}^{D} \left(2 - e^{-ip_j a} - e^{ip_j a}\right) e^{-ipx_n}$$

Choosing $x_m = 0$ and using that the Fourier transform of $\delta(x_n)$ is 1, we can write

$$(\Delta_L + m^2) \int_{-\frac{\pi}{a}}^{\frac{\pi}{a}} \frac{dp}{(2\pi)^D} e^{-ipx_n} G(p) = \int_{-\frac{\pi}{a}}^{\frac{\pi}{a}} \frac{dp}{(2\pi)^D} e^{-ipx_n} 1$$

or

$$\int_{-\frac{\pi}{a}}^{\frac{\pi}{a}} \frac{dp}{(2\pi)^D} e^{-ipx_n} \left(\left[\frac{2}{a^2} \sum_{j=1}^{D} (1 - \cos(p_j a)) + m^2\right] G(p) - 1\right) = 0$$

valid for all x_n, from which we conclude that $(\cdot) = 0$.

4.

$$\left(\frac{2}{a^2} \sum_{j=1}^{D} (1 - \cos(p_j a)) + m^2\right) G(p) = 1 \Rightarrow G(p) = \frac{a^2}{\left(\sum_{j=1}^{D} 4 \sin^2 \frac{ap_j}{2}\right) + m^2 a^2}$$

5.

$$G(p) = \frac{1}{p^2 + m^2 + O(a^2)} \qquad a \to 0.$$

6. It follows from the very definition of the discretized version of Δ_L that we have the matrix elements:

$$-a^2(\Delta_L)_{nm} = 2D\delta_{nm} - Q_{nm}$$

where Q_{mn} is 1 if n and m label neighboring sites and zero for all choices of n and m.

7. Δ_L is an operator on $\ell^2(\mathbb{Z}^D)$, the sequences $f(x_n)$ which are square summable, and by Parseval's and Plancherel's theorems Fourier transformation is a unitary map from $\ell^2(\mathbb{Z}^D)$ to $L^2([-\pi,\pi]^D)$. This map conserves the norm of operators and we can thus analyze the Fourier transformed operator which we have already found. By Fourier transformation we obtain

$$-a^2(\Delta_L + m^2) \to (2D + m^2 a^2)I - 2\sum_{j=1}^{D}\cos(p_j a).$$

Thus it is a simple multiplication operator and we have already found the inverse, namely $G(p)$. We also see that the Fourier transform of Q is a multiplication operator

$$Q \to \hat{Q} = 2\sum_{s3j=1}^{D}\cos(p_j a), \quad \text{i.e} \quad ||\hat{Q}\hat{f}|| \leq 2D||\hat{f}||.$$

The norm of $\hat{Q}$ is thus less than or equal 2D (in fact it is easily seen to be 2D) and the norm of $\hat{Q}/(2D + m^2 a^2)$ correspondingly less than 1. (The norm of a bounded operator A is defined as $\sup_f ||Af||/||f||$). An operator $I - A$ with $||A|| < 1$ has an inverse operator, which has a convergent expansion in powers of A (the Neumann series):

$$\frac{1}{I-A} = \sum_{n=0}^{\infty} A^n$$

Writing

$$-a^2(\Delta + m^2) = (2D + m^2 a^2)\left[I - \frac{Q}{2D + m^2 a^2}\right]$$

leads to the asked for Neumann series:

$$-(\Delta + m^2)^{-1} = \frac{a^2}{2D + m^2 a^2}\frac{1}{I - \frac{Q}{2D+m^2 a^2}} = \frac{a^2}{2D + m^2 a^2}\sum_{n=0}^{\infty}\frac{Q^n}{(2D + m^2 a^2)^n}.$$

8. From the definition of Q it follows that $(Q^k)_{mn}$ is the number of connected lattice paths of length k which connect site m to site n and the representation asked for follows.

9. Comparing formulas for $G_a(x_n,x_m)$ and $(-\Delta_L+m^2)^{-1}$ as a power series in Q one obtain that
$$a m_0(a) = \log(2D+m^2 a^2)$$

10.
$$m_0(a) = \frac{\log(2D+m^2 a^2)}{a} = \frac{\log 2D}{a} + m^2 a + O(a^3).$$

The power dependence on a for the two first terms is universal, but the coefficients are not.

11. The number of connected paths on the lattice, starting at a given point x_n and made of ℓ links, is
$$\ell^{2D} = e^{\ell \log 2D}.$$

The formula for $G_a(x_n,x_m)$ can be written as
$$G_a(x_n,x_m) = \sum_\ell e^{-m_0 a \ell} \mathcal{N}(\ell,x_n,x_m),$$

where $\mathcal{N}(\ell,x_n,x_m)$ are the number of connected lattice paths from x_n to x_m. It turns out that this number is
$$\mathcal{N}(\ell,x_n,x_m) \propto \frac{1}{\ell^{D/2}} \ell^{2D} \quad \text{for} \quad \ell \gg 1.$$

The contraint that the path should be from x_n to x_m rather the just being an arbitrary path starting at x_n only results in a subleading correction to the exponential growth of the number of paths. It is thus seen that the term in $m_0(a)$, which is divergent for $a \to 0$ is precisely cancelling the exponential growing number of paths of length ℓ, i.e. the *entropy* of paths. *This will be a universal feature of all the geometric systems we will consider: the number of geometric objects of a certain kind will grow exponentially with length, area, volume or whatever we consider, and in order to have a well defined partition function for these objects, we have to adjust the "bare" coupling constants (**renormalise the coupling constants**) such that this exponential growth is cancelled.*

C.4　SOLUTIONS TO PROBLEM SET 4

1. The measure is not affected by the redefinition $S_i = \langle S(h) \rangle + \delta S_i$:

$$dS_i = d(\langle S(h) \rangle + \delta S_i) = d\delta S_i.$$

We now insert $S_i = \langle S(h) \rangle + \delta S_i$ into the expression for the partition function:

$$\begin{aligned}Z &= \int \prod_i dS_i \, \exp\left[-\sum_i \left(\kappa_0 S_i^2 + \lambda_0 S_i^4 - \beta h S_i\right) + \beta J \sum_{\langle ij \rangle} S_i S_j\right] \\ &= \int \prod_i d(\delta S_i) \, \exp\left[-\sum_i \left(\kappa_0 (\langle S(h) \rangle + \delta S_i)^2 + \lambda_0 (\langle S(h) \rangle + \delta S_i)^4 \right.\right. \\ &\quad \left.\left. - \beta h (\langle S(h) \rangle + \delta S_i)\right) + \beta J \sum_{\langle ij \rangle} (\langle S(h) \rangle + \delta S_i)(\langle S(h) \rangle + \delta S_j)\right]\end{aligned}$$

The sum $\langle ij \rangle$ indicates that we sum over all the nearest neighbors. Next, we expand to quadratic order in δS_i:

$$\begin{aligned}Z &= \int \prod_i d(\delta S_i) \, \exp\left[-\sum_i \left[\{\kappa_0 \langle S(h) \rangle^2 + \lambda_0 \langle S(h) \rangle^4 - \beta h \langle S(h) \rangle\} + \right.\right.\\ &\quad \{2\kappa_0 \langle S(h) \rangle + 4\lambda_0 \langle S(h) \rangle^3 - \beta h\} \delta S_i + \{\kappa_0 + 6\lambda_0 \langle S(h) \rangle^2\} \delta S_i^2\Big] \\ &\quad \left.+ \beta J \sum_{\langle ij \rangle} \left(\langle S(h) \rangle^2 + \langle S(h) \rangle (\delta S_i + \delta S_j) + \delta S_i \delta S_j\right)\right] \\ &= Z_{\text{mf}}^V \int \prod_i d(\delta S_i) \exp\left[-\sum_i \Big[\{2(\kappa_0 - DJ\beta)\langle S(h)\rangle + 4\lambda_0 \langle S(h)\rangle^3 - \beta h\}\delta S_i \right.\\ &\quad \left.+ \{\kappa_0 - DJ\beta + 6\lambda_0 \langle S(h) \rangle^2\}\delta S_i^2\Big] - \frac{1}{2}\beta J \sum_{\langle ij \rangle}(\delta S_i - \delta S_j)^2\right]\end{aligned}$$

Here we have defined Z_{mf} as the part of Z independent of δS_i:

$$Z_{\text{mf}} = \exp\left[-(\kappa_0 - DJ\beta)\langle S(h)\rangle^2 - \lambda_0 \langle S(h)\rangle^4 + \beta h \langle S(h) \rangle\right]$$

and used the fact that $\sum_{\langle ij \rangle} \langle S(h) \rangle = D \sum_i \langle S(h) \rangle$ and $\sum_{\langle ij \rangle} \delta S_i = D \sum_i \delta S_i$ as well as

$$\sum_{\langle ij \rangle} \delta S_i \delta S_j = -\frac{1}{2}\sum_{\langle ij \rangle}(\delta S_i - \delta S_j)^2 + D\sum_i \delta S_i^2$$

If we now let $\langle S(h) \rangle$ satisfy

$$2(\kappa_0 - D\beta J)\langle S(h) \rangle + 4\lambda_0 \langle S(h) \rangle^3 = \beta h \tag{C.4.1}$$

we see that the term linear in δS_i in the exponential drops out. The remaining terms are quadratic in the δS_i, which implies that calculating $\langle \delta S_i \rangle$ we will indeed obtain $\langle \delta S_i \rangle = 0$.

2. We have

$$\beta f_{\text{mf}} = (\kappa_0 - DJ\beta)\langle S(h)\rangle^2 + \lambda_0 \langle S(h)\rangle^4 - \beta h \langle S(h)\rangle \qquad \text{(C.4.2)}$$

and differentiating wrt h we obtain

$$\beta \frac{\partial f_{\text{mf}}}{\partial h} = \left[2(\kappa_0 - DJ\beta)\langle S(h)\rangle + 4\lambda_0 \langle S(h)\rangle^3 - \beta h\right] \frac{\partial \langle S(h)\rangle}{\partial h} - \langle S(h)\rangle \qquad \text{(C.4.3)}$$

and from eq. (C.4.1) then

$$\frac{\partial f_{\text{mf}}}{\partial h} = -\langle S(h)\rangle \qquad \text{(C.4.4)}$$

3. Just substitute h_{eff} for h in f_{free}.

4. For $\kappa_0 > DJ\beta$ the solution is $\langle S(h)\rangle|_{h=0} = 0$, and for $\kappa_0 < DJ\beta$ we have

$$\langle S(h)\rangle|_{h=0} = \sqrt{\frac{2(DJ\beta - \kappa_0)}{4\lambda_0}} = c\sqrt{\beta - \beta_c}, \quad c = \sqrt{\frac{DJ}{2\lambda_0}}, \quad \beta_c = \frac{\kappa_0}{DJ} \qquad \text{(C.4.5)}$$

5. Follows from (C.4.5)

6. If we differentiate eq. (C.4.1) wrt h we obtain

$$2\left[(\kappa_0 - DJ\beta) + 6\lambda_0 \langle S(h)\rangle^2\right] \frac{\partial \langle S(h)\rangle}{\partial h} = \beta$$

i.e.

$$\frac{\partial \langle S(h)\rangle}{\partial h} = \frac{\beta}{2\left[(\kappa_0 - DJ\beta) + 6\lambda_0 \langle S(h)\rangle^2\right]}$$

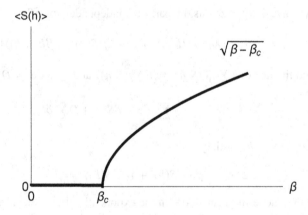

Figure C.4.1 The graph $\langle S(h)\rangle|_{h=0}(\beta)$. It is zero for $\beta < \beta_c$ and proportional to $\sqrt{\beta - \beta_c}$ for $\beta > \beta_c$.

For $\beta < \beta_c$ we have $\langle S(h) \rangle |_{h=0} = 0$ and this leads to

$$\left.\frac{\partial \langle S(h) \rangle}{\partial h}\right|_{h=0} = \frac{\beta}{2(\kappa_0 - D\beta J)} \qquad (C.4.6)$$

while for $\beta > \beta_c$ we have from (C.4.5): $6\lambda_0 \langle S(h) \rangle^2 |_{h=0} = 3(DJ\beta - \kappa_0)$ and thus

$$\left.\frac{\partial \langle S(h) \rangle}{\partial h}\right|_{h=0} = \frac{\beta}{4(D\beta J - \kappa_0)} \qquad (C.4.7)$$

7. This follows since $|\beta - \beta_c| = |\beta - \frac{\kappa_0}{D} J|$.

8. Differentiating wrt J_i and J_j in the integral we obtain $\langle x_i x_j \rangle$, while differentiating instead $e^{s4 J_i A_{ij}^{-1} J_j / 2}$ (and setting $J = 0$) we obtain A_{ij}^{-1}.

9. This follows directly from the expression for $H_F(\delta S_i)$ and the definition of the discrete Laplace operator.

10. This follows from the expression for m^2 given by (B.4.17) and the definition of T_c.

11. Clear again from from (B.4.17).

C.5 SOLUTIONS TO PROBLEM SET 5

Rooted planar trees

1. Assume $h(L) = L^{\gamma-2}$. We have (with notation $\Delta\mu := \mu - \mu_c$)

$$z(\mu) = \sum_L e^{-\mu L} e^{\mu_c L} L^{\gamma-2} = (\Delta\mu)^{1-\gamma} \sum_L \Delta\mu\, e^{-\Delta\mu L} (\Delta\mu L)^{\gamma-2}.$$

Assume now that $\gamma > 1$. When we take the limit $\mu \to \mu_c$ the sum turns into a finite integral (with value $\Gamma(\gamma-1)$). Thus we have for $\gamma > 1$:

$$z(\mu) \to \frac{\Gamma(\gamma-1)}{\Delta\mu^{\gamma-1}}, \quad \Delta\mu \approx \left(1 - \frac{g_c}{g}\right), \quad \text{i.e.} \quad z(g) \propto \left(1 - \frac{g_c}{g}\right)^{1-\gamma}$$

If $0 < \gamma \leq 1$ then differentiate $z(\mu)$ wrt μ. Then the argument is true for $z'(\mu)$, and integrate to obtain $z(\mu) = z(\mu_c) - c(\mu - \mu_c)^{1-\gamma}$.

Recall that for rooted branched polymers the susceptibility is $\chi(\mu) \propto -z'(\mu)$ and by definition $\chi(\mu) \sim (\mu - \mu_c)^{-\gamma}$. Thus $\gamma = 1/2$ corresponds to $L^{-3/2}$.

Now let us assume that $z(g) - z(g_c) \propto (1 - g_c/g)^{1-\gamma}$. We can now Taylor expand and obtain

$$z(g) - z(g_c) \propto \left(1 - \frac{g_c}{g}\right)^{1-\gamma} = \sum_L c_L \left(\frac{g_c}{g}\right)^L, \quad c_L = \frac{\Gamma(L-(1-\gamma))}{\Gamma(L+1)\Gamma(\gamma-1)}$$

More precisely we find (writing $s = 1 - \gamma$ and assuming for simplicity that $s < 0$, but the arguments are also correct for $s > 0$)

$$c_L = \frac{(-1)^L s(s-1)\cdots(s-L+1)}{L!} = \frac{(L-s-1)\cdots(-s)}{L!} \frac{\Gamma(-s)}{\Gamma(-s)} = \frac{\Gamma(L-s)}{\Gamma(L+1)\Gamma(-s)}.$$

Here we have used the definition of the Gamma function $s\Gamma(s) = \Gamma(s+1)$, $s! = \Gamma(s+1)$. Finally, the property

$$\frac{\Gamma(L-s)}{\Gamma(L+1)} \to \frac{1}{L^{s+1}} \quad \text{for} \quad L \to \infty$$

leads to the desired power dependence. On can prove the asymptotic behavior using Stirling's formula for the asymptotic behavior of the Γ-function for large argument, as well as $(1 - s/L)^L \to e^{-s}$ for $L \to \infty$.

2. In order to determine g_c, first find z_c by solving $\dfrac{dg}{dz} = 0$. We see that

$$g(z) = \frac{1 + \sum_{n=2}^{\infty} w_n z^{n-1}}{z} = \frac{\sum_{n=0}^{\infty} z^n}{z} = \frac{1}{z(1-z)} = \frac{1}{z} + \frac{1}{1-z}.$$

$$\frac{dg}{dz} = -\frac{1}{z^2} + \frac{1}{(1-z)^2} = 0 \quad \Rightarrow \quad z_c = \frac{1}{2} \quad \Rightarrow \quad g_c = g(z_c) = 4$$

Since

$$\left.\frac{d^2g}{dz^2}\right|_{z_c} = \frac{2}{z_c^3} + \frac{2}{(1-z_c)^3} > 0,$$

we know that for z close to z_c we have $g(z) - g(z_c) \approx c(z_c - z)^2$. Thus we obtain $z(g) = z_c - \tilde{c}\sqrt{g - g_c}$ and thus $\gamma = 1/2$ and $\mathcal{N}(L) \approx 4^L L^{-3/2}$.

3.
$$g = \frac{1}{z(1-z)} \quad \Rightarrow \quad z^2 - z + \frac{1}{g} = 0 \quad \Rightarrow \quad z = \frac{1 - \sqrt{1 - \frac{4}{g}}}{2}.$$

4. If we take $g(z) = 1/z + z$ we can repeat the steps in question 3.:

$$g = 1/z + z \quad \Rightarrow \quad z^2 - gz + 1 = 0 \quad \Rightarrow \quad z(g) = \frac{g - \sqrt{g^2 - 4}}{2}$$

It follows that $g_c = 2$ and $z_c = 1$. Introducing $\Delta g = g - g_c$ we have:

$$z(g) = z_c - \sqrt{2}\sqrt{\Delta g}\left(1 + O(\Delta g)\right), \qquad \gamma = \frac{1}{2}$$

5. Now $g(z) = 1/z + z^{n-2}$ and $n > 2$ by assumption.

$$\frac{dg}{dz} = \frac{-1}{z^2} + (n-2)z^{n-3}, \qquad \frac{d^2g}{dz^2} = \frac{2}{z^3} + (n-2)(n-3)z^{n-4}.$$

Thus we find from $g'(z_c) = 0$:

$$z_c = \frac{1}{(n-2)^{\frac{1}{n-1}}}, \quad g(z_c) = (n-2)^{\frac{1}{n-1}} + \frac{1}{(n-2)^{\frac{n-2}{n-1}}}, \quad g''(z_c) > 0.$$

This implies as before the $z(g) - z_c \approx c\sqrt{1 - g_c/g}$ and that $\gamma = 1/2$. We have

$$g_c(n) \approx 1 + \frac{\ln n}{n} \quad \text{for} \quad n \gg 1.$$

This choice of weights corresponds to only allowing extremely high branching. This means that for $L < n$ the only BP will be the root connected to a vertex of order 1, and for $L < 2n$ there will only be one more BP, namely the one where the root is connected to a vertex of order n. It is thus not surprising that exponential growth g_c^L is slow, i.e. $g_c(n)$ is close to 1 (although this argument is *not* a proof).

Thus we see explicitly in the examples that the critical value of z_c (and therefore g_c) will change when choosing different branching weight ratios, but that the power law behavior so far has been universal, corresponding to $\gamma = 1/2$. We will study under what conditions one gets the 'standard' value of $\gamma = 1/2$, and what one can do to change it.

6. Now $g(z) = 1/z + 1$, which we can solve for z as

$$z(g) = \frac{1}{g-1} = \frac{1}{g}\frac{1}{1-\frac{1}{g}} = \frac{1}{g}\sum_{n=0}^{\infty}\left(\frac{1}{g}\right)^n = \sum_{n=1}^{\infty}(e^{-\mu})^n, \quad g_c = 1, \quad \gamma = 2.$$

Recall that we assigned a factor $e^{-\mu}$ for every unit length 'link' of the BP, so this expression for z simply sums over all possible BPs without branching (since $w_2 = 1$ and all other weights are 0). Since these "BPs" are not embedded in a target space, the only property of such a BP is its length, and each length is counted exactly once. It is thus not a "real" BP and we have a corresponding non-standard γ which is obtained for $z_c = \infty$.

Criticality of BPs

7. We see that z is an expansion in terms of $e^{-\mu}$, i.e. in $1/g$, the first term being $1/g$. Therefore, higher terms in the expansion are higher powers of $1/g$, and in the limit $g \to \infty$ all terms will vanish, giving $z = 0$ (assuming we have an absolute convergent power series).

8. We expand $zg(z)$ around the point z_c. The first two derivatives are then

$$\frac{d}{dz}(zg(z)) = g(z) + zg'(z),$$

$$\frac{d^2}{dz^2}(zg(z)) = g'(z) + g'(z) + zg''(z) = 2g'(z) + zg''(z).$$

Continuing this procedure, we see that

$$\frac{d^n}{dz^n}(zg(z)) = ng^{(n-1)}(z) + zg^{(n)}(z).$$

The Taylor expansion is then

$$(zg(z))|_{z_c} = z_c g(z_c) + \sum_{m=1}^{\infty}\left(mg^{(m-1)}(z_c) + z_c g^{(m)}(z_c)\right)\frac{(z-z_c)^m}{m!}.$$

However, using the fact that $zg(z)$ is a polynomial in z of order n, we know that all terms with $m > n$ vanish. Furthermore, by the assumption that $g^{(m)}(z_c) = 0$ for all $1 \leq m < n$, we see that all terms drop out except the one of order n and the $g^{(0)}$ for $m = 1$. The term of order n can be determined from the fact that $g(z)$ is the sum of $1/z$ and a polynomial of order $n-1$. It is clear that $ng^{(n-1)}(z_c) = 0$ (by our assumption), so we only need to compute $z_c g^{(n)}(z_c)$. The nth derivative of $g(z)$ can only receive a contribution from the $1/z$ part, since the nth derivative of an order $n-1$ polynomial is zero. Therefore

$$zg(z)) = z_c g(z_c) + g(z_c)(z-z_c) + z_c g^{(n)}(z_c)\frac{(z-z_c)^n}{n!}$$

$$= zg(z_c) + z_c(-1)^n n! z_c^{-1-n}\frac{(z-z_c)^n}{n!} = zg(z_c) + \left(1 - \frac{z}{z_c}\right)^n.$$

$w_2 = 0$ and we have

$$zg(z) = 1 + f(z) = 1 + w_2 z + w_3 z^2 + w_4 z^2 \cdots = 1 + w_3 z^2 + w_4 z^3 \cdots$$

$$zg(z) = zg(z_c) + \left(1 - \frac{z}{z_c}\right)^n = 1 + \left(g(z_c) - \frac{n}{z_c}\right) z + \frac{n(n-1)}{2z_c^2} z^2 + \cdots$$

where we have just expanded the bracket. Therefore $z_c g(z_c) - n = 0$.

9. We simply find the weights by using $f(z) = zg(z) - 1$ and by expanding $(1 - z/z_c)^n$: the coefficient to z^{m-1} is the weight w_m:

$$w_m = \frac{(-1)^{m-1}}{z_c^m} \binom{n}{m-1} = \frac{(-1)^{m-1}}{z_c^{m-1}} \frac{\Gamma(n+1)}{\Gamma(n-m+2)\Gamma(m)}, \quad 3 \le m \le n+1. \tag{C.5.1}$$

and by definition $w_1 = 1$ and $w_2 = 0$.

10. We solve $z_c^n (g - g_c) z = (z_c - z)^n$ iteratively wrt z:

$$z = z_c - z_c (g - g_c)^{1/n} z^{1/n} = z_c - z_c (g - g_c)^{1/n} \left(z_c - z_c (g - g_c)^{1/n} z^{1/n}\right)^{1/n}$$
$$= z_c - z_c^{1+1/n} (g - g_c)^{1/n} + O((g - g_c)^{2/n})$$

We have $z = z_c - c(g - g_c)^{1-\gamma}$ for the rooted BPs. Thus $\gamma = 1 - 1/n$.

11. If all weights are positive, the second derivative of $g(z)$ is

$$g''(z) = \frac{2}{z^3} + \text{positive terms} > 0 \quad \text{for all } z > 0.$$

Since z by definition is larger than zero if the weights are positive and $g(z)$ has a minimum if we assume that the weights $w_m = 0$ for $m > n+1$, expansion around this minimum at z_c leads to the standard result

$$g(z) = g(z_c) + c_1 (z - z_c)^2 + \cdots \quad \text{i.e.} \quad z(g) = z_c - c_2 (g - g_c)^{1/2} + \cdots$$

Clearly $\gamma = 1/2$ if and only if the second derivative of $g(z)$ is positive at the critical point, and one way to achieve this is to have a finite number of branching weights, which in addition are all positive.

12. Again compute the second derivative of $g(z)$. One can then notice that

$$g''(z) = \frac{f''(z)}{z} - \frac{2}{z} g'(z)$$

after collecting terms. At the critical point, $g'(z) = 0$, so positivity of $f''(z)$ at the critical point implies positivity of $g''(z)$ there. From our previous discussion, this again leads to $\gamma = 1/2$, provided there *is* a critical point. If we add the assumptions that $f(0), f'(0) = 0$ and $f''(x) > 1/x^2$ for large x this is ensured since we can write

$$g'(z) = -\frac{1}{z^2}(1 + f(z) - z f'(z)) = -\frac{1}{z^2}\left(1 - \int_0^z dx \, x f''(x)\right).$$

The assumptions ensure that $g'(z)$ is negative for small z and positive for large z.

13. The two given examples of functions with the desired property show that not all w_m need to be positive, as long as $f''(z) > 0$ (at z_c). Working out the w_m is straightforward from the power series of the functions:

$$w_m = \frac{1}{(m-1)!}, \quad m > 2, \quad w_{2k-2} = \frac{2^{2k}(2^{2k}-1)B_{2k}}{(2k)!}, \quad w_4 = 1, \, w_6 = -\frac{1}{3}, \ldots$$

The Bernoulli numbers B_{2k} enter in the weights w_{2m}, $m \geq 2$ defined by the power series of $z^2 \tanh z$, and result in oscillating signs, and a radius of convergence $r = \frac{\pi}{2}$, but the function is perfectly regular along the real axis (there are poles on the imaginary z-axis at $\pm i\pi/2$).

14. The problem of extracting the coeficients w_m from $(1-z/z_c)^s$ is identical to the problem of finding the Taylor coefficient c_L in $(1-g_c/g)^{1-\gamma}$, $s = 1-\gamma$, which we addressed in the first question in this Problem Set. We can then directly use these results with $L = m-1$.

$$\begin{aligned} w_m &= \frac{(-1)^{m-1}}{z_c^{m-1}} \frac{s \cdot (s-1) \cdot (s-2) \cdots (s-(m-1)+1)}{(m-1)!} \\ &= \frac{1}{z_c^{m-1}} \frac{(-1)^{m-1}\Gamma(s+1)}{\Gamma(m)\Gamma(s-m+2)} = \frac{\Gamma(m-1-s)}{\Gamma(m)\Gamma(-s)}. \end{aligned} \quad (C.5.2)$$

The first expression in the second line is simply the generalization of what we already derived in question 9 for s being an integer n, and the second expression is the same as we derived in question 1 with $L = m-1$. The second formula is well suited to find the asymptitic behavior of w_m for large m, as we did in question 1, and we find:

$$w_m \to \frac{(-1)^n}{|\Gamma(-s)|} \frac{1}{m^{s+1}} \quad \text{for} \quad m \to \infty, \quad (C.5.3)$$

The factor $(-1)^n$ comes from $\Gamma(-s)$.

15. The oscillating sign of w_m for $m < s+2$ follows from the first expression in the second line of eq. (C.5.2) and the constancy for $m > s+2$ follows from the second expression in second line of eq. (C.5.2).

16. We have directly from (C.5.2) that w_m is positive for $1 < s < 2$ and $m > 2$ (by construction $w_1 = 1$ and $w_2 = 0$ since we have fixed $g(z_0) = s/z_c$).

17. The proof is identical to that in the case $s = n$, considered in question 10.

C.6 SOLUTIONS TO PROBLEM SET 6

1. The figures are clearly the only possibility if we have the so-called *hard dimers*. That leads to, in terms of equations

$$z = e^{-\mu}\left(1 + z^2 + 2z\tilde{z}\right), \qquad \tilde{z} = e^{-\mu}\xi\left(1 + z^2\right) \qquad \text{(C.6.1)}$$

or the ones given in the problem set.

2. Eliminating $\tilde{z}$ from (C.6.1) leads to

$$g = \frac{1+z^2}{z} + \frac{2\xi}{g}(1+z^2). \qquad \text{(C.6.2)}$$

This is a second-order equation in g and we find

$$g = \frac{1}{2}\left[\frac{1+z^2}{z} + \sqrt{\frac{(1+z^2)^2}{z^2} + 8\xi(1+z^2)}\right] \qquad \text{(C.6.3)}$$

where one has to choose the plus sign for square root since we know from (C.6.1) that $g \to 1/z$ for $z \to 0$.

3. Differentiate g given by eq. (C.6.2) wrt z:

$$g'(z) = -\frac{1}{z^2} + 1 + \frac{4\xi z}{g} + F_1(z, g, g'), \qquad \text{(C.6.4)}$$

$$g''(z) = \frac{2}{z^3} + \frac{4\xi}{g} + F_2(z, g, g') - \frac{2\xi(1+z^2)}{g^2}g''(z) \qquad \text{(C.6.5)}$$

where F_1 and F_2 are functions which vanish if $g'(z) = 0$. Thus we have for the value of z where $g'(z) = g''(z) = 0$:

$$0 = -\frac{1}{z^2} + 1 + \frac{4\xi}{g}z, \qquad 0 = \frac{1}{z^3} + \frac{2\xi}{g}, \qquad \text{(C.6.6)}$$

4. From this we can find $1/z^2$ and ξ/g, and then from (C.6.2) the values of g and ξ:

$$\frac{1}{z_c^2} = \frac{1}{3}, \quad \frac{2\xi_c}{g_c} = \frac{-1}{3\sqrt{3}}, \quad \text{i.e.} \quad z_c = \sqrt{3}, \quad g_c = \frac{8}{3\sqrt{3}}, \quad \xi_c = \frac{-4}{27}. \qquad \text{(C.6.7)}$$

Finally, differentiating $g''(z)$ in (C.6.5) one more time and using $g'(z_c) = g''(z_c) = 0$ we obtain

$$g'''(z_c) = -\frac{6}{z_c^4} - \frac{2\xi_c(1+z_c^2)}{g_c^2}g'''(z_c) \qquad \text{(C.6.8)}$$

and for the values in (C.6.7) we conclude $g'''(z_c) \neq 0$.

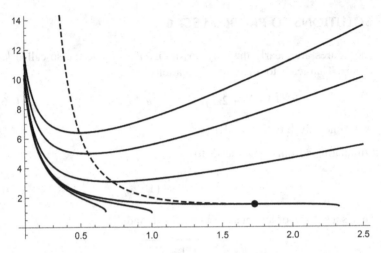

Figure C.6.1 The various curves $z \to g(z,\xi)$ for a number of values of ξ in the range from 10 (top curve) to -0.4 (bottom curve). The third curve from the bottom corresponds to $\xi = -4/27$ which has the lowest value of the curve minimum. The dashed curve is the curve of the minima for the curves $g(z,\xi)$ for $\xi > -4/27$, i.e. the part of the curve $g_k(z_k)$ given by eq. (C.6.9) until its minimum, i.e. for $0 < z < \sqrt{3}$. An extended range of $g_k(z_k)$ is shown in Fig. C.6.2. The two lowest curves have $\xi < -4/27$ and no local extrema. For $\xi < -1/8$ the square root in the expression (C.6.3) for $g(z,\xi)$ can become negative for sufficient large z and the function is only defined up to this point in the graph.

5. We write now $g(z,\xi)$ for the function (C.6.3). As for ordinary BP, for a fixed $\xi > \xi_c$ we have now a critical point $z_k(\xi) < z_c$ where $g'_z(z_k,\xi) = 0$, and the corresponding $g_k(\xi) = g(z_k,\xi)$. Thus we have a critical curve $\xi \to (z_k(\xi), g_k(\xi))$ for $\xi > \xi_c$ (see Fig. C.6.1). Rather than finding the parametric form of the curve, we can directly find the form by solving $g'_z(z,\xi) = 0$ from the first eq. in (C.6.6) (finding ξ/g) and inserting in (C.6.2):

$$\frac{2\xi}{g_k} = \frac{1-z_k^2}{2z_k^3}, \quad \text{i.e.} \quad g_k = (1+z_k^2)\left(\frac{1}{z_k} + \frac{2\xi}{g_k}\right) = \frac{(1+z_k^2)^2}{2z_k^3}. \quad (C.6.9)$$

6. However, the curve $g_k(z_k)$ seemingly continues happily for $z_k > z_c$. This can be understood from the explicit solution for $g(z,\xi)$ given in (C.6.3), the behavior of which are shown in Fig. C.6.2 for various values of ξ. We are (from the point of view of physics of BPs) only interested in the first minimum of $g(z,\xi)$ for a given value ξ. This first minimum we meet precisely on the first part of the curve $g_k(z_k)$. As long as ξ decreases, starting at ∞, $z_k(\xi)$ will increase and $g_k(z_k)$ will decrease *until* $dg_k/dz_k = 0$. This point is exactly $(z_k, g_k) = (z_c, g_c)$. For $z_k > z_c$ $g_k(z_k)$ will start to increase again. This new point (z_k, g_k) is associated with a second extremum, a local

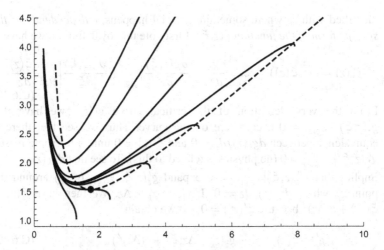

Figure C.6.2 The various curves $z \to g(z, \xi)$ for a number of values of ξ. For $\xi \in\,]-4/27, -1/8]$ the curves have a local maximum to the right of the local minimum and the curve $g_k(z_k)$ given by eq. (C.6.9) passes through these local maxima for $z_k > \sqrt{3}$, as seen in the figure.

maximum for the function $g(z, \xi)$ for a given value of ξ. From the explicit form of $g(z, \xi)$ one can show that for $\xi \geq -1/8$ there is only one extremum, the minimum of $g(z, \xi)$ at $z_k(\xi)$. For $-4/27 < \xi < -1/8$ there are *two* values z_k for which $g'_z(z, \xi) = 0$. Fig. C.6.2 shows how the part of the curve $g_k(z_k)$ for $z_k > z_c$ passes through the maxima of the curves $g(z, \xi)$ for $-4/27 < \xi < -1/8$. The point $z_k(\xi_c = -4/27) = \sqrt{3}$ is precisely the point where $dg_k/dz_k = 0$. Finally, for $\xi < \xi_c = -4/27$ there is no local minimum, $g'(z, \xi) < 0$ for all z where $g(z, \xi)$ are defined.

7. We can find the function $z_k(\xi)$, or more easily the inverse function $\xi = \xi(z_k)$. From (C.6.9) we obtain

$$\xi(z) = \frac{(1-z^4)(1+z^2)}{8z^6}. \qquad (C.6.10)$$

8. We now want to show that

$$\left.\frac{d\xi(z)}{dz}\right|_{z=z_c} = 0 \qquad (C.6.11)$$

One can of course directly differentiate (C.6.10) and insert $z_c = \sqrt{3}$. Also, taking the double derivative one finds that $\xi''(z_c) \neq 0$.

However it is more in the spirit of critical phenomena to show that the point z_k^c where $dg_k(z_k)/dz_k = 0$ is also the point where both $\xi'(z_k) = 0$ and $g'_z(z, \xi)|_{\xi=\xi(z_k)} = g'''_{zz}(z, \xi)|_{\xi=\xi(z_k)} = 0$, i.e. this point, z_k^c, is naturally identified with z_c where something special happens, *independently of the specific form of the function $g(z,\xi)$*. First note that by definition we have

$$g_k(z) = g(z,\xi(z)) \quad \text{i.e.} \quad \frac{dg_k}{dz} = \frac{\partial g(z,\xi)}{\partial z}\bigg|_{\xi(z)} + \frac{\partial g(z,\xi)}{\partial \xi}\bigg|_{\xi(z)} \frac{d\xi(z)}{dz} \tag{C.6.12}$$

From the very definition of the critical curve $g_k(z_k)$ we have that $g'_z(z,\xi)|_{\xi=\xi(z)} = 0$ everywhere on the curve. Thus we see that there is equivalence between $dg_k(z)/dz = 0$ and $\xi'(z) = 0$ unless for some reason $g'_\xi(z,\xi)|_{\xi=\xi(z)} = 0$ (no physics is related to this). To see that $dg_k(z)/dz = 0$ implies that $g''_{zz}(z,\xi)|_{\xi=\xi(z)} = 0$ expand $g'_z(z,\xi)|_{\xi=\xi(z)}(=0)$ around the point z_k^c where $dg_k(z)/dz = 0$. Let $z = z_k^c + \Delta z$. We then have $\xi(z) = \xi(z_k^c) + c(\Delta z)^2$ because $\xi'(z_k^c) = 0$ and we obtain

$$\begin{aligned}g'_z(z,\xi(z)) &= g'_z(z_k^c+\Delta z, \xi_k^c + c(\Delta z)^2) &\text{(C.6.13)}\\ &= g'(z_k^c,\xi_k^c) + g''_{zz}(z_k^c,\xi_k^c)\Delta z + O((\Delta z)^2). &\text{(C.6.14)}\end{aligned}$$

Since, by definition, $g'_z(z,\xi(z)) = g'(z_k^c,\xi_k^c) = 0$, we conclude that $g''_{zz}(z_k^c,\xi_k^c) = 0$ and we have shown that we indeed can identify z_c with z_k^c.

9. The two terms

$$\frac{\partial^2 g}{\partial \xi \partial z}\bigg|_{z_c,\xi_c}(z-z_c)(\xi-\xi_c) + \frac{1}{6}\frac{\partial^3 g}{\partial z^3}\bigg|_{z_c,\xi_c}(z-z_c)^3 \tag{C.6.15}$$

in the expansion (B.6.11) of $g(z,\xi)$ around (z_c,ξ_c) both behave like $(\xi-\xi_c)^{3/2}$ when one uses eq. (B.6.10) from the problem sheet.

10. Clear, differentiating twice one obtains the desired behavior of $g''_k(\xi)$. Recall that the free energy $f(\xi) = -\log g_k(\xi)$. Differentiation twice we find the that singular behavior of f'' is the same as that of g''_k and we have finally

$$f''(\xi) \propto \frac{1}{(\xi-\xi_c)^{1/2}} \propto (\xi-\xi_c)^{\sigma-1}, \quad \text{i.e.} \quad \sigma = \frac{1}{2}. \tag{C.6.16}$$

C.7 SOLUTIONS TO PROBLEM SET 7

BPs with infinite Hausdorff dimension

1.
$$g(z) = \frac{1+f(z)}{z} \quad (C.7.1)$$

$g(z) \to \infty$ for $z \to 0$. The radius of convergence for $f(z)$ is $z=1$. If $s \leq 1$ then the power series of the derivative of f is infinite at $z=1$ ($\sum_n n^{-s}$ is divergent for $s \leq 1$). Thus also the derivative of $g(z)$ will go to infinity at $z=1$ for these values of s. The value of z, z_c, for which $g(z)$ assumes the minimum is thus $0 < z_c < 1$, and it is a simple minimum: $g''(z_c) > 0$. This argument does not require $g'(1) = \infty$, only $g'(1) > 0$, which can be shown to lead to $s < 1.59....$

2. We have to cancel an nth-order polynomial in $g(z)$ to obtain $g(z) - g_c \sim c(1-z)^s$. This can clearly be done by adding a suitable term $\sum_{k=2}^{n+2} \tilde{w}_k z^{k-1}$ to $f(z)$.

3. Solve by iteration:

$$\Delta z = \frac{\Delta g}{c_1} - \frac{c_2}{c_1}(\Delta z)^2 - \frac{c_3}{c_1}(\Delta z)^3 + \cdots \quad (C.7.2)$$

$$(\Delta z)_1 = \frac{\Delta g}{c_1} \quad (C.7.3)$$

$$(\Delta z)_{1,2} = \frac{\Delta g}{c_1} - \frac{c_2}{c_1}\left(\frac{\Delta g}{c_1}\right)^2 \quad (C.7.4)$$

$$(\Delta z)_{1,2,3} = \frac{\Delta g}{c_1} - \frac{c_2}{c_1}\left(\frac{\Delta g}{c_1} - \frac{c_2}{c_1}\left(\frac{\Delta g}{c_1}\right)^2\right)^2 \Bigg|_{2,3} - \frac{c_3}{c_1}\left(\frac{\Delta g}{c_1}\right)^3 \quad (C.7.5)$$

4. We have

$$G_\mu^{(l)}(r) = \frac{(1+f(z))^2}{f'(z)} e^{-m_I(\mu)r}, \quad m_I(\mu) = -\log(e^{-\mu} f'(z)). \quad (C.7.6)$$

where $g = e^\mu$. Recall from the Chapter 4 (differentiating $e^\mu z = 1+f(z)$ wrt μ)

$$e^{-\mu} f'(z) = 1 + \frac{z}{z'} = 1 + z\frac{d\mu}{dz}, \quad \text{i.e.} \quad m_I(\mu) = -\log\left(1+z\frac{d\mu}{dz}\right). \quad (C.7.7)$$

Thus, if $d\mu/dz \neq 0$ for $z=z_c$ we obtain $m_I(\mu_c) \neq 0$. The reason for this is simply that if $s > 2$ then $g(z)$ (or $\mu(z)$) is a decreasing function of z all the way to $z_c = 1$ and thus $d\mu/dz < 0$ also at z_c. This is contrary to the situation for $\gamma = 1/2$ where $\mu = \mu_c - k(z_c - z)^2$ and $d\mu/dz \to 0$ for $z \to z_c$.

Ising model coupled to BPs

5. Assume the spin of the root is +. Depending on whether the first vertex after the root has spin + or spin -, we get factors

$$e^{-\mu}e^{\beta+h}(1+f(Z_+)), \quad \text{or} \quad e^{-\mu}e^{-\beta-h}(1+f(Z_-)), \quad \text{(C.7.8)}$$

and similarly, if the spin of the root is -, we obtain

$$e^{-\mu}e^{-\beta+h}(1+f(Z_+)), \quad \text{or} \quad e^{-\mu}e^{+\beta-h}(1+f(Z_-)). \quad \text{(C.7.9)}$$

6. If $h = 0$ then clearly it is consistent to choose $Z_+ = Z_-$ and solve the equation for $Z = Z_+ = Z_-$. Assuming there is only one solution to the equations, $Z_- = Z_+$ is justified. The rest of the questions are easily answered.

7.
$$\frac{d}{dZ}\frac{1+f(Z)}{Z}\bigg|_{Z_c} = 0 \implies \frac{1+f(Z_c)}{Z_c^2} = \frac{f'(Z_c)}{Z_c}. \quad \text{(C.7.10)}$$

8. The log 2 comes from the fact that for $\beta = 0$ the action $e^{\beta\sum_{<i,j>}\sigma_i\sigma_j} = 1$ and in $Z(\beta)$ we thus get $\sum_{\sigma_i} = 2^V$ for the spin contribution. This is precisely the number of spin configurations, and the classical entropy is k_B log(number of configurations). The entropy density is thus log 2 in our units where $k_B = 1$.

9. The equations are readily obtained by expanding the defining equations to linear order in h, writing $\mu(\beta,h) = \mu_c(\beta) + \Delta\mu$, where we assume $\Delta\mu = k \cdot h + O(h^2)$, and then adding and subtracting the resultant equations.

10. We clearly obtain $\Delta\mu = 0$ from

$$(\Delta Z_+ + \Delta Z_-)\left(e^{\mu_c(\beta)} - 2\cosh\beta f'(Z_c)\right) + 2Z_c e^{\mu_c(\beta)}\Delta\mu = 0$$

since $e^{\mu_c(\beta)} - 2\cosh\beta f'(Z_c) = 0$ and this means (since we are expanding only to linear order in h, that $\Delta\mu = O(h^2)$, which implies that $d\Delta\mu/dh \to 0$ for $h \to 0$. Thus the spontaneous magnetization is zero.

11. Using the information that $\Delta Z_+ = 0$, we can write

$$\Delta Z_-\left(e^{\mu_c(\beta)} - 2\cosh\beta f'(Z_c)\right) + 2Z_c e^{\mu_c(\beta)}\Delta\mu = 0 \quad \text{(C.7.11)}$$

$$-\Delta Z_-\left(e^{\mu_c(\beta)} - 2\sinh\beta f'(Z_c)\right) = 4h(1+f(Z_c))\sinh\beta. \quad \text{(C.7.12)}$$

It follows from the equations that $\Delta Z_- \propto \Delta\mu$ and $\Delta Z_- \propto h$, thus $\Delta\mu = k \cdot h$, where $k \neq 0$.

Solutions to Problem Sets 1–13

12. From the definition $\langle m \rangle = d\Delta\mu/dh\big|_{h=0} = k$ and from (C.7.11) and (C.7.12) we find explicitly, using $e^{\mu_c(\beta)} = 2\cosh\beta\,(1+f(Z_c))/Z_c$ as well as $Z_c = 1$:

$$\langle m \rangle = \tanh\beta \,\frac{1-\tanh\beta\,\frac{f'(1)}{1+f(1)}}{1-\frac{f'(1)}{1+f(1)}}, \qquad f(1) = \mathrm{Li}_{s+1}(1), \quad f'(1) = \mathrm{Li}_s(1).$$

(C.7.13)

It is seen that $\langle m \rangle \to 0$ for $\beta \to 0$ and $\langle m \rangle \to 1$ for $\beta \to \infty$, as one would expect.

Ising model and dimers

13. Use the suggested formula for each term in the action and extract cosh-factors.

14. The lowest order $\tanh\beta$ term is obtained by having a link $\sigma_i\sigma_j\tanh\beta$ anywhere on the lattice. However we have to "close" the link with two terms $\sigma_i\tanh h$ and $\sigma_j\tanh h$ in order that the summations over σ_i and σ_j do not give zero. Similarly when we put down two links. If they do not touch (hard dimers) they have to be "closed" by four terms of the form $\sigma_k\tanh h$. A given term $\sigma_{k_1}\cdots\sigma_{k_n}\tanh^n h$ has to meet links at the vertices k_i in order that the sum over σ_{k_i} does not give zero. Thus n has to be even (since links bring an even number of σs). Next, the *smallest* number of links one can use is obtained if the links can be put down as hard dimers (and it is $n/2$). There are many other ways one can dress the $\tanh^n h$ term with links but they always involve more links and thus higher powers of $\tanh\beta$ and thus higher powers of β.

15. Follows from the expansion given for $Z(\beta, h)$.

C.8 SOLUTIONS TO PROBLEM SET 8

Asymptotic expansions

1. Write

$$f(x) = xh(x), \qquad B(h)(x) = \sum_{n=0}^{\infty}(-1)^n x^n = \frac{1}{1+x}. \qquad (C.8.1)$$

$$h(x) = \int_0^{\infty} dt\, e^{-t} B(h)(xt), \qquad f(x) = x\int_0^{\infty} dt\, e^{-t}\frac{1}{1+tx}. \qquad (C.8.2)$$

2. Differentiate the formal power series for $f(x)$ to obtain another formal power series

$$f'(x) = \sum_{n=0}^{\infty}(n+1)!(-1)^n x^n = \frac{1}{x^2}\left(-\sum_{k=0}^{\infty} k!(-1)^k x^{k+1} + x\right) = -\frac{1}{x^2}f(x) + \frac{1}{x}$$

3. The general solution to an inhomogeneous linear differential equation:

$$f' + a(x)f = b(x), \quad f(x) = e^{-A(x)}\int^x dy\, e^{A(y)} b(y), \quad A(x) = \int^x dy\, a(y).$$

Applied to our differential equation we obtain

$$f(x) = e^{1/x}\int_0^x dy\, e^{-1/y}\frac{1}{y} = \int_0^{\infty} dt\, \frac{xe^{-t}}{1+xt} \qquad \left[t = \frac{1}{y} - \frac{1}{x}\right] \qquad (C.8.3)$$

4. Differentiating $e^{1/x}\text{Ei}(-1/x)$ it is easily seen that it satisfies the differential equation. Also, changing variables as above ($t = \frac{1}{y} - \frac{1}{x}$) it is clear that it is the Borel sum of the original formal power series.

5. Successive partial integrations: ($y = 1/x$)

$$e^y\left(\int_y^{\infty} dt\, \frac{e^{-t}}{t}\right) = -\frac{e^{-t}}{t}\bigg|_y^{\infty} + \frac{e^{-t}}{t^2}\bigg|_y^{\infty} - 2\frac{e^{-t}}{t^3}\bigg|_y^{\infty} + \cdots + (-1)^n n!\int_y^{\infty} dt\, \frac{e^{-t}}{t^{n+1}}$$

$$(C.8.4)$$

6. Formally we have from the power series that $g(x) = -f(-x)$. This clearly leads to both the integral and the differential equation. The reason that we have changed the definition of $\text{Ei}(u)$ to $\text{Ei}_c(u)$, is that we do not want 0 to be part of the integration interval, since the integral then is ill defined (one can include it by a so-called principle value prescription, but it is easier to avoid 0)

To find the asymptotic expansion of $e^{-1/x}\text{Ei}_c(1/x)$, we perform the partial integration as in (C.8.4)

$$\int_{-1/x}^{-c} dt\, \frac{e^{-t}}{t} = -\frac{e^{-t}}{t}\bigg|_{\frac{-1}{x}}^{-c} + \frac{e^{-t}}{t^2}\bigg|_y^{\infty} - 2\frac{e^{-t}}{t^3}\bigg|_{\frac{-1}{x}}^{-c} + \cdots + (-1)^n n!\int_{\frac{-1}{x}}^{-c} dt\, \frac{e^{-t}}{t^{n+1}}$$

$$= e^{1/c}\left(\frac{1}{c} + \frac{1}{c^2} + 2!\frac{1}{c^3} + \cdots\right) - e^{1/x}\left(x + x^2 + 2!x^3 + \cdots\right)$$

and thus

$$-e^{-1/x}\operatorname{Ei}_c(1/x) = \left(x+x^2+2!x^3+\cdots\right)+F(c)e^{-1/x}$$

Note that $F(c)e^{-1/x}$ is a solution to the homogeneous differential equation. This is why we have a solution for any positive c and they all have the same asymptotic expansion since $F(c)e^{-1/x}$ *does not contribute to the asymptotic series*.

C.9　SOLUTIONS TO PROBLEM SET 9

Branched polymers with loops

1.
$$z^2 - gz + 1 + j = 0, \quad \text{i.e.} \quad z(g,j) = \frac{g}{2} - \sqrt{\frac{g^2}{4} - (j+1)} = \frac{g}{2} - \sqrt{\Delta(g,j)}$$
(C.9.1)

We have to choose the minus sign in front of the square root since $z \to 0$ for $g \to \infty$, from the very definition of the partition function.

2.
$$g = \frac{1+j+z^2}{z}, \quad \frac{dg}{dz} = -\frac{1+j}{z^2} + 1, \quad \frac{d^2g}{dz^2} = \frac{2(1+j)}{z^3} \quad \text{(C.9.2)}$$

Thus
$$\frac{dg}{dz} = 0 \Rightarrow z_c = \sqrt{1+j}, \quad g_c = 2\sqrt{1+j}, \quad \Delta(g_c, j) = 0, \quad \text{(C.9.3)}$$

and since $g''(z_c) > 0$ we have $\gamma = 1/2$.

3. From the figure it follows that
$$\chi = \frac{1}{g} + \frac{2z}{g^2} + \frac{2^2 z^2}{g^3} + \cdots = \frac{1}{g-2z} = \frac{1}{2\sqrt{\Delta}} = \frac{1}{g} + 2\frac{1+j}{g^3} + 6\frac{(1+j)^2}{g^5} + \cdots$$
(C.9.4)

4. The expansion in loops of the two first of the hierarchial equations for Z and $\chi^{(k)}$ can be written
$$Z_0 + \frac{Z_1}{\Lambda} + \frac{Z_2}{\Lambda^2} + \cdots = \frac{1}{g}\left(1 + g + \left[Z_0 + \frac{Z_1}{\Lambda} + \frac{Z_2}{\Lambda^2} + \cdots\right]^2 + \frac{1}{\Lambda}\left[\chi_0^{(2)} + \frac{\chi_1^{(2)}}{\Lambda} + \cdots\right]\right)$$
(C.9.5)

$$\chi_0^{(2)} + \frac{\chi_1^{(2)}}{\Lambda} + \cdots = \frac{1}{g}\left(1 + 2\left[Z_0 + \frac{Z_1}{\Lambda} + \cdots\right]\left[\chi_0^{(2)} + \frac{\chi_1^{(2)}}{\Lambda} + \cdots\right] + \frac{1}{\Lambda}\left[\chi_0^{(3)} + \cdots\right]\right)$$
(C.9.6)

From this we find
$$Z_1 = \frac{1}{g}\left(2Z_0 Z_1 + \chi_0^{(2)}\right) \Rightarrow Z_1(g - 2Z_0) = \chi_0^{(2)} \Rightarrow Z_1 = \left(\chi_0^{(2)}\right)^2$$
(C.9.7)

and
$$\chi_1^{(2)} = \frac{1}{g}\left(2Z_0 \chi_1^{(2)} + 2Z_1 \chi_0^{(2)} + \chi_0^{(3)}\right), \quad \text{(C.9.8)}$$

i.e.
$$\chi_1^{(2)}(g - 2Z_0) = 2Z_1 \chi_0^{(2)} + \chi_0^{(3)} = 4\left(\chi_0^{(2)}\right)^3 \Rightarrow \chi_1^{(2)} = 4\left(\chi_0^{(2)}\right)^4.$$
(C.9.9)

From the $1/\Lambda^2$ term in (C.9.5) we obtain

$$Z_2 = \frac{1}{g}\left(2Z_0 Z_2 + Z_1^2 + \chi_1^{(2)}\right) \Rightarrow Z_2(g-2Z_0) = Z_1^2 + \chi_1^{(2)} \Rightarrow Z_2 = 5\left(\chi_0^{(2)}\right)^5. \tag{C.9.10}$$

5. In any one-loop graph of the kind we discuss, the root is unique, the vertex where the loop starts is unique, the shortest path connecting the root to the vertex is unique and the shortest loop-line is unique. The shortest path between the root and the marked vertex is dressed with all kind of outgrowths and can be used to represent all BPs where the shortest path between the marked vertices has a fixed length. When we then sum over the length of these paths we obtain all BPs connecting the root and the vertex where the loop starts, i.e $\chi_0^{(2)}$. Similar arguments apply to the loop. The vertex where the loop starts, seen from the root, was unique and can be labelled a new root. We now open the loop by splitting this vertex in two. One part is the root, the other vertex will act as the marked vertex in a new BP, before forming the loop, but now a BP of the kind belonging to $\chi_0^{(2)}$. This makes sense since the vertex we split was of order 3 and had thus no j attached. After the split it becomes a root and a vertex of order 1, i.e. precisely the two vertices of order 1 which have no j attached in a BP belonging to $\chi_0^{(2)}$. The shortest path between these two vertices is exactly the shortest path mentioned before in the loop. Summing over such graphs we obtain again all BPs with the marked points separated a given distance and summing over the length we obtain again $\chi_0^{(2)}$. In total thus $(\chi_0^{(2)})^2 = 1/(4\Delta)$.

$$Z_1(g,j) = \frac{1}{4\Delta} = \frac{1}{g^2-4(j+1)} = \frac{1}{g^2} + 4\frac{s9j+1}{g^2} + 4^2\frac{(1+j)^2}{g^4} + \cdots \tag{C.9.11}$$

These are thus 1 one-loop diagram with two lines, 4 one-loop diagrams with 4 lines and 16 one-loop diagrams with 6 lines, see Fig. C.9.1

6. First, the 5 "skeleton" graphs shown in Fig. B.9.5 in the Problem Set are precisely the two-loop graphs generated by iterating the graphical Fig. B.9.3 in the Problem Set to two loops, assuming that all vertices are of order 3 except for the root which is of order 1. The graphical iteration is more or less identical to the algebraic iteration we performed above, which gave us Z_2 (eq. (C.9.10)). The middle graph in Fig. B.9.5 corresponds to the term Z_1^2, while the four other terms come from the fact that the one-loop propagator can be decomposed in four components, which according to eq. (C.9.8) can be written as $2Z_1\chi_0^{(2)}$ (leading to the two graphs to the left in in Fig. B.9.5) plus $\chi_0^{(3)}$ which leads to the two graphs to the right in Fig. B.9.5 (graphically $\chi_0^{(3)}$ is 2 times a ϕ^3 vertex connected to three external points if we only allow graphs with internal vertices of order 3).

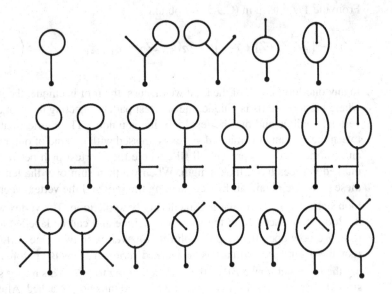

Figure C.9.1 The rooted BPs with one loop and two, four and six links.

Second, whenever one draws a ϕ^3 graph where the only vertex of order 1 is the root, one can "extend" the lines (i.e. the links in the graph) to a full BP with two marked points, i.e. to $\chi_0^{(2)}$ (which we in the following just denote χ). Also, given such a ℓ-loop BP-graph, one can, starting from the root, in a unique way identify the vertices which constitute the vertices in a "skeleton" ϕ^3 graph where the only vertex of order one is the root. Thus the total contribution is χ^L, where L is the number of links in the skeleton ϕ^3-graph. Let now G be a ϕ^3-graph with V_{ex} external vertices (i.e. vertices of order 1) and V_I internal vertices (i.e. vertices of order 3) and L links and ℓ loops. We then have[1]

$$L - (V_I + V_{ex}) + 1 = \ell \quad 3V_I + V_{ex} = 2L \quad \text{i.e.} \quad 3\ell - 1 = L, \quad \text{(C.9.12)}$$

in the case of tadpoles where $V_{ex} = 1$. Thus the total BP contribution coming from BPs with skeleton graph G with ℓ loops is $\chi^{3\ell-1}$, and the BP partition function with ℓ loops will be the sum over all such skeleton graphs, i.e. all "tadpole" ϕ^3-graphs with ℓ loop

$$Z_\ell(g,j) = C_\ell \, \chi^{3\ell-1}, \quad C_\ell = \# \text{ tadpole-}\phi^3 \text{ graphs.} \quad \text{(C.9.13)}$$

---

[1] The first equation defines the number of loops in the way we meet them in a Feynman diagram: we have to integrate over the momentum for each line (propagators), but for each vertex we have momentum conservation, except for allover momentum conservation. The ℓ is then the independent momenta we have to integrate over, i.e. the number of loops associated with the Feynman integral.

It is clear that $Z_\ell = C_\ell \chi^{3\ell-1}$ is precisely what we algebraically proved above for Z_1 and Z_2, and it is not too difficult to extend this algebraic proof to all orders in ℓ.

Note that we are not really specifying in a precise way what we mean by the *number* of tad-pole ϕ^3 graphs. It will not be important for us. The important point is that $Z_\ell \propto \chi^{3\ell-1}$ since this is what determines the singular behavior of $Z_\ell(g)$ when $g \to g_c$.

7. We simply insert

$$Z(g,j,\Lambda) = \frac{g}{2} - \sqrt{\Delta}F(t), \quad \Lambda\Delta^{\frac{3}{2}} = \frac{3}{2}t, \quad \frac{d}{dj} = -\frac{d}{d\Delta} = -\sqrt{\Delta}\Lambda\frac{d}{dt}$$

into the equation

$$gZ = (1+j) + Z^2 + \frac{1}{\Lambda}\frac{dZ}{dj} \qquad (C.9.14)$$

and obtain

$$g\left(\frac{g}{2} - \sqrt{\Delta}F\right) = 1+j+\left(\frac{g^2}{4} - g\sqrt{\Delta}F + \Delta F^2\right) + \frac{1}{\Lambda}\left(\frac{F}{2\sqrt{\Delta}} + \Lambda\Delta\frac{dF}{dt}\right)$$

or

$$\left(\frac{g^2}{4} - (1+j)\right) = \Delta F^2 + \Delta\frac{F}{2\Lambda\Delta^{3/2}} + \Delta\frac{dF}{dt},$$

i.e.

$$1 = F^2 + \frac{F}{3t} + \frac{dF}{dt}. \qquad (C.9.15)$$

8. Inserting the asymptotic expansion in the differential equation leads to a power expansion in $1/t$, where the coefficient multiplying $1/t^{n+1}$ has to be zero. Expressed in terms of the coefficients c_n of the asymptotic expansion of $F(t)$ we then obtain the equations:

$$1 = c_0^2 \qquad (C.9.16)$$

$$0 = \sum_{k=0}^{n+1} c_k c_{n+1-k} + (\frac{1}{3} - n)c_n = 0, \quad n \geq 0. \qquad (C.9.17)$$

The first few equations are (starting with $c_0 = 1$)

$$2c_1 = -\frac{1}{3}$$

$$2c_2 = c_1 - \frac{1}{3}c_1 - c_1^2$$

....

$$2c_{n+1} = nc_n - \left(\frac{1}{3}c_n + c_1 c_n + c_2 c_{n-1} \cdots + c_{n-1}c_2 + c_n c_1\right)$$

The two first equations give $c_1 = -1/6$ and $c_2 = -5/72$.

9. If we can ignore the bracket in the equation for c_{n+1}, a consistent solution for large n is clearly $c_n = -k\Gamma(n)/2^n$, where k is a constant. It is a consistent solution for large n up to power corrections $n^{-\alpha}$, since assuming it, one has

$$\frac{c_{n+1}}{c_n} = \frac{n}{2}\left(1 - \frac{1}{n}\left[\frac{1}{3} + \frac{c_1 c_n + \cdots + c_n c_1}{c_n}\right]\right)$$

and one can check (numerically) that $[\cdot] < 1/3 + 3k$.

10. We write the definition of γ_ℓ as $Z_\ell \sim (g - g_c)^{-\gamma_\ell + 1}$ and since $\Delta \sim g - g_c$ for $g \to g_c$ we have (question (5)): $Z_\ell \sim \Delta^{-\frac{3}{2}\ell + \frac{1}{2}}$, and we conclude $\gamma_\ell = \frac{3}{2}\ell + \frac{1}{2}$.

11. We know that the scaling limit of the BPs is universal, independent of the weights $w_3, w_4, \ldots$ as long as there are only a finite number of them and they are positive. Thus the susceptibility without loops behaves as $\chi_0(g) \sim 1/\sqrt{g - g_c}$. If we have $w_4, w_5, \ldots$ different from zero we can form many more skeleton graphs, involving vertices of order 4,5 etc. Note also that even if $w_3 = 0$ we have not problem constructing skeleton graphs with ϕ^3 vertices. Given the number of loops ℓ, the question is: which skeleton graph is most singular. Each link in the skeleton graph is represented by a BP propagator $\chi_0(g) \sim 1/\sqrt{g - g_c}$, so we simply want the tadpole graphs with ℓ loops and the maximal number of links L. Let V_n, $n = 3, 4, \ldots$ denote the number of vertices of order n. For a tadpole skeleton graph G with one "external" link and one "external" vertex, L links and ℓ loops we have

$$L = (V_3 + V_4 + \cdots) + \ell, \qquad 1 + 3V_3 + 4V_4 + \cdots = 2L.$$

Thus it is seen that L becomes maximal if all vertices (except the root) are order 3 vertices. (if we only have vertices of order n: $L = (n\ell + 1)/(n - 2)$.)

C.10 SOLUTIONS TO PROBLEM SET 10

A general even potential V(x)

1. Contracting the contour as mentioned (using that the contour integral does not change), and writing $1/\sqrt{\omega^2-a^2} = 1/(\sqrt{\omega-a}\sqrt{\omega+a})$ we obtain

$$\oint_C \frac{d\omega}{2\pi i} \frac{f(\omega)}{\sqrt{\omega^2-a^2}} = \int_a^{-a} \frac{dx}{2\pi i} \frac{-i}{\sqrt{a^2-x^2}} f(x) + \int_{-a}^a \frac{dx}{2\pi i} \frac{i}{\sqrt{a^2-x^2}} f(x)$$

$$= \int_{-a}^a \frac{dx}{\pi} \frac{f(x)}{\sqrt{a^2-x^2}} = \int_{-1}^1 \frac{dy}{\pi} \frac{f(ay)}{\sqrt{1-y^2}} \quad \text{(C.10.1)}$$

2. The formula for $W(z)$ in Chapter 6 can, for an even potential where $c_+ = a$ and $c_- = -a$, be written as

$$\oint_C \frac{d\omega}{2\pi i} \frac{f(\omega)}{\sqrt{\omega^2-a^2}}, \quad f(\omega) = \frac{V'(\omega)\sqrt{z^2-a^2}}{2(z-\omega)} = \frac{(z+\omega)V'(\omega)\sqrt{z^2-a^2}}{2(z^2-\omega^2)}. \quad \text{(C.10.2)}$$

Thus, for z outside the contour we can directly apply (C.10.1). Since $V'(\omega)$ is an odd function, the integral with $zV'(\omega)$ is zero and only the contribution with $\omega V'(\omega)$ survives, leading to the wanted formula, using that $\omega V'(\omega)$ is an even function.

3. Expanding the expression for $W(z)$ in powers of $1/z$ we obtain

$$W(z) = \frac{1}{g} \int_0^a \frac{dx}{\pi} \frac{x\tilde{V}'(x)}{\sqrt{a^2-x^2}} \frac{1}{z} + O\left(\frac{1}{z^2}\right) \quad \text{(C.10.3)}$$

which leads to the determination of g as a function of a.

4. The t_n term will lead to the following integral

$$\int_0^a \frac{dx}{\pi} \frac{2n\,x^{2n}}{\sqrt{a^2-x^2}} = \frac{2n}{\pi} \int_0^{\frac{\pi}{2}} d\theta \sin^{2n}\theta = \frac{1}{2} \cdot \frac{(2n-1)!!}{(2n-2)!!} \quad \text{(C.10.4)}$$

if we set $x = \sin\theta$ and use the hint. We now use

$$(2n-1)!! = 2^{n-1}(n-\tfrac{1}{2})(n-\tfrac{3}{2})\cdots\tfrac{1}{2} = 2^{n-1}\frac{\Gamma(n+\tfrac{1}{2})}{\Gamma(\tfrac{1}{2})}$$

$$(2n-2)!! = 2^{n-2}(n-1)(n-2)\cdots 1 = 2^{n-2}\Gamma(n).$$

This provides the formula.

5. We have $B(1,\tfrac{1}{2})=2$ and $B(2,\tfrac{1}{2})=4/3$, and thus

$$g(a^2) = \frac{1}{4}a^2 - \frac{3}{16}a^4 \quad \text{(C.10.5)}$$

The critical point is where $g'(a^2)=0$, i.e. $a_c^2 = \tfrac{2}{3}$ and thus $g_c = g(a_c^2) = \tfrac{1}{12}$.

6. Assume $t_1 > 0$, and the other $t_n \leq 0$ and that $t_n = 0$ for $n > N$. $g'(a^2) = 0$ leads to

$$\frac{t_1}{B(1,\frac{1}{2})} + \sum_{n>1} \frac{nt_n}{B(n,\frac{1}{2})} (a^2)^{n-1} = 0 \qquad (C.10.6)$$

which clearly has only one (positive) a_c solution. Furthermore we have

$$g''(a^2) = \sum_{n>1} \frac{n(n-1)t_n}{B(n,\frac{1}{2})} (a^2)^{n-2} < 0 \qquad (C.10.7)$$

so expanding around a_c we have

$$g(a^2) = g(a_c^2) + \frac{1}{2}g''(a_c^2)(a_c^2-a^2)^2 + O((a_c^2-a^2)^3), \qquad (C.10.8)$$

7. We know by now that if we have an even polynomial $V(x)$ of order $2m$, then $g(a^2)$ will also be an even polynomial of order $2m$ and we know the relations between t_n and the coefficients g_n in the polynomial

$$g(a^2) = \sum_{n=1}^{m} g_n a^{2n}. \qquad (C.10.9)$$

From the assumptions it is clear, by Taylor expanding around a_c^2 that $g(a^2)$ can be written as stated. We only need to determine $g(a_c^2)$ and c. For this we use the expansion around $a^2 = 0$:

$$g(0)=0, \quad g(a^2)=g_1 a^2 + O(a^4) = \frac{t_1}{B(1,\frac{1}{2})} a^2 + O(a^4) = \frac{t_1}{2}a^2 + O(a^4) \qquad (C.10.10)$$

The result now follows ($t_1 = 1/2$)

$$g(a_c^2) = c\, a_c^{2m}, \quad c\, m a_c^{2m-2} = \frac{1}{4}. \qquad (C.10.11)$$

8. For $a_c^2 = 1$ we have for the coefficient g_n in (C.10.9), expanding $-c(1-a^2)^m$

$$g_n = \frac{t_n}{B(n,\frac{1}{2})} = (-1)^{n-1} c \binom{m}{n}, \quad c = \frac{1}{4m}. \qquad (C.10.12)$$

9. The formula $\tilde{M}_0(a^2) = 2g(a^2)$ follows directly from the definitions of $\tilde{M}_0$ and g, and it is a replacement of $M_0(a^2) = 2$ discussed in Chapter 6.

10. We know from the definitions that

$$\tilde{M}_k(a^2) \propto \frac{d^k}{d(a^2)^k} \tilde{M}_0(a^2) \propto \frac{d^k}{d(a^2)^k} g(a^2) \propto \frac{d^k}{d(a^2)^k}(1-a^2)^m \qquad (C.10.13)$$

which gives the desired result.

11. In Problem Set 5 we saw that the coefficient c_n to a^{2n} in the power expansion of $(1-a^2)^s$ is:

$$c_n = \frac{\Gamma(n-s)}{\Gamma(-s)\Gamma(n+1)} \propto \frac{1}{n^{s+1}} \quad \text{for} \quad n \to \infty.$$

$$t_n = -\frac{c_n}{4s} B(n, 1/2) \propto \frac{1}{n^{s+3/2}} \quad \text{for} \quad n \to \infty.$$

12. Follows from the definition of $\tilde{M}_k$ by differentiation wrt a^2.
13. This relation is just as discussed in Chapter 6, the only difference being a factor $1/g$.
14. Using the hint, and the relations proven earlier we want to prove that

$$\left(\tilde{M}(z) - 2(z^2-a^2)\frac{d\tilde{M}(z)}{da^2}\right) = \tilde{M}_1. \quad \text{(C.10.14)}$$

We have (using $\tilde{M}_k = 0$ for $k > m$)

$$2(z^2-a^2)\frac{d\tilde{M}(z)}{da^2} = -2(z^2-a^2)\sum_{k=2}^{m}(k-1)(z^2-a^2)^{k-2}\tilde{M}_k +$$
$$+2(z^2-a^2)\sum_{k=1}^{m-1}(k+\frac{1}{2})(z^2-a^2)^{k-1}\tilde{M}_{k+1}$$
$$= (z^2-a^2)\sum_{k=2}^{m}(z^2-a^2)^{k-2}\tilde{M}_k = \tilde{M}(z)-\tilde{M}_1,$$

i.e. the wanted formula.

C.11 SOLUTIONS TO PROBLEM SET 11

Multiple Ising spins coupled to 2d quantum gravity

1. We can sum over the spin configurations in the following way: let T be a triangulation where all spins are aligned to the spins on the two boundary triangles who per definition have the same spin (e.g. +). Now take an arbitrary interior link (there are $N_{L(I)} = N_L(T) - 2$ of these). We either leave the link untouched, it gives a factor 1, or we can open this link into two, connected to the same vertices and glue a new universe with − spin at its boundary to close the surface. In this way we effectively add a factor $e^{-2\beta} G(\mu,\beta)$ to the link. In total we then associate a factor $(1 + e^{-2\beta} G(\mu,\beta))$ with each interior link. In this way we actually perform the sum over allowed spin configurations and it leads to the self-consistent equation for $G(\mu,\beta)$.

    Let T be a triangulation with $N_{L(ex)}$ boundary links, $N_{L(I)}$ intrinsic links and N_T triangles. Then we have

    $$2N_{L(I)} + N_{L(ex)} = 3N_T$$

    which results in the last equation since $N_{L(ex)} = 2$.

2. The equation should be clear: summing over $\mathcal{T}^{(2)}(2)$ with exponential weight is by definition G_0.

3. Just a rearrangement using $G(\mu,\beta) = G_0(\bar\mu)$.

4. Differentiate eq. (12) from the Problem Set wrt $\bar\mu$, using the definition

    $$\chi_0(\bar\mu) = -\frac{dG_0(\bar\mu)}{d\bar\mu}.$$

5. Differentiate $G(\mu,\beta) = G_0(\bar\mu)$ using the chain-rule on the rhs and the definition of χ.

6. we have

    $$G_0(\bar\mu) = \sum_T e^{-\bar\mu N_T}, \quad \chi_0(\bar\mu) = -\frac{dG_0(\bar\mu)}{d\bar\mu} = \sum_T N_T\, e^{-\bar\mu N_T}$$

    Thus

    $$\frac{3}{2}\chi_0(\bar\mu) - G_0(\bar\mu) = \sum_T \left(\frac{3}{2} N_T - 1\right) e^{-\bar\mu N_T}$$

    which is clearly a decreasing function of $\bar\mu$ (each term is..)

7. The obvious choice of β_0 is the value where

    $$e^{2\beta_0} = \frac{3}{2}\chi_0(\mu_0) - G_0(\mu_0), \quad \text{i.e.} \quad \beta_0 = \frac{1}{2}\log\left(\frac{3}{2}\chi_0(\mu_0) - G_0(\mu_0)\right). \tag{C.11.1}$$

since this is, from above arguments, the largest value the rhs can assume. Thus we know for sure that if $\beta > \beta_0$ then

$$e^{2\beta} - \left(\frac{3}{2}\chi_0(\bar{\mu}) - G_0(\bar{\mu})\right) > 0 \quad \text{for} \quad \bar{\mu} \geq \mu_0,$$

and from $G(\mu,\beta) = G_0(\bar{\mu})$ we know that for $\mu > \mu_0(\beta)$ also $\bar{\mu} > \mu_0$ (else both sides of the equation could not exist).

8. The above considerations show that all the way down to $\bar{\mu} = \mu_0$ there is a simple linear relationship between $\bar{\mu}$ and μ for small changes. The derivatives $\partial \bar{\mu}/\partial \mu$ and $\partial \mu/\partial \bar{\mu}$ are finite as long as $\beta > \beta_0$ all the way down to and including $\bar{\mu} = \mu_0$. Thus the only source of non-analyticity in the relation (B.11.16) in the Problem Set can come from $\chi_0(\bar{\mu})$ when $\bar{\mu} \to \mu_0$. Since $\partial \mu/\partial \bar{\mu}$ is finite at that point and a non-singular function of $\chi_0(\bar{\mu})$ and $G_0(\bar{\mu})$, the non-analyticity of $\chi(\mu,\beta)$ must be the same as that of $\chi_0(\bar{\mu})$, and thus $\gamma(\beta) = \gamma_0$ as long as $\beta > \beta_0$.

9. The first of the relations

$$\bar{\mu}_0(\beta_0) = \mu_0, \qquad \bar{\mu}_0(\beta) > \mu_0 \quad \text{for} \quad \beta < \beta_0.$$

follows from the definition of β_0 given above. The other relation is also a consequence of that definition: when $\beta < \beta_0$

$$e^{2\beta} = \frac{3}{2}\chi_0(\bar{\mu}) - G_0(\bar{\mu})$$

has a solution $\bar{\mu} > \mu_0$ simply because the rhs is an increasing function when $\bar{\mu}$ decreases toward μ_0 and $\beta < \beta_0$

10. Eq. (B.11.20) in the Problem Set clearly implies

$$\mu - \mu_0(\beta) = c(\bar{\mu} - \bar{\mu}_0(\beta))^2 + O((\bar{\mu} - \bar{\mu}_0(\beta))^3). \qquad \text{(C.11.2)}$$

unless, for some reason, the second derivative $\dfrac{\partial^2 \mu(\bar{\mu},\beta)}{\partial \bar{\mu}^2} = 0$, which we will assume is not the case.

11. Since $\beta < \beta_0$ we have that $\bar{\mu}_0(\beta) > \mu_0$. Thus $\chi_0(\bar{\mu})$ and $G_0(\bar{\mu})$ are analytic around that point. The source of singularity in $\chi(\mu,\beta)$ in the expression (B.11.16) in the Problem Set:

$$\chi(\mu,\beta) = \chi_0(\bar{\mu}(\mu,\beta)) \frac{\partial \bar{\mu}(\mu,\beta)}{\partial \mu}$$

therefore cannot come from $\chi_0(\bar{\mu})$ or $G_0(\bar{\mu})$ and has to comes from the denominator in $\frac{\partial \bar{\mu}(\mu,\beta)}{\partial \mu}$, which goes to zero for $\bar{\mu} \to \bar{\mu}_0(\beta)$. However, since

$\bar{\mu}_0(\beta) > \mu_0$ we can Taylor expand the denominator around $\bar{\mu}_0(\beta)$ and we obtain, using (C.11.2), the desired result

$$\chi(\mu,\beta) \sim \frac{1}{\bar{\mu}-\bar{\mu}_0(\beta)} \sim \frac{1}{\sqrt{\mu-\mu_0(\beta)}} \qquad \beta < \beta_0.$$

12. we have for $\bar{\mu} \to \mu_0$ by definition (see expansion (B.11.7) in the Problem Set)

$$\frac{3}{2}\chi_0(\bar{\mu}) - G_0(\bar{\mu}) = \frac{3}{2}\chi_0(\mu_0) - G_0(\mu_0) - c(\bar{\mu}-\mu_0)^{-\gamma_0} + O(\bar{\mu}-\mu_0),$$
(C.11.3)

and thus

$$e^{2\beta_0} - \left(\frac{3}{2}\chi_0(\bar{\mu}) - G_0(\bar{\mu})\right) = c(\bar{\mu}-\mu_0)^{-\gamma_0} + O(\bar{\mu}-\mu_0). \qquad \text{(C.11.4)}$$

This implies

$$\frac{\partial\mu}{\partial\bar{\mu}} \sim (\bar{\mu}-\mu_0)^{-\gamma_0} + O(\bar{\mu}-\mu_0), \qquad \frac{\partial\bar{\mu}}{\partial\mu} \sim \frac{1}{(\bar{\mu}-\mu_0)^{-\gamma_0}} \qquad \text{(C.11.5)}$$

13. and by integration of (C.11.5)

$$\mu - \mu_0(\beta_0) = c(\bar{\mu}-\mu_0)^{1-\gamma_0} + O((\bar{\mu}-\mu_0)^2) \qquad \text{(C.11.6)}$$

14. Finally from

$$\chi(\mu,\beta) = \chi_0(\bar{\mu}(\mu,\beta)) \frac{\partial\bar{\mu}(\mu,\beta)}{\partial\mu}$$

we obtain, using (C.11.5)

$$\chi(\mu,\beta_0) \sim \frac{1}{(\bar{\mu}-\mu_0)^{-\gamma_0}}$$

and using (C.11.6)

$$\chi(\mu,\beta_0) \sim \frac{1}{(\mu-\mu_0(\beta_0))^{-\gamma_0/(1-\gamma_0)}}.$$

This is the desired result.

C.12 SOLUTIONS TO PROBLEM SET 12

The purpose of this Problem Set is to derive the the multiloop formulas (6.73), (6.77) and (6.78) using (6.72). We will simply use the representation (6.61) for the loop insertion operator and act on the disk function $w(\vec{g},z)$ written in the form (6.52), using the results (6.57)–(6.60). Let us for convenience write the two-loop function (6.73) in the following way

$$w(\vec{g},\omega,z) = \frac{1}{(z^2-\omega^2)^2}\left(-2z\omega + \frac{2z^2\omega^2-c^2(z^2+\omega^2)}{(z^2-c^2)^{1/2}(\omega^2-c^2)^{1/2}}\right) \quad \text{(C.12.1)}$$

1. *Show that*

$$\frac{2}{\tilde{M}_1}\frac{d}{dc^2}\sum_{k=1}^{\infty}\tilde{M}_k(\omega^2-c^2)^{k-1/2} = -\frac{1}{(\omega^2-c^2)^{1/2}}. \quad \text{(C.12.2)}$$

We have

$$\frac{d}{dc^2}\sum_{k=1}^{\infty}\tilde{M}_k(\omega^2-c^2)^{k-\frac{1}{2}} = \sum_{k=1}^{\infty}\frac{d\tilde{M}_k}{dc^2}(\omega^2-c^2)^{k-\frac{1}{2}} + \tilde{M}_k\frac{d(\omega^2-c^2)^{k-\frac{1}{2}}}{dc^2}$$

$$= \sum_{k=1}^{\infty}\left((k+\frac{1}{2})\tilde{M}_{k+1}(\omega^2-c^2)^{k-\frac{1}{2}} - (k-\frac{1}{2})\tilde{M}_k(\omega^2-c^2)^{k-\frac{3}{2}}\right)$$

$$= -\frac{1}{2}\frac{\tilde{M}_1}{(\omega^2-c^2)^{1/2}} \quad \text{(C.12.3)}$$

leading to (C.12.2)

2. *Show that*

$$\frac{\partial}{\partial V(z)}\sum_{k=1}^{\infty}\tilde{M}_k(\omega^2-c^2)^{k-1/2} = \frac{d}{dz}\left[\left(\frac{\omega^2-c^2}{z^2-c^2}\right)^{1/2}\frac{z}{z^2-\omega^2}\right]. \quad \text{(C.12.4)}$$

We have

$$\sum_{k=1}^{\infty}\frac{\partial \tilde{M}_k}{\partial V(z)}(\omega^2-c^2)^{k-\frac{1}{2}} = \frac{d}{dz}\sum_{k=1}^{\infty}\frac{z(\omega^2-c^2)^{k-\frac{1}{2}}}{(z^2-c^2)^{k+\frac{1}{2}}}$$

$$= \frac{d}{dz}\left(\frac{z(\omega^2-c^2)^{\frac{1}{2}}}{(z^2-c^2)^{\frac{3}{2}}}\sum_{l=0}^{\infty}\left(\frac{\omega^2-c^2}{z^2-c^2}\right)^l\right)$$

$$= \frac{d}{dz}\left(\frac{z(\omega^2-c^2)^{\frac{1}{2}}}{(z^2-c^2)^{\frac{3}{2}}}\frac{z^2-c^2}{z^2-\omega^2}\right) = \frac{d}{dz}\left(\frac{z(\omega^2-c^2)^{\frac{1}{2}}}{(z^2-c^2)^{\frac{1}{2}}}\frac{1}{z^2-\omega^2}\right)$$

3. *Use now (6.57) to write*

$$\frac{dw(\vec{g},\omega)}{dV(z)} = \frac{-2\omega z}{(z^2-\omega^2)^2} + \qquad\qquad\text{(C.12.5)}$$
$$\frac{c^2}{(z^2-c^2)^{3/2}(\omega^2-c^2)^{1/2}} - \frac{d}{dz}\left[\left(\frac{\omega^2-c^2}{z^2-c^2}\right)^{1/2}\frac{z}{z^2-\omega^2}\right].$$

This is simple consequence of eq. (6.57) and the form of the loop insertion operator given by (6.52).

and show that the last two terms, after differentiation, can be reorganized in the following way:

$$\frac{1}{(z^2-\omega^2)^2}\left(\frac{(z^2-\omega^2)\omega^2}{(z^2-c^2)^{1/2}(\omega^2-c^2)^{1/2}} + \frac{(\omega^2-c^2)(z^2+\omega^2)}{(z^2-c^2)^{1/2}(\omega^2-c^2)^{1/2}}\right)$$
(C.12.6)

Here we just have to perform the differentiation wrt z which leads to the term

$$\frac{(\omega^2-c^2)^{\frac{1}{2}}}{(z^2-c^2)^{\frac{3}{2}}}\frac{z^2}{z^2-\omega^2} - \left(\frac{\omega^2-c^2}{z^2-c^2}\right)^{\frac{1}{2}}\left(\frac{1}{z^2-\omega^2} - \frac{2z^2}{(z^2-\omega^2)^2}\right) \qquad\text{(C.12.7)}$$

and combining the first term in this expression with the second term in (C.12.5) we obtain

$$\frac{1}{(z^2-c^2)^{\frac{3}{2}}(\omega^2-c^2)^{\frac{1}{2}}}\left(c^2 + \frac{z^2(\omega^2-c^2)}{z^2-\omega^2}\right) = \frac{1}{(z^2-c^2)^{\frac{1}{2}}(\omega^2-c^2)^{\frac{1}{2}}}\frac{\omega^2}{z^2-\omega^2}$$
(C.12.8)

This provides us with the first term in (C.12.6). The second term in (C.12.7) is just the second term in (C.12.6).

4. *Use the above to prove formula (C.12.1).*

It is just trivial algebra in the numerator of (C.12.6) and the use of (6.57).

We now turn to the proof of the three-loop formula (6.77). Since the two-loop function only depends on the coupling constants $\vec{g}$ via the position of the cut, $c(\vec{g})$, the loop insertion operator becomes very simple in the form (6.52) when acting on the two-loop function.

(5) *Prove that*

$$\frac{d}{dc^2}\left(\frac{2z^2\omega^2-c^2(z^2+\omega^2)}{(z^2-c^2)^{1/2}(\omega^2-c^2)^{1/2}}\right) = \frac{1}{2}\frac{c^2(z^2-\omega^2)^2}{(z^2-c^2)^{3/2}(\omega^2-c^2)^{3/2}} \qquad\text{(C.12.9)}$$

Just differentiate and use some simple algebra.

Solutions to Problem Sets 1–13   265

(6) *Use this to prove formula (6.77) for the three-loop function*

We have

$$w(u,z,\omega) = \frac{d}{dV(u)} w(z,\omega) = \frac{2}{\tilde{M}_1(u^2-c^2)^{\frac{3}{2}}} \frac{d}{dc^2} w(z,\omega) \quad (C.12.10)$$

$$= \frac{c^2}{\tilde{M}_1(u^2-c^2)^{\frac{3}{2}}} \frac{1}{(z^2-\omega^2)^2} \frac{d}{dc^2}\left(\frac{2z^2\omega^2 - c^2(z^2+\omega^2)}{(z^2-c^2)^{1/2}(\omega^2-c^2)^{1/2}}\right)$$

and the result now follows from (C.12.9):

$$w(u,z,\omega) = \frac{1}{2c^2\tilde{M}_1} \frac{c^2}{(u^2-c^2)^{\frac{3}{2}}} \frac{c^2}{(z^2-c^2)^{\frac{3}{2}}} \frac{c^2}{(\omega^2-c^2)^{\frac{3}{2}}} \quad (C.12.11)$$

Let us next prove the 4-loop formula. What we have to show is that

$$\frac{d}{dV(z)} \frac{f(c)}{\tilde{M}_1} = \frac{2}{\tilde{M}_1} \frac{d}{dc^2} \frac{f(c)}{\tilde{M}_1(z^2-c^2)^{3/2}}. \quad (C.12.12)$$

7. *Show that*

$$\frac{\partial \tilde{M}_1}{\partial V(z)} = -\frac{d}{dc^2} \frac{2c^2}{(z^2-c^2)^{3/2}} \quad (C.12.13)$$

and use this to show (C.12.12).

From (6.58) we have

$$\frac{\partial \tilde{M}_1}{\partial V(z)} = \frac{d}{dz} \frac{z}{(z^2-c^2)^{3/2}} = -\frac{2z^2+c^2}{(z^2-c^2)^{5/2}} = -\frac{d}{dc^2} \frac{2c^2}{(z^2-c^2)^{3/2}} \quad (C.12.14)$$

Next, we have

$$\left(\frac{\partial}{\partial V(z)} + \frac{2c^2}{\tilde{M}_1(z^2-c^2)^{3/2}} \frac{d}{dc^2}\right) \frac{f(c)}{\tilde{M}_1} = \quad (C.12.15)$$

$$-\frac{1}{\tilde{M}_1^2}\left(\frac{\partial \tilde{M}_1}{\partial V(z)} + \left(\frac{d}{dc^2}\frac{2c^2}{(z^2-c^2)^{3/2}}\right)\right) f(c) + \frac{2}{\tilde{M}_1} \frac{d}{dc^2} \frac{f(c)}{\tilde{M}_1(z^2-c^2)^{3/2}}$$

and eq. (C.12.13) then leads to formula (C.12.12).

Finally, let us turn to the n-loop formula, which we have just proven for $n = 3, 4$. Assume it is correct up to $n-1 \geq 3$.

8. *prove the following*

$$\left[\frac{d}{dV(z)}, \frac{2}{\tilde{M}_1}\frac{d}{dc^2}\right] = 0 \quad (C.12.16)$$

Using (6.61) for the loop insertion operator, the commutator can be written

$$\left[\frac{\partial}{\partial V(z)}, \frac{2}{\tilde{M}_1}\frac{d}{dc^2}\right] + \left[\frac{c^2}{(z^2-c^2)^{\frac{3}{2}}}\frac{2}{\tilde{M}_1}\frac{d}{dc^2}, \frac{2}{\tilde{M}_1}\frac{d}{dc^2}\right] \qquad (C.12.17)$$

The first commutator is

$$\left(\frac{\partial}{\partial V(z)}\frac{2}{\tilde{M}_1}\right)\frac{d}{dc^2} = -\frac{2}{\tilde{M}_1^2}\left(\frac{d}{dz}\frac{z}{(z^2-c^2)^{\frac{3}{2}}}\right)\frac{d}{dc^2} \qquad (C.12.18)$$

The second commutator is (using $[AB,C] = A[B,C] + [A,C]B$):

$$\left[\frac{c^2}{(z^2-c^2)^{\frac{3}{2}}}, \frac{2}{\tilde{M}_1}\frac{d}{dc^2}\right]\frac{2}{\tilde{M}_1}\frac{d}{dc^2} = -\frac{4}{\tilde{M}_1^2}\left(\frac{d}{dc^2}\frac{c^2}{(z^2-c^2)^{\frac{3}{2}}}\right)\frac{d}{dc^2} \qquad (C.12.19)$$

Thus the sum of the commutators in (C.12.17) is zero when using (C.12.13) and we have proven (C.12.16).

9. *Use this to prove the multiloop formula (6.78)*
   Formulas (C.12.16) and (C.12.12) show that we have

$$\frac{d}{dV(z_n)}\left(\frac{2}{\tilde{M}_1}\frac{d}{dc^2}\right)^{n-4}\left(\frac{1}{2c^2\tilde{M}_1}\prod_{k=1}^{n-1}\frac{c^2}{(z_k^2-c^2)^{3/2}}\right) \qquad (C.12.20)$$

$$= \left(\frac{2}{\tilde{M}_1}\frac{d}{dc^2}\right)^{n-4}\frac{d}{dV(z_n)}\left(\frac{1}{2c^2\tilde{M}_1}\prod_{k=1}^{n-1}\frac{c^2}{(z_k^2-c^2)^{3/2}}\right)$$

$$= \left(\frac{2}{\tilde{M}_1}\frac{d}{dc^2}\right)^{n-3}\left(\frac{1}{2c^2\tilde{M}_1}\prod_{k=1}^{n}\frac{c^2}{(z_k^2-c^2)^{3/2}}\right)$$

i.e. the multiloop formula.

C.13 SOLUTIONS TO PROBLEM SET 13

The characteristic function and the two point function

1.
$$\frac{dY}{dt} = -\hat{W}(Y) \Rightarrow dt = -\frac{dY}{\hat{W}(Y)} \Rightarrow T_2 - T_1 = -\int_{Y(T_1)}^{Y(T_2)} \frac{dy}{\hat{W}(y)}. \qquad (C.13.1)$$

2. We obtain

$$\frac{d\sinh^{-1}\sqrt{F(x)}}{dx} = \frac{1}{\sqrt{1+F(x)}} \frac{F'(x)}{2\sqrt{F(x)}}, \qquad \left(F'(x) = \frac{-\sqrt{\frac{g_s}{2\alpha}}\sinh^2\beta}{(x-\alpha)^2}\right)$$

$$= \frac{-\sqrt{\frac{g_s}{2\alpha}}\sinh^2\beta/(2(x-\alpha))}{\sqrt{\cosh^2\frac{\beta}{2}(x-\alpha)+\sinh^2\beta\sqrt{\frac{g_s}{2\alpha}}}\sqrt{\sinh^2\frac{\beta}{2}(x-\alpha)+\sinh^2\beta\sqrt{\frac{g_s}{2\alpha}}}}$$

$$= \frac{-\sqrt{\frac{g_s}{2\alpha}}\sinh\beta}{(x-\alpha)\sqrt{(x-\alpha)^2+4\cosh\beta\sqrt{\frac{g_s}{2\alpha}}(x-\alpha)+4\sinh^2\beta\frac{g_s}{2\alpha}}}$$

$$= \frac{-\Sigma}{\sqrt{(x-\alpha)^2+4\alpha(x-\alpha)+4\Sigma^2}} = \frac{-\Sigma}{\sqrt{(x+\alpha)^2-\frac{2g_s}{\alpha}}} \qquad (C.13.2)$$

3. Eq. (B.13.4) in the Problem Set implies that
$$F(\bar{X}(T)) = \sinh^2\left(\Sigma T + \sinh^{-1}\sqrt{F(X)}\right) \qquad (C.13.3)$$

and thus

$$\bar{X}(T) - \alpha = \frac{\sqrt{g_s/2\alpha}\sinh^2\beta}{\sinh^2\left(\Sigma T + \sinh^{-1}\sqrt{F(X)}\right) - \sinh^2(\beta/2)} \qquad (C.13.4)$$

$$= \frac{\Sigma^2}{\sqrt{g_s/2\alpha}} \frac{1}{\sinh^2\left(\Sigma T + \sinh^{-1}\sqrt{F(X)}\right) - \sinh^2(\beta/2)}.$$

4. For T large we have from (C.13.4)

$$\bar{X}(T) - \alpha \to \frac{4\Sigma^2}{\sqrt{g_s/2\alpha}} e^{-2\Sigma T} e^{-2\sinh^{-1}\sqrt{F(X)}}$$

$$= \frac{4\Sigma^2}{\sqrt{g_s/2\alpha}} e^{-2\Sigma T} \left(\cosh\left(\sinh^{-1}\sqrt{F(X)}\right) - \sinh\left(\sinh^{-1}\sqrt{F(X)}\right)\right)^2$$

$$= \frac{4\Sigma^2}{\sqrt{g_s/2\alpha}} e^{-2\Sigma T} \left(\sqrt{1+F(X)} - \sqrt{F(X)}\right)^2 \qquad (C.13.5)$$

5. Let us now take $X \to \infty$. Then

$$\sqrt{F(X)} \to \sinh(\beta/2) \quad \text{and} \quad \sinh^{-1}\sqrt{F(X)} \to \beta/2. \qquad (C.13.6)$$

$$\sinh^2(\Sigma T + \frac{\beta}{2}) - \sinh^2\frac{\beta}{2} = \sinh(\Sigma T + \beta)\sinh\Sigma T \qquad (C.13.7)$$

and we obtain the desired formula by using

$$\sinh(\Sigma T + \beta) = \sinh\beta\cosh\Sigma T + \cosh\beta\sinh\Sigma T, \qquad (C.13.8)$$

as well as $\sqrt{\frac{g_s}{2\alpha}}\sinh\beta = \Sigma$ and $\sqrt{\frac{g_s}{2\alpha}}\cosh\beta = \alpha$.

6. For $g_s \to 0$ we have $\alpha = \Sigma = \sqrt{\Lambda}$ and eq. (B.13.7) in the Problem Set then reads:

$$\bar{X}(T) = \sqrt{\Lambda} + \frac{2\sqrt{\Lambda}}{e^{2\sqrt{\Lambda}T} - 1} = \sqrt{\Lambda}\coth\sqrt{\Lambda}T, \qquad (C.13.9)$$

which is also (8.23) for $X \to \infty$.

7. The first equation in question 7 in the Problem Set is a trivial consequence of the definitions and that $\hat{W}(X(T)) = -\frac{d\bar{X}(T)}{dT}$. The second line follows by differentiating $\bar{X}(T)$ (do not do it by hand). Finally $4(\alpha^2 - \Sigma^2) = 2g_s/\alpha$.

The average shape of CDT and GCDT universes

8. By assumption $Y > -\alpha$ and for $T = \infty$ we have $\bar{X}(T;X) = \alpha$. This implies that $\hat{W}(\bar{X}(T)) = \hat{W}(\alpha) = 0$. Thus eq. (B.13.13) in the Problem Set becomes

$$\frac{\hat{W}'(\bar{X}(t)) - \hat{W}'(\alpha)}{(\bar{X}(t) - \alpha)/\tilde{W}(\bar{X}(t))} \to -\frac{2\tilde{W}'(\alpha)}{\tilde{W}(\alpha)} \quad \text{for } t \text{ large.} \qquad (C.13.10)$$

We have here used

$$\hat{W}'(x) = \frac{1}{\tilde{W}(x)} - \frac{(x-\alpha)\tilde{W}'(x)}{\tilde{W}^2(x)}, \quad \hat{W}'(\alpha) = \frac{1}{\tilde{W}(\alpha)} \qquad (C.13.11)$$

$$\hat{W}''(x) = -\frac{2\tilde{W}'(x)}{\tilde{W}^2(x)} - (x-\alpha)\left[\frac{\tilde{W}''(x)\tilde{W}(x) - 2(\tilde{W}'(x))^2}{\tilde{W}^3(x)}\right]. \qquad (C.13.12)$$

and thus

$$\hat{W}'(\bar{X}(t)) - \hat{W}'(\alpha) = \hat{W}''(\alpha)(x-\alpha) + O((x-\alpha)^2), \qquad (C.13.13)$$

where

$$\hat{W}''(\alpha) = \frac{-2\tilde{W}'(\alpha)}{\tilde{W}^2(\alpha)} \qquad (C.13.14)$$

For GCDT we have

$$-\frac{2\tilde{W}'(x)}{\tilde{W}(x)}\bigg|_{x=\alpha} = \frac{2(x+\alpha)}{(x+\alpha)^2 - 2g_s/\alpha}\bigg|_{x=\alpha} = \frac{\alpha}{\Sigma^2}. \quad (C.13.15)$$

og since we have $\bar{X}(t) = \alpha + O(e^{-2\Sigma t})$ this is also the order of the correction to $\langle L(t) \rangle$.

9. Inserting $Y = -\alpha$ in eq. (B.13.13) in the Problem Set we obtain

$$\langle L(t) \rangle_{X,Y=-\alpha} = \frac{\hat{W}'(\bar{X}(t)) - \hat{W}'(\alpha)}{\hat{W}(\bar{X}(t))} + \frac{\tilde{W}(\alpha)}{\hat{W}(\bar{X}(t))}$$

$$= \frac{1}{\bar{X}(t) - \alpha} - \frac{\tilde{W}'(\bar{X}(t))}{\tilde{W}(\bar{X}(t))} \quad (C.13.16)$$

where we have used (C.13.11)–(C.13.13).

Now assume we have a finite T, but both t and T are large and $T \gg t$. Since in this limit the leading corrections are $\bar{X}(T) - \alpha \propto e^{-2\Sigma T}$ and $\bar{X}(t) - \alpha \propto e^{-2\Sigma t}$, and since the leading correction to formula (B.13.20) in the Problem Set when expanding (B.13.13) in the Problem Set around $\bar{X}(T = \infty) = \alpha$ (always assuming $Y = -\alpha$) is of the form

$$\Delta(\langle L(t) \rangle_{X,Y=-\alpha}) = O\left(\frac{\bar{X}(T) - \alpha}{\bar{X}(t) - \alpha}\right) = O(e^{-2\Sigma(T-t)}), \quad (C.13.17)$$

we have obtained the desired estimate.

10. The CDT solution (8.23) can be written as

$$\bar{X}(t;X) = \sqrt{\Lambda} \, \frac{1 + e^{-2\sqrt{\Lambda}t} \frac{X - \sqrt{\Lambda}}{X + \sqrt{\Lambda}}}{1 - e^{-2\sqrt{\Lambda}t} \frac{X - \sqrt{\Lambda}}{X + \sqrt{\Lambda}}} = \sqrt{\Lambda} \coth \sqrt{\Lambda}(t + t_0) \quad (C.13.18)$$

where $t_0(X)$ is defined by

$$\frac{X - \sqrt{\Lambda}}{X + \sqrt{\Lambda}} = e^{-2\sqrt{\Lambda}t_0} \quad \text{i.e.} \quad X = \sqrt{\Lambda} \coth \sqrt{\Lambda} t_0. \quad (C.13.19)$$

Inserting this $\bar{X}(t;X)$ in (C.13.16) and using $\tilde{W}(x) = 1/(X + \sqrt{\Lambda})$ we obtain, using for convenience the notation $\tilde{t} := t + t_0$:

$$\langle L(t) \rangle_{X,Y=-\sqrt{\Lambda}} = \frac{1}{\sqrt{\Lambda}} \left(\frac{1}{\coth \sqrt{\Lambda}\tilde{t} - 1} + \frac{1}{\coth \sqrt{\Lambda}\tilde{t} + 1} \right) \quad (C.13.20)$$

$$= \frac{1}{\sqrt{\Lambda}} \frac{2 \sinh \sqrt{\Lambda}\tilde{t} \cosh \sqrt{\Lambda}\tilde{t}}{\cosh^2 \sqrt{\Lambda}\tilde{t} - \sinh^2 \sqrt{\Lambda}\tilde{t}} = \frac{\sinh 2\sqrt{\Lambda}\tilde{t}}{\sqrt{\Lambda}}.$$

11. We have

$$G(X,Y;T) = \frac{\hat{W}(\bar{X}(T;X))}{\hat{W}(X)} \frac{1}{\bar{X}(T;X)+Y}, \qquad (C.13.21)$$

$$G(X,L;t) = \frac{\hat{W}(\bar{X}(t;X))}{\hat{W}(X)} e^{-\bar{X}(t;X)L}. \qquad (C.13.22)$$

One can write

$$\int_0^\infty dL\, e^{-\bar{X}(t;X)L} L\, G(L,Y;T-t)$$
$$= -\frac{d}{d\bar{X}(t;X)} \int_0^\infty dL\, e^{-\bar{X}(t;X)L} G(L,Y;T-t)$$
$$= -\frac{d}{d\bar{X}(t;X)} G(\bar{X}(t;X),Y;T-t) \qquad (C.13.23)$$

and we have

$$G(\bar{X}(t;X),Y;T-t) = \frac{\hat{W}(\bar{X}(T-t;\bar{X}(t;X)))}{\hat{W}(\bar{X}(t;X))} \frac{1}{\bar{X}(T-t;\bar{X}(t;X))+Y}. \qquad (C.13.24)$$

Using eqs. (C.13.21) - (C.13.24) in eq. (B.13.12) one obtains

$$\frac{(Y+\bar{X}(T;X))\hat{W}(\bar{X}(t;X))}{\hat{W}(\bar{X}(T;X))} \frac{d}{d\bar{X}(t;X)} \left[\frac{-\hat{W}(\bar{X}(T-t;\bar{X}(t;X)))}{\hat{W}(\bar{X}(t;X))(\bar{X}(T-t;\bar{X}(t;X))+Y)} \right] \qquad (C.13.25)$$

12. We have, differentiating wrt X:

$$t = \int_{\bar{X}(t;X)}^{X} \frac{dy}{\hat{W}(y)} \implies 0 = \frac{1}{\hat{W}(X)} - \frac{d\bar{X}(t;X)}{dX} \frac{1}{\hat{W}(\bar{X}(t;X))}. \qquad (C.13.26)$$

13. When solving the differential equation with the specific boundary condition $\bar{X}(t=0)=X$ we can stop at any time t where we have reached $\bar{X}(t;X)$, and then continue after the coffee break for the remaining $T-t$ time, reset to new starting time 0, provided we start out with the value $\bar{X}(t;X)$ we reached at time t. Thus $\bar{X}(T-t;\bar{X}(t;X))=\bar{X}(T,X)$, the result we would have obtained in one go, without the coffee break. It is seen explicitly from our solution

$$T = \int_{\bar{X}(T;X)}^{X} \frac{dy}{\hat{W}(y)} = \left[\int_{\bar{X}(t;X)}^{X} + \int_{\bar{X}(T-t,\bar{X}(t;X))}^{\bar{X}(t;X)} \right] \frac{dy}{\hat{W}(y)} = t+(T-t) \qquad (C.13.27)$$

14. We are now ready to perform the differentiation, obtaining

$$\frac{d}{d\bar{X}(t;X)}\left[\frac{-\hat{W}(\bar{X}(T-t;\bar{X}(t;X))}{\hat{W}(\bar{X}(t;X))(Y+\bar{X}(T-t;\bar{X}(t;X)))}\right] \quad \text{(C.13.28)}$$

$$= \frac{\hat{W}(\bar{X}(T-t;X))}{\hat{W}^2(\bar{X}(t;X))}\left[\frac{\hat{W}'(\bar{X}(t;X))-\hat{W}'(\bar{X}(T-t;\bar{X}(t;X)))}{Y+\bar{X}(T-t;\bar{X}(t;X))}\right.$$

$$\left.+\frac{\hat{W}(\bar{X}(T-t;X)}{(Y+\bar{X}(T-t;\bar{X}(t;X)))^2}\right]$$

In this formula we can now use $\bar{X}(T-t;\bar{X}(t;X))=\bar{X}(T,X)$ and go back and insert (C.13.28) in (C.13.25). This will produce the wanted formula for $\langle L(t)\rangle_{X,Y}$.

References

1. J. Ambjørn, B. Durhuus and T. Jonsson. *Quantum Geometry: A Statistical Field Theory Approach.* Cambridge University Press, Cambridge, UK, 1997.
2. H. Kleinert. *PATH INTEGRALS in Quantum Mechanics, Statistics, Polymer Physics and Financial Markets.* World Scientific Publishing Co. Pte. Ltd. Singapore, 2009.
3. J. Zinn-Justin. *Path Integrals in Quantum Mechanics.* Oxford University Press, UK, 2005.
4. J. Ambjørn, B. Durhuus and T. Jonsson. *Statistical Mechanics of Paths With Curvature Dependent Action.* J. Phys. A **21** (1988), 981–996.
5. J. Ambjørn, B. Durhuus and T. Jonsson. *A Random Walk Representation of the Dirac Propagator.* Nucl. Phys. B **330** (1990), 509–522.
6. R, Lyons and Y. Peres. *Probabilities on Trees and Networks.* Cambridge University Press, UK, 2016.
7. M. Luscher and P. Weisz. *Scaling Laws and Triviality Bounds in the Lattice phi**4 Theory. 1. One Component Model in the Symmetric Phase.* Nucl. Phys. B **290** (1987), 25–60.
8. J. Frohlich, *On the Triviality of Lambda (phi**4) in D-Dimensions Theories and the Approach to the Critical Point in $D \geq$ Four-Dimensions,* Nucl. Phys. B **200** (1982), 281–296.
9. Russell Lyons and Yuval Peres, *Probabilities on Trees and Networks,* Cambridge University Press, 2016.
10. D. Aldous. *Continuum Random Trees III.* The Annals of Probabilities **21** (1993), 248–289.
11. I. Aniceto, G. Basar and R. Schiappa. *A Primer on Resurgent Transseries and Their Asymptotics.* Phys. Rept. **809** (2019), 1–135. arXiv:1802.10441 [hep-th].
12. M. Mariño, *Lectures on Non-perturbative Effects in Large N Gauge Theories, Matrix Models and Strings.* Fortsch. Phys. **62** (2014), 455-540. arXiv:1206.6272 [hep-th].
13. J. Ambjørn and Y. Makeenko. *String Theory as a Lilliputian World.* Phys. Lett. B **756** (2016), 142–146. arXiv:1601.00540 [hep-th].
14. J. Ambjørn and Y. Makeenko. *Scaling Behavior of Regularized Bosonic Strings.* Phys. Rev. D **93** (2016) no.6, 066007. arXiv:1510.03390 [hep-th].
15. J. Ambjørn, T. Budd and Y. Makeenko. *Generalized Multicritical One-Matrix Models.* Nucl. Phys. B **913** (2016), 357–380. [arXiv:1604.04522 [hep-th]].
16. J. Ambjørn, J. Jurkiewicz and Y. M. Makeenko. *Multiloop Correlators for Two-dimensional Quantum Gravity.* Phys. Lett. B **251** (1990), 517–524.
17. J. Ambjørn, L. Chekhov, C. F. Kristjansen and Y. Makeenko. *Matrix Model Calculations Beyond the Spherical Limit.* Nucl. Phys. B **404** (1993), 127–172. arXiv:hep-th/9302014 [hep-th].
18. B. Eynard and N. Orantin. *Topological Recursion in Enumerative Geometry and Random Matrices.* J. Phys. A **42** (2009) no.29, 293001.
19. E. Gwynne. *Random surfaces and Liouville quantum gravity.* https://arxiv.org/abs/1908.05573
20. G. Miermont. *Aspects of random maps.* http://perso.ens-lyon.fr/gregory.miermont/coursSaint-Flour.pdf

21. J. Miller. *Random planar geometry.* https://www.dpmms.cam.ac.uk/ jpm205/teaching/lent2020/rpg notes.pdf
22. J. Ambjørn and T. Budd. *Multi-point Functions of Weighted Cubic Maps.* Ann. Inst. H. Poincare Comb. Phys. Interact. **3** (2016), 1–44. arXiv:1408.3040 [math-ph].
23. Jean-François Le Gall. *Brownian Geometry.* Survey paper written for the 21st Takagi lectures. arXiv:1810.02664 [math.PR]
24. H. Kawai, N. Kawamoto, T. Mogami and Y. Watabiki. *Transfer Matrix Formalism for Two-dimensional Quantum Gravity and Fractal Structures of Space-Time.* Phys. Lett. B **306** (1993), 19–26. arXiv:hep-th/9302133 [hep-th].
25. J. Ambjørn and Y. Watabiki. *Scaling in Quantum Gravity.* Nucl. Phys. B **445** (1995), 129–144. arXiv:hep-th/9501049 [hep-th]
26. V. G. Knizhnik, A. M. Polyakov and A. B. Zamolodchikov. *Fractal Structure of 2D Quantum Gravity.* Mod. Phys. Lett. A **3** (1988), 819.
27. F. David. *Conformal Field Theories Coupled to 2D Gravity in the Conformal Gauge.* Mod. Phys. Lett. A **3** (1988), 1651.
28. J. Distler and H. Kawai. *Conformal Field Theory and 2D Quantum Gravity.* Nucl. Phys. B **321** (1989), 509–527.
29. J. Bouttier, P. Di Francesco and E. Guitter. *Geodesic Distance in Planar Graphs.* Nucl. Phys. B **663** (2003), 535–567. arXiv:cond-mat/0303272 [cond-mat].
30. B. Durhuus, T. Jonsson and J. F. Wheater. *On the Spectral Dimension of Causal Triangulations.* J. Statist. Phys. **139** (2010), 859. arXiv:0908.3643 [math-ph].
31. J. Ambjørn and R. Loll. *Nonperturbative Lorentzian Quantum Gravity, Causality and Topology Change.* Nucl. Phys. B **536** (1998), 407–434. arXiv:hep-th/9805108 [hep-th].
32. J. Ambjørn, R. Loll, Y. Watabiki, W. Westra and S. Zohren. *A New Continuum Limit of Matrix Models.* Phys. Lett. B **670** (2008), 224–230 arXiv:0810.2408 [hep-th]
33. J. Ambjørn, T. Budd and Y. Watabiki. *Scale-Dependent Hausdorff Dimensions in 2d Gravity.* Phys. Lett. B **736** (2014), 339–343. arXiv:1406.6251 [hep-th].
34. J. Ambjørn and T. G. Budd. *Trees and Spatial Topology Change in CDT.* J. Phys. A: Math. Theor. **46** (2013), 315201. arXiv:1302.1763 [hep-th].
35. J. Ambjørn, R. Janik, W. Westra and S. Zohren. *The Emergence of Background Geometry from Quantum Fluctuations.* Phys. Lett. B **641** (2006), 94–98. arXiv:gr-qc/0607013 [gr-qc].
36. J. Ambjørn, S. Arianos, J. A. Gesser and S. Kawamoto. *The Geometry of ZZ-Branes.* Phys. Lett. B **599** (2004), 306–312. arXiv:hep-th/0406108 [hep-th].
37. J. Ambjørn, R. Loll, Y. Watabiki, W. Westra and S. Zohren. *A String Field Theory Based on Causal Dynamical Triangulations.* JHEP **05** (2008), 032. arXiv:0802.0719 [hep-th].
38. J. Ambjørn, R. Loll, Y. Watabiki, W. Westra and S. Zohren. *A Matrix Model for 2D Quantum Gravity Defined by Causal Dynamical Triangulations.* Phys. Lett. B **665** (2008), 252–256. arXiv:0804.0252 [hep-th].

Index

(p,q) conformal theory, 209
$G(\Lambda;R)$, definition, 115
$W(V;L_1,\ldots,L_n)$, formula, 101
$W(V;L_1,\ldots,L_n)$, definition, 82
$W(\Lambda;L_1,\ldots,L_n)$, formula, 101
$W(\Lambda;L_1,\ldots,L_n)$, definition, 82
$W(\Lambda;Z_1,\ldots,Z_n)$, definition, 82
$W(\Lambda;Z_1,\ldots,Z_n)$, formula, 101
$\mu_c = -f(\text{matter})$, 183
$i\varepsilon$ prescription, 149
m^{th}-multicritical point, 207

Γ-function, 158
$\tilde{M}_k(\vec{g})$, definition, 94
M_k, definition, 91

action, 1
additive renormalization, 100
advanced Green function, 150
Airy functions, 141, 199
algebraic equations for $\chi^{(k)}(g,j)$, 197
annealed average, 179
anomalous geodesic dimension, 112
arbitrary high branching, 177
asymptotic expansions, 191

baby universes, 131, 141
bare cosmological constant, 100
BCH formula, 10
bending invariant, 55
beta-function, 205
bipartite graphs, 93
Borel summable, 141
Borel summation, 65, 191
Borel transformation, 191
bosonic string action, 49
bosonic strings, 49
boundaries of a surface, 49
boundary condition for two-loop propagator, 113
boundary conditions, 145
boundary cosmological constant, 81, 125

boundary curve, 58
branched polymers, 35, 86
branched polymers as a ϕ^3 field theory, 202
branched polymers with loops, 67, 141, 195
branched polymers with matter, 178
branching, 35
Brillouin zone, 167

Cartesian coordinates, 26
Causal Dynamical Triangulations, 121
causal Green function, 150
CDT critical exponents, 129
CDT Hamiltonian, 129
CDT partition function, 124
central charge of matter fields, 104
central limit theorem, 22
characteristic equation, 113
characteristic function, 22, 219
class $\mathcal{T}^{(2)}$ of triangulations, 68
class $\mathcal{T}^{(3)}$ of triangulations, 67
classical limit of CDT, 140
closed string, 50
combinatorial Laplacian, 62
conformal anomaly, 105
conformal field theories coupled to gravity, 104
continuity equation, 158
continuum limit of $w(\mu,\lambda_1,\ldots,\lambda_n)$, 99
contraction of loops to points, 108
convolution of functions, 162
correlation length, 107
cosmological coupling constant, 26
Coulomb's law, 152
counting triangulations of the disk, 83
critical exponent, 3
critical exponents α, β and γ, 4
critical exponents η_l, ν_l and γ_l, 44
critical exponents η_{bp}, ν_{bp} and γ_{bp}, 42
critical exponents of $G_\mu(r)$, 116
curved spacetime, 25

cut-off, 19, 29

deficit angle, 56
degenerate triangulations, 83
determination of $c_\pm(\vec{g})$, 91
diffeomorphism, 28
differential equation for $Z(g,j)$, 196
diffusion equation, 154
dimers on a lattice, 178
discrete convolution, 111
discrete inverse Laplace
    transformation, 111
discrete Laplace transformation, 111
disk amplitude, 127
double expansion in g and Λ, 197
double scaling limit, 66, 201
double-line notation, 83
dual ϕ^3 graph, 138
dynamical triangulations (DT), 82
Dyson-Schwinger equation, 203

eigenfunction of operator $\hat{D}$, 146
eigenfunctions, CDT, 130
eigenvalues, CDT, 130
Einstein-Hilbert action, 81
electrodynamics, 153
electrostatics, 152
endpoint of phase transition line, 181
engtrance loop, 108
entropy of path, 19
equidistance mass excitations, 115
equilateral triangulations, 57, 82
Euclidean action, 13
Euclidean Dynamical Triangulations, 121
Euclidean Green function, 150
Euler characteristic, 58
Euler's relation, 138
exit loop, 108
exponential growth, 88
exponential growth of number
    of surfaces, 52
exponential-integral function $\text{Ei}(u)$, 192

Fekete's lemma, 53
ferromagnetic classical spin model, 170

Feynman graph, 202
Feynman Green function, 17, 149
Feynman path integral, 7
Feynman propagator, 149
Feynman-Kac path integral, 12
Fisher's scaling relation, 5, 44, 46, 116
foliation preserving diffeomorphisms, 140
Fourier transformation, 162
fractal structure of 2d gravity, 107
Free energy, 54
free non-relativistic particle, 161
Fresnel integrals, 162
fugacity, 178

Gauss-Bonnet theorem, 58
Gaussian action of scalar field, 26
Gaussian distribution, 22
Gaussian integrals, 161
general critical behavior for 2d gravity,
    208
general loop equation, 90
Generalized CDT, 131
generalized eigenfunctions, 147
generating function, 84, 92
generating functions, 165
geodesic curvature, 58
geodesic distance between loops, 110
geometric action, 17
geometric interpretation of $G_L^{(l)}(r)$, 46
geometric interpretation of $G(V;R)$, 118
global Hausdorff dimension, 46
global Hausdorff dimension of 2d gravity, 117
graph distance, 108
graphic BP equation, 38
gravitational coupling constant, 26
Green function, 15, 16, 145

Hamilton function, 1
Hamiltonians unbounded from
    below, 141
handles, 54
handles of a surface, 49
hard dimers, 178
harmonic oscillator, 164

Hartle-Hawking wave function, 102, 127, 142
Hausdoff dimension, 13
Hausdorff dimension, 30, 31, 74
Hausdorff dimension $d_H^{(bp)}$, 42
heat conduction equation, 154
height of a vertex, 87
Hermitean operator, 145
high temperature phase, 216
Hilbert space $\mathcal{H} = L^2(\mathbb{R})$, 6
Hořava-Lifshitz gravity, 140
hyperbolic plane, 222
hyperscaling relations, 5

imaginary time, 12
indeterminate, 85, 125
infinite Hausdorff dimension, 185
integration over geometries, 55
interacting scalar fields, 33
intersection indices of Riemann surfaces, 96
intrinsic area of a surface, 74
intrinsic curvature, 28, 55
intrinsic link distance, 43
intrinsic properties of branched polymers, 42
Ising model and hard dimers, 189
Ising model coupled to gravity, 104
Ising model coupled to branched polymers, 187
Ising model on a regular lattice, 210

j-polygon, 92

Kato-Trotter theorem, 9
Kazakov potential, 207
Kirchoff's matrix-tree theorem, 67

Lagrange function, 1
Laplace operator, 167
Laplace-Beltrami operator, 60
lattice delta function, 167
lattice propagator, 167
lattice propagator in x-space, 168
lattice spin Hamiltonian, 170
Lebesque measure, 11
Lee-Yang edge singularity, 183
Lie-Trotter formula, 9
link distance, 87
Liouville quantum gravity, 121
local Hausdorff dimension, 46
local Hausdorff dimension of 2d gravity, 117
loop equation, 88, 89
loop-insertion operator, 92
lowest mass excitation, 78

macroscopic spin clusters, 211
magnetic critical exponent β, 172
magnetic susceptibility exponent, 173
magnetization m, 3
magnetized baby universes, 216
marked boundary, 85
mass renormalization, 21
Maxwell's equations, 153
mean field Ising model on DT, 211
mean field partition function, 171
mean fields exponents, 33
mean-field exponents, 5
metric tensor $g_{ab}(\xi)$, 25
Minkowskian spacetime, 13
modified Bessel function, 16
Multi-Ising spins coupled to 2d gravity, 210
multicritical exponents, 46
multicritical branched polymers, 44, 103
multicritical gravity models, 103
multicritical point, 206
multiloop, 92
multiloop formulas, 217

n-loop function, 49, 61
n-loop function $w(\vec{g}, z_1, \ldots, z_n)$, 96
n-point function, 36, 49, 61
Nambu-Goto action, 49
Nash's theorem, 60
negative frequencies, 151
non-critical strings, 55
number of GCDT baby universes, 136
number of geometries with V and L, 82

number of geometries with V, L, 101
number of handles h, 82
number of path, 19
number of surfaces in $\mathbb{R}^D$, 51

one-dimensional gravity, 27
one-point function, 36, 37, 70
Onsager exponents, 104, 211
Onsager solution, 104
order of a vertex, 35

parallel transportation, 56
partition function, 2, 87
path integral of the closed string, 50
path integral, relativistic particle, 17
peeling procedure, 110
phase transition, 3
physical length, 29
physical mass, 20
piecewise linear curve, 20
piecewise linear paths, 10
piecewise linear surface, 55
planar branched polymers, 35
planar maps, 139
planar trees, 35
Poisson's equation, 152
Polyakov action, 49
polylogarithm function, 185
positive frequences, 151
potentials with infinite power series, 208
primary operator, 209
projectable Hořava-Lifshitz gravity, 140
propagator, 16
proper time, 140
pseudosphere, 222

quadrangulations, 139
quantum geometry, 131
quantum Hamiltonian, 6
quantum Liouville theory, 105, 209
quantum partition function, 7
quenched average, 179

random surface action, 49
random surfaces, 49, 50

random walks, 21
recursion relations, 165
reduced one-point function, 37
regular triangulations, 83
regularization of geometries, 82
regularized propagator, 30
relativistic massive particle, 151
relativistic particle, 15
relativistic string, 50
renormalization of the gravitational
    coupling constant, 67
renormalized cosmological constant, 100
renormalized physical mass, 73
reparametrization invariance, 18
residue calculus, 149
residue theorem, 157
retarded wave function, 153
Riccati equation, 199
Riemann zeta function, 185
Riemannian geometry, 25
rooted branched polymers, 37
rooted branched polymers of
    height t, 123
rooted planar trees, 175
rotating integration contour, 203

scalar curvature, 26
scalar field, 26
scaling limit, 21
scaling limit of graphs, GCCDT, 137
scaling of the bosonic string mass, 68
scaling parameter, 31
scaling parameter a, 22
scaling relations, 30
Schrödinger equation, 7, 155
Schwinger proper-time, 29
self-intersecting surfaces, 51
shape of CDT and GCDT universes, 220
shortest link-path, 39
single atom spin partition function, 170
singular behavior of $c^2(g)$, 97
skeleton graphs, 139
solution to loop equation, 91
spatial diffeomorphisms, 140
specific heat c_V, 3

spin clusters, 211
spin correlation length $\xi(T)$, 174
spin partition function, 170
spin system, 2
spin-spin correlation function, 173
spin-spin correlator, 4
statistical system, 2
string coupling constant, 65
string field theory, 143
string perturbation theory, 65
string tension, 54, 76
stringy feature of two-point function, 115
Sturm-Liouville boundary conditions, 146
subadditive function, 53
subleading power of number of surfaces, 52
summation over topologies, 64
susceptibility, 3, 21, 41, 195
susceptibility function $\chi^{(k)}(\mu)$, 41
susceptibility function $\chi^{(n)}(\kappa)$, 51
Suzuki-Trotter formula, 10

tachyon, 78
Theorema Egregium, 55
three-loop function $w(\vec{g}, z_1, z_2, z_3)$, 96
Time-ordered Green function, 150
topology of a triangulation, 56
two-dimensional quantum gravity, 55, 81
two-loop function $w(\vec{g}, z_1, z_2)$, 95
two-loop GCDT function, 134
two-loop propagator, 107
two-point function, 36, 39, 70, 114, 219
two-point GCDT function, 135

uniform random rooted trees, 175
universal disk amplitude, 208
universal two-loop function $w(\vec{g}, z_1, z_2)$, 96
universality, 5, 21, 37, 102, 107, 206
universality class, 135
universality class, CDT, 129
universality classes of 2d quantum gravity, 102
universality wrt branching, 37
univesal behavior of $w_{k,l_1,...,l_n}$, 98

unrestricted triangulations $\mathscr{T}^{(0)}$, 83

variance σ, 22

wave function of the free particle, 163
Wiener measure, 13
wormholes, 141
Wronskian, 146

Printed in the United States
by Baker & Taylor Publisher Services

Printed in the United States
by Baker & Taylor Publisher Services